The future is open. It is not predetermined and thus cannot be predicted—except by accident. The possibilities that lie in the future are infinite.

—Karl Popper (1995)

The Myth of the Framework, Routledge, London

THE COMPUTABLE CITY

THE COMPUTABLE CITY

HISTORIES, TECHNOLOGIES, STORIES, PREDICTIONS

MICHAEL BATTY

The MIT Press
Cambridge, Massachusetts
London, England

The MIT Press would like to thank the anonymous peer reviewers who provided comments on drafts of this book. The generous work of academic experts is essential for establishing the authority and quality of our publications. We acknowledge with gratitude the contributions of these otherwise uncredited readers.

This book was set in Stone Serif and Stone Sans by Westchester Publishing Services. Printed and bound in the United States of America.

Library of Congress Cataloging-in-Publication Data

Names: Batty, Michael, author.
Title: The computable city : histories, technologies, stories, predictions / Michael Batty.
Description: Cambridge, Massachusetts : The MIT Press, [2024] | Includes bibliographical references and index.
Identifiers: LCCN 2023019243 (print) | LCCN 2023019244 (ebook) | ISBN 9780262547574 | ISBN 9780262377843 (epub) | ISBN 9780262377850 (pdf)
Subjects: LCSH: Smart cities.
Classification: LCC TD159.4 .B38 2024 (print) | LCC TD159.4 (ebook) | DDC 307.1/416--dc23/eng/20231031
LC record available at https://lccn.loc.gov/2023019243
LC ebook record available at https://lccn.loc.gov/2023019244

ISBN: 978-0-262-54757-4

10 9 8 7 6 5 4 3 2 1

CONTENTS

PREFACE

Go back to the earliest days of digital computing and you would have encountered arcane environments where electrical engineers, physicists, and mathematicians were embarked upon building electrical switching machines while at the same time involving themselves in computing problems that could ultimately be represented by different sequences of zeros and ones. As these environments developed, those who built and operated such machines quickly became separate from their users, who devoted themselves ever more exclusively to the way they wrote programs to achieve predictions to problems only soluble using these same computers. Users would physically take their data and programs to the machine and often wait for hours, even days, for outputs from their computations. The original assumption then was that computers would primarily be used to simulate models of some phenomena, abstracted from the real system itself. The idea that computers could be the very systems that were being simulated was far-fetched, and in the context of cities, it was way beyond our imagination that the very thing we were simulating might itself be eventually composed of computers. But, in fact, this was what one of the pioneers of digital computing, Alan Turing, used to demonstrate the very idea of computation from the very beginning. There were still glimpses of this world when I became involved with computers as a graduate student in the mid-1960s.

As the world moved on and computers became smaller and faster at exponential rates, the idea that computers might move out of their locations into the wider environment of the city gained ground. But even by the time the personal computer had become widely established in the 1980s and the internet was consolidating itself in the 1990s, there was considerable skepticism over whether the city itself might ever be considered as a computer, despite the long-standing notion that the computer was a universal machine. It took the invention of the smartphone to really bring home the notion of global connection, universal computation, and the fact that were we to count computers and measure the amount of information being transmitted, we would soon figure out that a city is now composed of a multitude of digital information processors. From our understanding of the city using computer models of their form and function to the very notion that their form and function might be built from as many computers as there are city dwellers, implies that we can now say that the city is computable. This is more than simply a cliché that defines what has also been called the "smart city"; it implies a paradox and indeed a puzzle that we use the same machines to simulate systems composed of those same machines. In this context, it is a recursion that dominates every application of computers that we can envisage. This is what the history underlying this thesis in this book is all about.

The computable city began in the 1950s, with the first models of land use and transportation developed in the US in large cities such as Chicago and Detroit and urban data systems in places such as Coventry in the UK. These computer models slowly developed, building on rudimentary theories that fashioned social physics, urban economics, and transportation into comprehensive frameworks relevant to urban planning. This was the world in which I cut my teeth with respect to how we got such models to work and make predictions. Research and development in this area has now evolved to the point where there are many new initiatives in urban science that continue to enrich this kind of understanding, but they inevitably date back to these early models. Much of this early work was about what we call here the "low-frequency city," where our focus tends to be on how cities change over years and decades. What has become significant since the millennium is the high-frequency city, which is largely a consequence of the scaling down of computers to individual devices, their embedding into the built environment, and their focus on data that is streamed in real time. Together, low- and high-frequency cities define the richness of the idea that we will call here the "computable city."

The other key theme in this book is based on the idea that technological revolutions, since the first, are intrinsically unpredictable. This also makes our science of the city very different from the nonhuman sciences, and as new technologies have been invented over the last 250 years, the city itself has become ever more complex. This is reflected in the inability of our theories to evolve at the same rate that the complexity of our cities has, notwithstanding the difficulties of generating good robust theory in an urban world that is forever changing. Unpredictability has become more and more significant as we have learned about the modern world and begun to accept the notion that there are many kinds of urban science and many kinds of urban model needed to get to grips with the problems of our cities. Although we will argue here that a good deal of the future is unpredictable, not everything is like this. There is no complete sea of ignorance about the future, and we are able to disperse some of the fog, despite this being a continuing challenge. Although our models of the city are somewhat limited and simply a basis for initial understanding, the evolution of urban theories so far provides a credible if minimal basis for speculating about their future.

I have written this book in the hope that it is accessible to many audiences, not just urbanists, planners, and engineers but also those who are interested in the evolution of technologies and the development of theories and models and computation in systems where there is a high degree of unpredictability and complexity. The arguments, I hope, also provide some insights into how social-science perspectives enrich systems where hard and soft technologies, particularly communications, are central to their functioning.

I need to thank a number of my colleagues for their help on various topics in the book. For myself, it all began in Manchester, where I trained as a town planner and urban designer. George Chadwick and Brian Mcloughlin, my erstwhile mentors from those far-off days, set me on the path to the computable city by opening up ideas that led to computers, systems theory, models, and complexity science, although I doubt they would now recognize much of what is contained in this book. Dave Foot taught me about computers and how to program (in Atlas Autocode) in Manchester and (Fortran) then in Reading, while Peter Hall, forever an influence, cemented his enthusiasm for urban modeling that he promoted at Reading before he was instrumental in helping set up my center, CASA (the Centre for

Advanced Spatial Analysis) at University College London (UCL). Alan Wilson remains a great advocate for our field since I first met him in 1969, while Paul Longley, my colleague from Cardiff days and now at UCL, continues to support my quests with a healthy skepticism, necessary in everything we do, for we can always do better. During my five-year sojourn in the US, Helen Couclelis from UC Santa Barbara continually provided me with various perspectives on how we might use computer models in many dimensions associated with cities. Mike Woldenberg from SUNY-Buffalo was a constant source of inspiration relating to scaling, morphology, and complexity that echo down the years and are implicitly reflected in the pages that follow.

Andy Hudson-Smith began our visualization focus at CASA, and he continues to pursue this, and some of his graphics are included here. This is increasingly important to the new science of cities and to disseminating and using our ideas in practical contexts that involve the design of more livable and sustainable cities. David O'Sullivan, Paul Torrens, Naru Shiode, and Andrew Crooks pushed my work hard toward urban models that were based on something more than social physics, while Martin Dodge and Rob Kitchin opened up new avenues to cyberspace and thence the smart city in work that they initiated at CASA and which has influenced some of the ideas reported here. Elsa Arcaute runs our networks and complexity science group at CASA, and her work and those of others is reflected in these pages. I need to thank Beth Clevenger, my editor at the MIT Press, who championed the book despite its length, and Anthony Zannino, who helped enormously advising on the graphics.

I must also thank my wife Sue and my son Dan who have helped me in ways too numerous to mention. I dedicate this work to them.

Michael Batty
Little Britain
London, EC1A 7BX
June 10, 2023

PREAMBLE

This book tells the story of how the digital computer and telecommunications have come to dominate the contemporary city since the middle of the twentieth century. Cities are one of humankind's greatest achievements, for they provide the crucible for processing information, economic innovation, and social interaction. Since the beginnings of the first industrial revolution some 250 years ago, cities have been dominated by new technologies, first mechanical, then electrical, and now digital. These have transformed the city from a physical space into one that is, in many senses, virtual, from a place where we move ourselves and our products physically across great distances to one where these movements are increasingly digital, dominated by information flows across communications networks. The role of distance in the city is thus being transformed, and in this, information is the key.

In chapter 1, we begin by sketching how digital technologies are currently changing the form and function of the city. We argue that such technologies are intrinsically unpredictable in that at each stage in their evolution, it has been impossible to anticipate the next stage in their development. The impact of computers on cities has been equally unpredictable, and to explain this, in part I, we begin by outlining the successive stages in the development of these technologies. In part II, we change tack and present the standard model that we use to explain how cities function and evolve. We focus on the role of distance and the ways we can look at cities over different time horizons, providing a pointer to a new urban science that is rapidly emerging. In part III, we illustrate how different simulation models of the city imply different kinds of computation and representation where these are focused on explaining how cities function. This takes us to part IV where we examine contemporary developments in the digital transformation that focus on the organization of cities and agencies involved in exploring urban futures.

These four parts emphasize computers, cities, models, and planning, which run alongside information, urbanization, computation, and organization. We array these as computers and information, cities and urbanization, models and computation, and planning and organization, weaving these themes in and out of one another to present a picture of how new technologies generate increasingly unpredictable urban futures.

1 THE UNPREDICTABLE TECHNOLOGY

It would appear that we have reached the limits of what it is possible to achieve with computer technology, although one should be careful with such statements, as they tend to sound pretty silly in 5 years.

—John von Neumann (1949)

IN THE BEGINNING

When the computer was first invented in the middle of the last century, no one ever anticipated that digital computation would become so all-pervasive that it would eventually underpin almost every activity we engage in. John von Neumann, one of the great minds of the twentieth century who peered into the future further than most,[1] was not able to predict that computers would get smaller and faster. This was the man who, along with Alan Turing, the other great sage of digital computation, clearly saw that the digital computer was a universal machine that could process any information that could ultimately be reduced to zeros and ones—the binary code. But none of this would have happened had the transistor not been invented quite independently at Bell Labs in downtown Manhattan in 1948, thus initiating the remorseless drive to miniaturization.

Many associated with this new technology, indeed those who were closest to digital computers and set to profit most from their development, did not anticipate its rapid evolution for many different kinds of information processing. The largest company, IBM, which specialized in how data might best be stored and manipulated using quasi-mechanical tabulators, was slow to realize the potential of these new digital equivalents. Indeed, Thomas J. Watson, who ran the company for forty years until 1956, is reported to have said, "I think there is a world market for maybe five computers," although he said this when the first prototypes were being developed. In fact, by the time Watson stepped down from running a booming IBM, it was the biggest commercial computer company in the world. This was notwithstanding that his statement is mirrored by many other fanciful quotes made about the origins of different types of computing,[2] versions of which rebound down the years, with every stage of the computer revolution being largely unanticipated by those closest to the machines.

SUCCESSIVE REVOLUTIONS

The industrial revolution had been in full pelt for at least 150 years when the digital computer was invented. From the fall of Rome around 400 CE and the emergence of the Dark Ages, classical knowledge with its own brand of science and society was just about kept alive in the Christian monasteries until the Western world slowly began to recover from the eleventh century onward. By the fifteenth century (1400 CE), the resurrection of Greek and classical philosophy and the Renaissance of this knowledge in the sixteenth and seventeenth centuries planted the seeds that led to the first industrial revolution that began in England in the middle of the eighteenth century. Although man's nomadic life had evolved to a settled agriculture more than ten thousand years ago, with the first cities appearing about 4000 BCE, it was not until the Renaissance in Europe that economic conditions appeared that favored large-scale technological innovations. These were first in farming and then in methods of manufacturing that spontaneously reinforced one another to lift living standards above the basic subsistence level for the first time in human history. Since then, this rise has shown little sign of faltering.

Several commentators on the future have characterized the industrial revolution as transitioning through at least four periods, each dominated by the invention and application of new technologies.[3] Alvin Toffler's "third wave" from pre- to post-industrial and Brynjolfsson and McAfee's "second machine age" define the emergence of computers starting during the last century.[4] These have sought to mark the continual increase in the amounts of energy devoted to moving and manipulating material resources, leading to a proliferation of machines and requisite organization, which, in turn, has led to a rapid rise in living standards globally. Each of these revolutions has built cumulatively on the preceding one, and although there is some sense in which old technologies are subsumed and made somewhat obsolete by new ones, much of what has been invented in the last 250 years remains in situ.

The first of these revolutions was devoted to the development of machines based on steam power, with the railways representing the quintessential imprint of this new technology on the way we were able to organize social and economic life. By the late nineteenth century, the revolution had begun to embrace electrical power that led to new forms of machine, in particular the automobile (and thence the airplane) that made new forms of mobility much more flexible and also introduced many new consumer aids into everyday life from mass automation. By the middle of the last century, the third revolution based on the invention of the digital computer—the focus of our history in this book—began. This involved automating production even further using electronics and information technology. For the first time, the great transition from a world based on energy to one based on information emerged, with computable devices rapidly becoming essential to mechanical and electrical technologies, representing a fusion of ideas from the earlier revolutions.[5]

The fourth industrial revolution, defined by Klaus Schwab in 2016 at the World Economic Forum, "is building on the Third, the digital revolution that has been occurring since the middle of the last century. It is characterized by a fusion of technologies that is blurring the lines between the physical, digital, and biological spheres."[6] In one sense, the fact that four revolutions build on one another makes it very probable that successive revolutions will continue to build in a similar cumulative fashion. The first revolution appeared to take at least a hundred years, the second fifty, and the third some twenty-five until the invention of the World Wide Web in 1990. The fourth presages a trajectory of successive revolutions that are taking place ever more rapidly. This makes it more difficult to define clearly separated temporal periods between each, with the waves of change that are now occurring, overlapping, merging, and coalescing.[7] The implication appears to be that as these waves occur, they will merge into one another, perhaps colliding into one another, adding to the complexity of everyday life and making it increasingly harder for populations to absorb their impact. It is no longer one generation thinking differently from a previous one, but rather each successive age cohort acquiring different abilities in making sense of our increasingly complex world.

In hindsight of course, all this might appear to have been predictable. The first industrial revolution led to mechanical technologies, the second to electrical, and the third to digital, while the fourth is another sea change, with digital merging with biomedical in a social context that is largely digital in terms of the way life is now organized. A fifth is under way based on many different forms of artificial intelligence (AI). The focus we have emphasized so far could be interpreted as a continuous evolution of technologies where energy and information represent different sides of the same coin, and thus the entire set of revolutions from the first onward define a continuing transformation toward a world that is largely underpinned by digital technologies. It is easy to see this when one examines any contemporary mechanical and electrical devices that do not throw away preexisting mechanics or electronics but rather reinterpret all of these using new digital forms. In this sense, our world is becoming populated by a multitude of digital twins that run alongside traditional technologies and social practices.

As we will argue throughout this book, every stage of the industrial-technological transformation has been largely unanticipated in its form and function. The mechanical was foisted on a world that, although prepared for a machine era, was largely unprepared for the machines that emerged. The automobile is the classic example. The famous quotation attributed to Henry Ford, who in the 1920s said, "If I had asked people what they wanted, they would have said faster horses," is surely apocryphal and unsubstantiated, but it does display the sentiment that we are never able to predict the future, even the short-term future, especially the short term.[8] We will develop the argument that the digital future is largely but never entirely unpredictable in the chapters that follow, notwithstanding there is still a degree of predictability in terms of the science that underpins the digital world.

We have largely assumed so far that the industrial revolution has proceeded through the invention of new technologies, largely in practice, while the science that lies behind these technologies has tended to explain these inventions in hindsight or in retrospect. In short, the science tended to lag rather than lead. In fact, an argument can be made that the science itself proceeded in parallel, developing increasingly important practical applications that fostered various technological inventions. But the science has also proceeded on a somewhat smoother trajectory than the technologies that it was able to explain. This is particularly so for mechanics and electronics, as we will note when we examine information in the next chapter, for these sciences tended to merge into one another and although the transitions in physical theory led to different paradigms, the scientific path has proven to be smoother.

To an extent, however, the third industrial revolution has become much more strongly dependent on the science of information, with its roots in statistical physics and its devices on the science of materials than on the technologies from the first two revolutions, while the fourth represents a synthesis, putting several basic sciences together. This reinforces the notion that a convergence of science and technologies is on the horizon—a singularity of sorts that has been anticipated by several writing about the extent to which our own physical nature can be infused with digital technologies. This is in essence Schwab's fourth industrial revolution, whose singularities were originally suggested by computer pioneers such as von Neumann and Ulam. It is also marked by those concerned with accelerating returns such as Ray Kurzweil, those such as Nick Bostrom who predict a possible evolution to a superintelligence, and the many avant-garde science fiction writers who have popularized the future in various forms of digital dystopia.[9]

THE PATH TO THE UNIVERSAL MACHINE

The idea of universality is the hallmark of physical theory in that the quest in physics is to discover theories that apply under all possible circumstances. In fact, even in physics, this quest has limits, as has been well documented by philosophers of science such as Thomas Kuhn, whose notion that science evolves through different paradigms is now the conventional wisdom. Paradigm shifts such as that which occurred a hundred years ago in the transition from Newtonian or classical physics to the physics of relativity and the quantum are akin to revolutions, although these are not quite comparable to the industrial revolutions or even the revolutions implicit in the development of new computer technologies that we will recount in this book.[10]

The closest comparable universality in the digital world is the notion that a computer is a universal machine where technology can be adapted to a very wide array of problems through its programming. As noted already, it was von Neumann and Turing who both argued that the idea of the computer was so general that it could compute anything that could be expressed in the binary code, in terms of problems that could be reduced to yes and no, black and white, zeros and ones. There were others at the beginning who had the same insights. Vannevar Bush, Roosevelt's science advisor during World War II, expressed the same sentiments and articulated this in his sketch for an all-purpose electronic office and digital science lab, which closely resembled a personal workstation nearly fifty years before such systems became commonplace.[11]

Yet, the idea that the core components of an electronic computer could be used to compute any problem that could be translated into the binary code was hard to accept. In 1945, this was the stuff of science fiction, and the notion that computers could be used to produce pictures and music as well as words seemed a long way off. In a world where those who had invented and worked with such machines considered the idea that there might be more than a handful of these machines, let alone machines that were small enough to be portable and therefore applicable in many different social and organizational contexts, seemed far-fetched. Even as late as 1975, computers struggled to deal with graphics, for they were either programmed as electronic drawing boards or organized so that external memories, called buffers, were attached to the machine to display pictures on crude grids or more usually as line drawings from storage tubes. The idea that one could use line printers to overprint characters to build up pictures in the manner of thematic maps is another example of the tortuous way the first computers attempted to embrace the world of graphics.

To reach the point where graphics really took off, it took the scaling down of computers to the point where parts of their memories could be given over to store and thence display graphics, and this meant that machines had to be dedicated to producing pictures either as part of their programming or as interfaces to their use. This was part and parcel of the development of personal computers that emerged quickly from the actions of amateur enthusiasts, developing devices quite separately from the large mainframe machines that dominated computer systems until the late 1970s. Many of the early graphics associated with such machines were motivated by games that still represent one of the cutting-edge functions of computation. The developments that followed broadened the range of universality through networking enabling computers to share data and evolve functions that were much closer to social, business, and government activities than their traditional applications in the physical sciences. This is the history we will recount in part I, and it is basic to the emergence of what we call here "the computable city."

We will describe this great range of functions as we develop our argument and applications throughout this book, but it is worth noting that the original idea of universality has gradually weakened as we have sought to qualify it. Although the extent to which computers have become all pervasive cannot be disputed, there are many areas where there are strict limits on their applicability. Many of these depend on the limits posed by computers to develop different forms of AI, and we will return to these below. But computers only in their most general sense were universal machines; to be useful, they needed to be coupled to the physical world, and this meant coupling to other external or networked objects as well as recognizing that the human always remained in the loop. The notion that a machine whose only logic consisted of manipulating zeros and ones being useful is only of theoretical value, and even Turing's original machine involved a tape, a head, a register, and a table of instructions.[12] Moreover, when digital and analog computing come together, this also launches a cornucopia of possibilities in the real world.

It is often forgotten that computers need to be switched on by their human operators, and they need to be configured according to various networking protocols if they are to act in concert. Only when this ensemble has been constructed can they begin to act in some specific way for some specific task, and even then, there are countless applications that are not feasible, given the state of the art. In time, many of these applications will be demonstrated, but as the future is unknowable, the range of possible applications is unlimited. In the euphoria of the post–World War II years, these limits were rarely considered. There were many overly ambitious speculations as to what computers might do. Key to these were the notions that the computer was an analog of the human mind. Even von Neumann suggested the computer might be used to simulate the brain,[13] and in 1950, Turing himself said, "The original question 'Can machines think!' I believe to be too meaningless to deserve discussion. Nevertheless, I believe that at the end of the century the use of words and general educated opinion will have altered so much that one will be able to speak of machines thinking without expecting to be contradicted."[14]

We now know that Turing's sentiment is far from what has and is likely to happen, for there is a sense in which the idea of the universal machine also implies that the world is intrinsically unpredictable. If you believe that the computer is that machine, then it can be developed for problems that have not yet emerged, for applications that no one has thought of so far. This assumes you have faith in such universality, but if you believe the world is intrinsically unpredictable, then even the idea that there is such a thing as a universal machine is problematic, contingent on one's faith in science, which time and again has been undermined. The revolutions that determine the way computation and information are continually being elaborated show no sign of ending as new applications emerge. In this book, we will chart some of these, showing how they build on one another from a succession of developments in computation over the last eighty years that have been impossible to predict. Of course, this means that even if we cannot predict the future of computation or the computer, all we can do is speculate, anticipate, and suggest trends whose trajectories we know will be wrong. In the following chapters, we will explore how this unpredictability can be explained and managed, showing how the ideas of computation have meshed with our concept of what a city is and continue to do so, and how the urban world that encompasses everything we do is becoming computable. Of course, it follows too that if the future of computation is intrinsically unpredictable, so indeed is the future city and even the future of cities.

MOBILITY, TECHNOLOGY, AND THE COMPUTABLE CITY

Arguably, the key theme that has run through the succession of industrial and scientific revolutions since the first involves our ability to control how we move, which we generically refer to as mobility. The concept of motion is central to Greek philosophy, defining movement in terms of changes in location, in size, and in the quality of any identifiable object, being, or even idea. When Greek science and philosophy was resurrected, rediscovered if you like, in the early Renaissance in Europe, motion was central to this quest. Leonardo da Vinci was the scholar who did most to set the wider scientific context, which culminated with Newton's laws of motion in the late seventeenth century.[15] The industrial revolution can be interpreted as the application of these laws to exercise control over nature, which, in its first instance, is control over how we move ourselves around and the resources that we wish to exploit.

Newton defined motion in terms of the inertia of an object, the force and acceleration of the mass associated with any object, and the interaction of forces between two objects.[16] His second law, based on force and acceleration, led to his law of gravitation, which described the attraction and deterrence between any two objects through the so-called inverse square law that lies at the basis of how any object relates to any other relative to their positions in space. We will elaborate on this in more detail in part II when we come to review location theories that lie at the heart of much of our thinking about how different activities in cities are distributed. The industrial revolution began with the invention of machines that could move heavy objects over increasing distances, harnessing the energy required to do this by its release from fossil fuels, usually by heat. The steam engine quickly became the fundamental device that enabled us to move objects over increasing distances and at increasing speeds. The result was that by the early nineteenth century, new forms of travel emerged, the railways for example, which could move people to live at greater distances from their work, all the way to enabling exploration and colonization of much bigger geographies, which changed the influence and size of nation-states.

The implications of this new technology for cities were profound. Before the first industrial revolution, the limits on movement were determined by how fast horses and coaches could travel, and to all intents and purposes, the world remained extremely local as it had done since prehistory. The invention of the wheel at much the same time as the first cities emerged around 4000 BCE did not make a massive difference to the impact that distance had on the internal structure of the city, although cities did become better defended once heavy objects could be moved more easily into place. Prior to the industrial revolution, the upper limits on the size of the biggest cities were little more than one million people, and only when the railways emerged in the 1820s did these limits on size collapse. In fact, arguably, the upper limits on size established in the first machine age still exist for most of the technologies we use for our daily movement, even though the automobile has made possible much more flexible forms of mobility. Only now at the time of writing is there a clear break with these limits. As we will explain below, this is due to a change in what we communicate and transmit, rather than any critical change in our own physiology of movement.

The second industrial revolution that led predominantly to electrical devices did not change physical travel from the first revolution that much either, but what it did do was enable electrical technologies to be used to communicate between one another at a distance. First, the telegraph emerged in the middle of the nineteenth century, then the telephone and the radio, and by the middle of the last century, the television. These technologies simply allowed information to be transmitted using sounds and pictures, but they did not allow the manipulation of information in any purposeful way, or if they did, this was largely accomplished offline, so-to-speak, using manual technologies. Information lies at the core of the third industrial revolution, as well as defining Toffler's third wave, which clearly enabled us to substitute quite a lot of physical movement for the transmission of information. There are of course many reasons for physical movement that do not necessarily involve human movement, but a good deal of economic life is based on communicating information at a distance, and much of this kind of activity is where physical travel can be substituted for digital interaction.

In the next chapter, we will explore what this meant in the nineteenth century and before for the preindustrial city with respect to the postal service. In our current century, the recent pandemic associated with COVID-19 has dramatically hastened these kinds of substitution with respect to work and entertainment, which have now become largely digital. The limits on cities are still largely based on how much time it takes to travel, with few wishing to travel more than ninety minutes each day to or from work. Although there are now at least thirty-five cities with more than ten million people and a handful with more than fifty million, travel within the biggest cities is still limited by basic constraints on the speed of travel and time available. It looks as if these limits will remain, even though the world is rapidly urbanizing and everyone will be living within cities of one size or another by the end of this century. Unless there are some radically new inventions that enable us to travel individually much more comfortably and quickly, it is likely that subsequent industrial revolutions from now on will involve an increasing amount of digital interaction and information rather than ever more stretching of the limits on physical travel.

Cities in 2100 will be linked together as they begin to fuse with one another.[17] But their functioning is still likely to be similar to cities of the industrial era. In fact, it is impossible to predict how much movement there will be relative to their current situation. Unless there are very radical changes in construction methods, they are likely to look pretty similar to the cities of today, while some would say that the cities of today are similar to yesteryear. But peel back the hood and examine what is underneath. Since the industrial revolutions began, network after network has been established as the hallmark of each wave of change, and this layered complexity with multiple cross links is increasingly invisible to our scrutiny. This is due not only to the relative invisibility of information technologies but also to their ownership as well as the privacy and confidentiality that control the use of such information. Although at an aggregate level when one looks at satellite photographs, at night-lights for example, cities look clustered and ordered worldwide, in fact, cities are now composed of layer upon layer of increasingly complex activities, built on top of one another, which pose very strict limits to our understanding. These layers pertain of course to the continued development of information technologies that are woven into one another and cannot be easily observed, while in a global world, everything appears to be related to everything else, and meaningful boundaries become harder to define.

The evolution of computation from its beginnings was initially focused on using computers to develop good science, to test abstracted hypotheses that gave rise to important technologies. But this soon changed, and by the mid-1950s, computers were being widely embedded into organizational contexts focused on better management and control of all kinds of human enterprise. Throughout the 1970s and 1980s with the development of networks and the scaling down of computers where they could be embedded in many micro contexts, computation has become an increasingly important part of the organizational environment that sustains contemporary societies. By the 1990s, as computers continued to transition down to laptops and phones, there were signs that these technologies were beginning to be embedded into the very fabric of cities, into their buildings, transport systems, environmental controls, and so on. In short, the city itself was becoming computable in that its functioning was being increasingly controlled by computers. This is loosely called "the smart city," and it must be contrasted with the use of often the same computers to understand the city's own functioning. The idea that cities could then be envisaged as computers in their own right with those same computers being used to analyze how the city actually functioned is the nexus that we call here "the computable city": an environment constructed to use computers to study and control systems that are composed of computers.[18] On top of this, we might use those same computers to generate future plans as well as manage the future city, with computation penetrating virtually every aspect of the city's functioning.

So, we define the computable city as the use of computers to both understand the city and control it: as ways in which we can simulate the city as well as affect the very functioning of the city by embedding computers into the elements that define it, its physical fabric, as well as the ways we can manipulate the fabric. The computable city is not simply the smart city, the information city, the virtual city, but rather all of these things, and in this context, it is simply a series of perspectives on the city that focus on the multitude of ways in which computers relate to cities. These foci have changed over the last eighty years from a concern for using computers to understand systems composed of non-computable objects to systems composed very often of those same computers.

In hindsight, the emergence of the computable city is fairly obvious when we take the long view on information technology, but it was never anticipated at any stage prior to its happening. In the 1960s, the notion that one could build useful computer models of the city came onto the agenda to enable us to understand the city better as well as to generate better plans. These models were symbolic, perhaps with some being demonstrated as analogs such as traffic-flow systems, but the idea that the geometry of the city could be made computable was hardly considered at all. There were glimmerings of such thinking in places such as the MIT's Media Lab,[19] but the notion that the very representation of the city might be computable was way off the agenda. But as soon as computers scaled down to the point where simulating such 3D geometries were possible, there were many who argued that such developments were obvious and inevitable. The prospect now beckons of course as to what is the next big thing. And before we begin to trace the origins of the computable city in subsequent chapters, we will speculate as to what the fifth and subsequent industrial revolutions might look like and how this might be consistent with evolving our cities into new forms of urban computation.

PREDICTING THE UNPREDICTABLE: THE FIFTH INDUSTRIAL REVOLUTION

The digital transformation that began in the middle of the last century first focused on the development of the computer itself in terms of its hardware, and this dominated computation for the first thirty years. During this time, computers became faster and smaller, and by the early 1980s, they had scaled to the point where small computers had emerged that one could interact with individually in person. The immediacy of this form of personal computing led to a much stronger focus on software engineering, which quickly became much more significant than the manpower devoted to the development of hardware. From that time on, the production and application of software became the dominant activity in computation. By the 1990s, computers were being widely networked and the development of software began to shift to web-based activities.

All of this history is more or less encapsulated in the third industrial revolution, which has established the digital domain as central to the way we organize contemporary societies. But as in the case of each revolution so far, it has been largely impossible to anticipate the next transition. Although Klaus Schwab defined the fourth industrial revolution as being organized around "a range of new technologies that are fusing the physical, digital and biological worlds, impacting all disciplines, economies and industries, and even challenging ideas about what it means to be human," its bare bones are still being established.[20] Although there is now a strong focus on merging digital and biological technologies, the application of computation to a wide range of economic and business activities is also part and parcel of this revolution.[21] What has been happening is that the economy, social life, and government are increasingly being developed on digital platforms where the focus is on new forms of economic marketplace. Production and consumption are being massively affected by providing producers and consumers with their own computable devices and enabling them to interact in ways that have taken us all by surprise. Federated sets of computable devices and their users are now being developed to provide new forms of group organization that are automating tasks traditionally handled by more centralized computing. Increasingly, storage of data and computation itself are being located in remote locations, in the so-called cloud. Platform urbanism[22] is the emerging term for this application domain of current information technologies.

This fourth revolution is still in its early stages, but no sooner than the fusion of these technologies began, the revolution appears to have been deepened and broadened away from its original focus to what we now consider might be a fifth revolution—impossible to see other than in outline at this stage but essential to thinking about the future. This fifth revolution is even less well formed than the fourth and harder to characterize, for like all previous revolutions, it is based on a spreading out and embedding of technologies into ever wider domains of social life. In terms of identifiable developments, AI, various virtual realities (VR) such as the metaverse, big data, cloud computing, and the Internet of Things (IoT) are among the many strands that define this emergent revolution. We might even think of the development of autonomous vehicles as being part of this, as well as the whole gamut of automated devices being used in medicine and education. As these phases of the digital transformation have occurred, what has been developed does not get abandoned but gets woven into the nexus of technologies so far, modified and adapted in ways that sometimes tend to disguise their origins but all the time increasing the complexity of the computational as well as social world. This type of entanglement is now the hallmark of the modernity that has evolved since the first industrial revolution began. As Mark Weiser remarked, "the most profound technologies are those that disappear. They weave themselves into the fabric of everyday life until they are indistinguishable from it."[23]

In many senses, all of these strands of technology can be traced back to the third revolution and, in the case of AI, which is currently the most hyped development to date, back to the very origins of computing itself. Although it is impossible to predict what new digital and biological technologies will evolve toward even over the shortest time horizons, there is a sense in which there is a recurrence of ideas that reflect the fact that new technologies are continually improved, adapted, and reinterpreted in terms of the impact that they make. We have already described how, in the beginning, computers were thought about as akin to the human brain, and by the end of the last century, it was assumed that AI would soon be writ large across the computational landscape. To an extent then, we were all mesmerized by the notion that the computer was a universal machine, a general-purpose technology, and the idea that we could replicate the way we think in an artificial nonbiological technology was high on the agenda. Back then, there was a real sense that through AI and computation, we might be on the threshold of producing a new kind of magic.

The first great wave of thinking about artificial intelligence began as attempts to program computers to replicate symbolic games such as chess, and this has remained the classic use case more or less until this day. The term "AI" was coined at a conference at Dartmouth, Connecticut, in 1956 by John McCarthy,[24] where those who led the field for the next twenty-five years developed a vibrant and optimistic research program into how computers could be programmed to simulate the way we as human beings think.[25] This was the original long-term goal of digital computation itself—to simulate the complexity of the human brain and, in this way, to simulate the kinds of problem solving, design, and perhaps creative tasks that we consider unique to ourselves as human beings. This was called strong AI, and it dominated the quest until the mid-1970s, when there was a great realization of how unrealistic the program that had been set out actually was. The notion that one could simulate a general intelligence came under severe scrutiny, and the claims for what it might do if it ever emerged began to seem increasingly far-fetched. But as the dominant field went into its nuclear winter, elements of it, particularly those associated with methods of searching for pattern in data underlying basic robotics, vision, and automated manufacturing, began to take over. This led to the kind of AI that now dominates the contemporary scene—weak AI—where the focus is no longer on finding the logic or the intuitions that we as human beings use in our problem solving but rather on searching for coherent patterns in data that provide it with structure, thus offering the possibility that such order might be useful for making certain forms of limited prediction.[26]

During the fourth industrial revolution, there have been great strides in applying weak AI to many routine tasks that involve searching for patterns in large databases and using these patterns not only to make better decisions but also to create new ways of producing objects that enable human problem solving to be improved. The current quest, which might ultimately come to be the hallmark of the fifth industrial revolution, may well be the extensive application of weak AI to routine tasks that can be automated. There is a renewed sense that this weak AI might also be on the threshold of reaching back for the magic that was sought at the beginning. Indeed, speculating about the future of technology, science fiction writer Arthur C. Clarke stated as part of his three laws of the future that "any sufficiently advanced technology is indistinguishable from magic."[27] Of course, in hindsight much of this magic might be explicable, but at any point in time, we will never know what the digital technologies of the future have in store for us.

The great obstacle, however, lies in the fact that most of this weak AI depends on real-time streamed data that is manipulated over and over again to discover patterns that replicate the system in question. When confronted with similar but separate systems, statistically robust AI can produce excellent predictions that are often highly applicable. But the problem is that this sort of AI treats the system as a black box, and that in situations where black swans can occur—completely unanticipated events that come out of the blue[28]—such systems fail. We have yet to acquire enough knowledge of how such systems can be built and how well they perform in general to be able to state with confidence that generic forms of weak AI would ever be as successful as we would like to assume they might be. Indeed, in the wider context, some argue that there is no such thing as AI, with the current dominant focus on machine learning being neither artificial nor intelligent. This is an argument that must clearly dominate our longer perspectives on the industrial revolutions.[29]

QUALIFYING THE UNPREDICTABLE

Not everything is unpredictable. It would be a mistake to assume that this was the central message of this book, for if collectively we believed this to be so, then there would be no motivation to explore the present and the past, to seek explanations, and to consider that these might be useful in predicting the future. We only have to consider our own internal logic that determines our personal behaviors to know that there are many, many small-scale individual events that can be predicted in routine fashion. In fact, as Tetlock and Gardner reveal in their work on superforecasting,[30] individuals can be trained to produce forecasts that are much better than random, and this depends on recognizing context. This view of prediction is rooted in the notion that as events become better defined, smaller scale, and nearer the present day, then it must be easier to predict them. Only when they are scaled up, aggregated, and examined in general, perhaps even across the long sweep of history, do they become intrinsically unpredictable. Complex systems with many parts whose constituents are often unknown are clearly unpredictable to an extent, although some of their elements might be quite predictable. Only when they are assembled into much bigger wholes do they often become unpredictable. To an extent then, complexity implies unpredictability.

This notion that the short term is likely to be more predictable than the long term can be easily demonstrated by many sequences of short-term events. As this book was being written, the COVID-19 pandemic swept the world. Pandemics are not new, for they dominate history, but this particular pandemic, like most, was simply not anticipated. It was outside our collective experience, even though all of us can imagine what a pandemic will do, how it spreads, and how devastating it might be. We have some experience of smaller-scale pandemics within the last hundred years, although the last major pandemic was a century ago at the end of World War I, at the edge of living memory, and the difference between now and then make comparisons difficult if not impossible to make. The economic consequences that have come from COVID-19 in the last three years (2020–2023) were not anticipated either; neither was the Russia–Ukraine war, which has thrown the world's energy systems into chaos. Combine all this with rapid aging of populations in developed societies, falling levels of productivity in many sectors, housing crises, and climate change, then immediate and medium-term futures are becoming entirely opaque. The future is more and more like a black box, at least a pretty gray box, but in many ways, it always has been, and as we acquire more experience of technological change, the unpredictability of the many elements defining complex systems, particularly social, economic, and city systems, is increasingly the new normal.

In this book, our focus begins with the intrinsic unpredictability of new technologies, but within this wider nexus, elements of these technologies have greater predictability than others. The applications here to cities also generate their own unpredictability, with the notion that cities are always in disequilibrium or far from equilibrium in the jargon of complexity theory, which is the dominant theme. The tools and methods that we have, even the data that define the contemporary city, are in themselves volatile in that as cities become ever more complex, as they are clearly doing, many of the theories and models that we have to make sense of them are ever more limited. In this sense, the future of our technologies and cities is unpredictable, but the models and tools that we have invented to make sense of present and future cities are always limited. This generates a tension between explanation and prediction that pervades our quest to produce more sustainable cities with a better quality of life. In the pages that follow, these will be themes that recur again and again as we recount the evolution of digital technologies over the last eighty years since their genesis and speculate on what the next eighty years might bring.

To an extent, the waves of change associated with the industrial revolutions have more or less now converged in digital technologies, and there is little by way of innovation that does not incorporate the digital world in some measure. If we take an even wider perspective on social change, it looks very much as though the world prior to the industrial revolution was one based on very limited forms of energy that were first unlocked on a massive scale during the first industrial revolution and have now transitioned to a world where energy is information. This is also a world that prior to the industrial revolution was largely nonurban, while by the end of this century, it will be entirely urban. Everywhere will be some sort of city, and when this occurs, we may no longer speak of cities in the same way as we have done for the last six thousand years. Information, as we will see, holds the key to thinking about this future. In the first part of this book, we will explore how computers evolved as information processors and how information can act as a substitute for energy, how cities emerged as the key processors of information in the modern world, and how cities are rapidly transforming into digital and informational functions that are changing their physical form. These are issues we will continue to speculate about in everything that follows.

I COMPUTERS AND INFORMATION

In part I, we illustrate the evolution of digital technologies—computers and information, emphasizing how these technologies have grown exponentially in power, have fallen rapidly in cost, and continue to get smaller to the point where we now compute and communicate with our own personal devices, which are increasingly embedded into the built form and social fabric that constitutes our cities. This is making the city computable in ways that are transforming it into a very different system from anything that has existed hitherto.

In chapter 2, we begin by examining the impact of an old communications technology, the postal service, on the form of cities and how new network technologies such as the telegraph began to annihilate distance. Chapter 3 charts a history of how the digital computer emerged, and we emphasize the role of the pioneers, Alan Turing and John von Neumann, who first proposed that the computer was a universal machine. In chapter 4, we outline how computers scaled down to the personal level, to the desktop and thence to handheld devices. But before we explore these later in this book, in chapter 5, we focus on how computers came to be linked together as networks, introducing the key infrastructure, the internet, and its popularization as the "World Wide Web." Without the "web," there would be no such thing as the computable city.

We argue here that this history is an essential construct in understanding the genesis of the computable city, and in this context, we show how important the roles of individual personalities are in fashioning the digital revolution.

2 THE INFORMATION MILE

It is vital to remember that information—in the sense of raw data—is not knowledge; that knowledge is not wisdom; and that wisdom is not foresight. But information is the first essential step to all of these.

—Arthur C. Clarke (2003)

THE GENERAL POST OFFICE

In January 1937, had you taken the Central London Railway from the West End at Tottenham Court Road three stations east, you would then have ascended in the lift to the street level and come face to face with the Central Telegraph Office. The tube station was then called "Post Office," and you would have been standing amid a large complex of buildings that were the heart not only of the British postal system but also an organization that was one of the biggest in the world, at the center of an empire that was still more or less intact and had been built up over the previous three hundred years.[1] In 1937, the Post Office was the biggest industry in Britain, employing nearly three hundred thousand, and it was to increase even further over the next thirty years, peaking in 1972 at close to half a million employees.[2] It was here in 1896 that Guglielmo Marconi demonstrated the first wireless signal, which was the forerunner of radio and one the main ways we now access computers. More than fifty years earlier in 1840, Roland Hill dramatically expanded the way information could be communicated through his introduction of the penny post, charging one penny for the delivery of a letter anywhere in the country.

However, 1937 was an auspicious year for many reasons, as we will emphasize throughout this book. The name the tube line had adopted by then was the "Central Line," and in February of that year, the London Passenger Transport Board decided to rename the tube station "St. Paul's." This was because the tube station of the same name was south of the cathedral, further away from it than Post Office, and it had been decided a more appropriate name for that station would be "Blackfriars."[3] But the name change was also in the vanguard of an enormous disruption in the role of the Post Office that would take place over the next half century. The Post Office was at its peak. The Digital Age was beckoning. In 1937, it was not quite there, but an imminent world war would propel us dramatically toward this future and change the role of communications in our cities and society in ways that would then and even now seem like science fiction.

The building complex that had developed since the Post Office moved north of the cathedral in the 1820s began with the construction of the general headquarters (General Post Office [GPO]) on the eastern side of St. Martin's Le Grand, the street running north of the tube station toward London Wall. This building was designed by Robert Smirke, who was also responsible for main facade of the British Museum to which it bore an uncanny resemblance. In fact, the GPO was demolished in 1912, which even at the time was greeted by many as sacrilege, but the building was cramped and infested, and in any case, the GPO was still in the process of expanding. In 1874, the Central Telegraph Office was built on the western side of the street. Then, in the 1890s, GPO North was constructed on an adjacent site, and in 1910, the King Edward Building, a little further west, was opened. South of the cathedral, GPO South had been built in Carter Lane, and it was to this location that Marconi directed his wireless signals in 1896. In fact, the Post Office continued to expand until 1933, with the construction of its last building, the Faraday Building, which housed the International Telephone Exchange on Queen Victoria Street. By then, the Post Office had consolidated its monopoly on all public telecommunications in Britain and its empire: the mail of course, but also the telegraph in the 1860s and, by the early 1900s, the telephone.[4]

The complex of buildings in St. Martin's Le Grand were at the core of a communications system that, in fact, had more or less dictated the evolution of the national road and even the rail system throughout the nineteenth century and before. When the Post Office was established during the reign of King Henry VIII, it first established a system of post roads, which were the main arteries linking London to the rest of the kingdom, some parts using roads built by the Romans. These were splayed out in the familiar spider's web that characterizes the geography of the country centered on the capital London—five roads that are more or less are coincident with the present-day trunk roads: the A1 to the east and north eventually to Scotland; the A2 to Kent and the Channel ports; the A3 to the southwest ultimately to Cornwall; the A4 to Bath, Bristol, and South Wales; the A5 to the Midlands, North Wales, the Irish ferries, and the northwest; and the A6 to Carlisle on the Anglo-Scottish border. The post riders and then the mail coaches that came to ply these routes increasingly focused the network on Central London, and the radial and more local routes that filled in this skeleton gradually established a complete network where ultimately everywhere and everyone came within reach.

This system established the universality of the communication network, and it was operated for many years as a highly routinized, disciplined means of moving information, albeit largely physical in the form of letters and parcels, until the telegraph and telephone were invented in the mid-to-late nineteenth century. In fact, it was a spectacle of great interest when the mail coaches would set off from St. Martin's Le Grand. In the 1820s and 1830s, people would gather to watch the coaches in their fine livery, with riders and guards ready to move at the due time, often in the early evening, the stagecoaches speeding off up Aldersgate—the Great North Road—where they would race against each other to establish a lead that translated into time and money saved.[5] From the 1850s, although not as spectacular, the railways came to play a similar role as letters and parcels shifted to the mail trains, even decentralizing the Post Office operations involving collection and sorting, which took place on the trains themselves.

If you now go to St. Paul's tube station, everything has changed. There is no longer a lift to the street but rather sets of escalators. The entrance has been moved slightly nearer to the cathedral, and doubtless you will exit the station using a credit or Oyster (smart) card, which is almost the only way of traveling on the tube today. On arriving at the street, the Central Telegraph Office is gone, and what is in its place is the only remnant of the postal system left in the area: the BT (British Telecom) Centre, which was their new headquarters, built in 1985 when the first echoes of privatization of the Post Office began, but even this building, sold off just before the pandemic began, is now being repurposed. GPO North was sold off to the Nomura Bank in 1984, and the King Edward Building, which was then the headquarters, became the property of Merrill Lynch in 1998. Rather quickly during the Great Recession of 2007–2008, it was acquired and merged with the Bank of America. More or less everything to do with the Post Office south of St. Paul's has disappeared but the area is still the domain of massive amounts of information and its technologies. The London Stock Exchange moved to this area in 2004, but this is now entirely virtual. As the financial quarter moved to occupy this area, which is at the western end of the City, barely a day goes by when the streets are not being disassembled yet again to plough in more fiber optics. The City has instituted public Wi-Fi on the streets. So, you can visit the plaque commemorating Marconi's first wireless transmission and send your own signal from the QR code using your smartphone, no longer to GPO South but to anywhere in the world.[6]

In many respects, this is now the reality of the smart city built of digital information. But lest one forget, the smart city has been there since the first cities emerged in ancient Sumeria more than five thousand years ago. This, in one respect, is the central message of this book. The smart city is not a new concept; only its current realization is new, and this realization is that the contemporary city is now computable in a multitude of ways, as we suggested in the first chapter. We will map out this picture in the rest of this book, elaborating and explaining the surprising unpredictability of these events through histories and stories that define the way our world of cities has changed since the industrial revolution began.

A WALK THROUGH THE SMART CITY

Cities are places where people come together to share their expertise and their labor, to trade the resources they control with those that they want to acquire, and to match demand with supply. Cities are thus places where materials as well as ideas are exchanged and where information to make this possible is generated.[7] The Post Office was one of the quintessential mechanisms for exchanging information in the Industrial Age, and its imprint on the city in terms of the physical channels that enable both materials and ideas to be communicated has left an indelible mark on the form of the city as we see it in the post Industrial Age. In short, information has been at the heart of the city ever since cities first appeared, with a continuing succession of innovations characterizing the city as being smart. Although the storage and movement of information through the post, the telegraph, and the telephone in and around St. Martin's Le Grand and St. Paul's is deeply embedded in the area, whichever way you walk away from this hub, information processing of one form or another appears to be the dominant activity. The main reason why Marconi demonstrated his wireless telegraphy, as it was called in the late nineteenth century, at St. Martin's was because his influential relatives in Italy assured him that he would get a better reception and interest from the British Post Office than from anyone in his home country. In Italy, he was dismissed as a crackpot, largely because even in the late nineteenth century, most people did not understand electricity. The telegraph and its successor, the telephone, were widely regarded as somewhat magical instruments, with their inventors being more likely charlatans than serious scientists. In fact, Marconi went on from strength to strength, but his experiments in 1896 were based on several emergent information technologies that were, perhaps coincidently, invented along the golden mile that lay to the west of the site of this first demonstration.

If you now head west from the old Central Telegraph Office, which became the BT headquarters, within a few yards, you cross the western edge of London's Roman wall, passing the site of Newgate Prison, which was the one of the most infamous places of execution until the mid-nineteenth century when public hangings were ended. Today, the Central Criminal Court, the Old Bailey, marks the spot where the prison was located. Another two hundred yards further west, the street that is now High Holborn crosses what was once one of London's largest rivers—the Fleet—over which a viaduct was built in the 1860s. Almost as soon as this street was improved and the viaduct finished, Thomas Edison, whose pioneering work in many aspects of electricity had just started in the US, somewhat remarkably, sought permission to build the first coal-fired power station in the world to light the street across the bridge in 1882. As he did not have to dig up the street but was simply able to hang the cables from the steel frame, he demonstrated the effect of public electric lighting immediately, and the scheme was soon extended to the entire street from Holborn Circus at the western end of the viaduct to St. Martin's Le Grand in the east.[8] But like so many schemes of its time, its economic value was not assured, and the street was converted back to gas lighting in 1886. At the beginning of any technological innovation, this kind of flip-flopping is not unusual, as we will see once we begin to explore the world of the digital computer.

Both Marconi and Edison's inventions, which developed quite independently of one another over a period of some fifteen years, were forerunners of the Digital Age that began in earnest in the mid-twentieth century. This required, of course, electricity to power the communication of information. This provided the essential juxtaposition between the physical media (devices) required, invariably built around electrical waves, and the content of the communication, which by and large involved written, spoken, and visual media as we now know it. In short, these kinds of technology require *a machine* to effect the transfer and *information* to be the subject of the transference, which in its most primitive form is a wire with some source of power and a message realized in the form of light or heat translatable into some meaning. Our information mile thus begins where Marconi sent his first radio wave, and today, as we head west toward the site of Edison's streetlights, we come across the world's biggest online retailer. One of Amazon's London offices, adjacent to the location of Edison's power station at 57 Holborn Viaduct, has been established at number 60. It is unlikely that Jeff Bezos and his Amazon company[9] are aware that they have located one of their major offices between Edison and Marconi's first demonstrations of information technology built around what was then our newfound knowledge of electric technology, but it is yet more evidence that information has always been the lifeblood of the contemporary city and will continue to be so. Although the term "smart city" is an icon of our modern world, throughout the earliest of the industrial revolutions, the smart city is a concept that has as much resonance to that bygone era as it has to thinking about how cities function now.

Instead of carrying on down High Holborn at the end of the viaduct, we can descend to what was once the bed of the River Fleet walking south to the Thames and Blackfriars. The river was covered over many years ago, the lower reaches first being culverted in the mid-eighteenth century, with the entire river up to Hampstead Heath being put underground by the late nineteenth century. Walking south a few hundred yards, Ludgate Hill peels off on the left up to St. Paul's, while on the right, Fleet Street, running east–west, begins. This is where our story goes back even further, for in this area where Blackfriars, Ludgate Hill, the Farringdon Road, and Fleet Street come together at what was once the Fleet Bridge (now Ludgate Circus), printing in England began in the early sixteenth century. The story is well known that Thomas Caxton brought the printing press to England after learning of Johannes Gutenberg's invention of moveable type, which he introduced in his press for the first time in 1439 in Mainz. The most widely read book in Europe during the Middle Ages was of course the Bible, and early printing was focused on printing as many copies as possible. Thus, it is no surprise that Thomas Caxton first set up his printing press in 1490 adjacent to Westminster Abbey, and it was only after his death that his associate, Wynkyn de Worde, and his followers decided to move their operations to the Ludgate Hill area of the City. This area soon became the focus of book publishing in and around St. Paul's churchyard. Throughout the sixteenth and seventeenth centuries, it became a hotbed for radical discussion of science, philosophy, and government in the many coffee houses that began to populate the area. It was here that newspapers began in Georgian times, while the sheer demand for news led to many new printing presses being established here, which evolved into the newspaper industry of contemporary times.

Wynkyn de Worde set up his printing press adjacent to St. Bride's Church in Ludgate Circus, which rapidly became the adopted sanctum for those who produced and printed the news that was devoured by the many who frequented the City's coffee houses. The first newspaper, *The Daily Courant*, was established here in 1702, and over the next fifty years, the newspaper industry grew westward along Fleet Street.[10] Unlike the Post Office hub centered on St. Martin's and St. Paul's, which was primarily focused on communicating a wide array of business, personal correspondence, and news, Ludgate Hill and Fleet Street were preoccupied with discovering, reporting, and inventing the news. During the eighteenth and nineteenth centuries, newspapers became big business, and the expansion up Fleet Street saw the establishment of one of the world's most significant locations for the daily press. But just as the Post Office began its rapid decline as the Digital Age kicked in, so too did the newspaper industry, first with respect to massive automation that meant that the presses needed more space for printing and distribution. Then, in more recent years, they decentralized with the rapid emergence of online publishing. But none of this began in earnest until the latter part of the twentieth century when the socialism of earlier times gave way to privatization and the rise of corporate society.

If we now make our way up Fleet Street, we pass many places where publishing, editing, and journalism are still the dominant industries. In turn, these are bolstered by the judiciary and the law—the so-called Inns of Court— which occupy much of the north side of the street, culminating in the Royal Courts of Justice at the end of our mile. There is much we might say about how the law represents another facet of the information society, but we will not digress, preferring to divert back up to High Holborn, which bounds the Inns on the north, a little west of Edison's power station. In Red Lion Square, another piece of information technology was invented in the eighteenth century that was key to Britain's naval power and its subsequent colonial empire. Over a period of fifty years, John Harrison worked away at perfecting a maritime clock—a chronometer—that would give the correct time at sea so that navigators would know exactly where they were with respect to their longitude.[11] Until this positioning problem was solved, any sea voyage was subject to extremely unreliable information usually taken from astronomical calculations. Harrison was spurred on by the fact that in 1714, Parliament chose to offer a considerable prize to the person who devised an accurate chronometer, and although he was never formally awarded the prize, the clocks and subsequently the watches that he produced were widely acknowledged, being used to solve the problem of determining longitude by dead reckoning by the time of his death in 1776.

This is perhaps a little far from our focus on information per se, but knowing where we are is, in fact, one of the key determinants in making us smarter. Harrison's clocks are the beginning of a long line of information technologies that culminate in the Global Positioning System (GPS)— geo-positioning satellites. These enable us to find out where we are, using a variety of digital maps—Google Maps for example—that on our walk along the information mile we can continually use to download map information from the free Wi-Fi that is available in the City. As we will find out as our story progresses, determining where we are in real time using wireless and related technologies is one of the triumphs of the Information Age. The annihilation of distance that it brings is absolutely central to our abilities to connect and compute with anyone, in any manner, at any point in time. Harrison's contributions were essential in enabling us to progress in this way. On our walk from St. Paul's to the end of Fleet Street, every plot, building, and street we encounter is packed with relevant detail about the way our society and cities can be seen as being instruments to process information to aid our quality of life and the human good, and although Harrison's clocks and watches mainly pertained to the sea not the land, they are instrumental in enabling us to figure out wherever we are located.

If we now retrace our steps across the Inns of Court to the Royal Courts of Justice and cross the road to King's College, we reach the place where the theory of how information is transmitted was first laid out in quite beautiful form in the mid-nineteenth century. It was in 1862 that James Clerk Maxwell, the greatest Scottish physicist, was appointed to the Chair of Natural Philosophy at King's College London.[12] Maxwell was a child prodigy, who had worked on many physical problems before he came to King's, particularly problems of geometry that also accord to ideas about geographical positioning,[13] but it was at King's that he produced the remarkable synthesis of how electricity, magnetism, and light moved like waves in the four famous equations that quickly came to be known as his—Maxwell's—equations. Maxwell only lived in London for five years, resigning his Chair in 1865 to return to his country house in Scotland, where he added the finishing touches to his equations. In 1870, he was appointed to lead the newly formed Cavendish Laboratory at Cambridge, but he was never awarded the ultimate prize, for the Nobel Prize was not established until 1885, and he passed away, still a comparatively young man, in 1879. His contribution echoes down the ages, with Albert Einstein remarking, "Without Maxwell, there would be nothing."[14] Richard Feynman put a similar gloss on it when he said, "From a long view of the history of mankind—seen from, say, ten thousand years from now—there can be little doubt that the most significant event of the 19th century will be judged as Maxwell's discovery of the laws of electrodynamics."[15]

We will end our walk here with this little drama, but were to we to continue it another mile west, we would reach the Royal Institution where, in the first half of the nineteenth century, Michael Faraday developed many experiments in electromagnetism that ultimately defined the field. Maxwell, although forty years his junior, knew Faraday, and it is tempting to ask if the other protagonists in this story ever met one another. Edison is sure to have known Marconi, but the link back to Maxwell and Faraday is perhaps a little too far.[16] In fact, it is likely that they met, for London was a small place in those days, and we can say with certainty that all those coming later knew of the work of those before, for these theories and innovations remain central to the idea of information and its impact on our modern

lives. If our walk were to continue—for we have only just reached the beginnings of the West End and Bloomsbury where the intellectual elite of the
nineteenth century gathered—then we would encounter Soho, where TV
was first demonstrated by John Logie Baird in 1926.[17] Another half mile or so
north, we would come across the site of the first internet connection outside
of North America established in 1973 at UCL.[18] Alan Turing, the pioneer who
began our story in chapter 1 where we noted his argument that the digital
computer was a universal machine, was born some three miles away to the
northwest in Maida Vale.[19] A little further on in north London's suburbs,
Tommy Flowers built what some consider to be the first digital computer (the
Colossus) at the Post Office Research Station[20] in Dollis Hill in 1943 prior to it
being shipped to Bletchley Park, the center of the UK's code-breaking operations during World War II. But these are other stories, and although we hint
at them here, for they are too good to miss, we will pick up their detail in
later chapters as we begin to fill in our picture of what the computable city
is all about.

THE FIRST ELECTRICAL COMMUNICATIONS

Prior to the industrial revolution, most communications involved shipping
products or disseminating information by manual means. Goods and letters were simply delivered by hand, using dispatch riders or stagecoaches
to speed up the process, or locally by a messenger walking. Signaling information using visible beacons that could be seen over long distances and
unpacked using simple codes that indicated the purpose of the information
in question were also used where social organization was sufficiently well
developed to make this kind of communications possible. More locally,
drums could be used to signal various events, and in many cities, drum
towers and bell towers provided information about what the time was.
Historically, until the early nineteenth century, the only way we were able
to communicate with one another over distances that were out of hailing
range—distances greater than a few yards—was to devise some system of signaling on the premise that one could see the signals visually. Implementing
signals using fire, smoke, and sound meant that distances up to twenty-five
miles could be traversed, but much depended on local conditions, climate,
and topography. Everything about these processes was physical, apart from
the occasional use of codes to enable the sender and receiver to translate

the physical signal into more complex forms of information. At sea, such signaling was often enabled using systems such as semaphore where the sender constructed words from an alphabet of different physical flag positions, with the receiver usually observing the sender using a telescope. Such systems reached their best developed form during the Napoleonic Wars in the early nineteenth century, just prior to the period where the great transition to the Industrial and thence Information Age began.[21]

Physical traces of such communications can still be seen in the form of watchtowers and beacons, lookout points high above the surrounding terrain from which signals could be sent. The Great Wall of China, which is one of the few man-made objects visible from space, consists of a long string of such towers, while beacons that could be lit to announce the impending arrival of the enemy from long distances away often dominated countrywide communications. Beacons along the south coast of England built to warn of the approach of the Spanish Armada in the late sixteenth century can still be seen. The dominant feature of these systems was that they were physical and visible, and apart from the codebooks that were sometimes required for the translation of more complex messages, everything about them was based on simple observation. As we shall see, as soon as we began to make the transition to a world where automated technologies—mechanical, electrical, digital—held sway, this visibility began to disappear. Thus, the great challenge of understanding information technologies in the modern day is that much of their structure and form are difficult to observe and measure, for it is essentially invisible.

The quintessential demonstration of the comparative invisibility of new technologies in cities came right at the beginning of this revolution in information transmission. It came with the telegraph, which was the first way in which we were able to send information over really long distances, which quite quickly became global. The steam engine that James Watt and Matthew Boulton worked to perfect from the 1770s quickly led to the locomotive and to the notion that such engines could be used to move people and goods rather than simply pump water, although it took fifty years or more for the first passenger railways to be built from the 1820s.[22] This was the first major innovation that led to a revolution in the way we could conquer distance, but it was largely restricted to physical motion. Yet, for years, experiments with electricity had suggested that you could get a signal to move some visible force apparently instantaneously along a wire if you attached a battery at one end and some means of detecting the signal at the other. Of course, the light bulb had not yet been invented, for it was eighty years in the future when Edison made that breakthrough, but you could connect something like a compass needle at the other end and this would provide sufficient motion for the signal to be recognized. No one really understood why this was possible, but the mechanism for doing it was robust, and like most of our understanding of the physical world, we simply have to accept that these forces exist. The key idea, however, was that as electricity flowed in a circuit around a wire, if you broke and then reconnected the circuit regularly, then you could send (and receive) a series of pulses, which would literally represent the circuit as being on or off. "On" or "off" is the same as "yes" or "no," which might be considered to be "one" or "zero." The rest, they say, is history, for our world is now built on this fundamental realization: the binary code.[23]

This invention was called the "telegraph," the term that had been used a few years earlier in France for the optical signaling with flag-like physical levers, but after several false starts, by the 1840s, it had been conclusively demonstrated that signals could be sent over a wire using electricity. Early on, there were lots of problems over the length of the wire and strength of the signal, but by 1850, these issues had been understood, with various fixes such as relays being introduced into the circuitry. Samuel Morse, the inventor of what became the preferred code for telegraphy (the Morse code), and William Cook and Charles Wheatstone developed extensive telegraphs beginning in the mid-1840s in the US and Britain, respectively.[24] Thus, quite suddenly, there was another technology to conquer distance, this time one dependent on coding information into a form that could be transmitted (and received) almost instantly and, of course, invisibly! It is worth noting that Wheatstone was also Professor of Experimental Philosophy at King's College before Maxwell, and we could easily have diverted our walk to take in the lecture hall where he demonstrated his first experiments.[25]

One of the features of new information technology that will recur over and over again as new ones are developed is that initially, there is widespread skepticism over whether the technology will work. Many think of such new technologies as being peddled by confidence tricksters, and this was all the more significant in the case of the telegraph, largely because the general population and even the intelligentsia of the time (and possibly even now!) did not understand electricity. This history is full of all sorts of amusing anecdotes about those wishing to use the telegraph to send not information but physical parcels such as cakes and even people. In fact, it is the first evidence that new information technology has a degree of invisibility in that it is hard to convince a public that cannot *see* electricity of its merits. In reference to Clarke's law, which we noted in chapter 1, the telegraph was indistinguishable from magic.[26] Another feature of such technologies is that the inventors themselves often do not see what quickly becomes the ultimate use of their inventions, and resistance to rolling them out for a very wide population often constrains their early adoption.

In fact, this was very much the case with the telegraph, for once it could be demonstrated, it became immediately obvious that it would initiate a revolution. Consider the time when it was first demonstrated: the 1840s. Railway building had just started in Britain and subsequently in Europe and America, and thus the physical conquest of distance had begun. The idea of a global world was in the ascendancy. The penny post had just been introduced to enable the population to communicate ever more easily. The telegraph would take this one stage further and convert such information into a form that could be sent and received almost instantly. Moreover, each of these technologies would reinforce one another. Visible would connect with invisible in new ways. The post would move from traditional horse-drawn technologies to the railway. The railways would be organized largely through the telegraphs. Local messaging would be revolutionized with some telegraphs even being placed in local homes and businesses, just as would happen at the beginning of the revolution in personal computing albeit 140 years in the future. Considered in this way, the rate of change in the 1840s appears to be as great as anything that has happened hitherto, and these changes are quite comparable with our own times. In fact, in the West, current rates of change in information technologies are substantial, but physically, the change is less obvious. Back in 1840, if you were in London, Manchester, Chicago, or New York, the rate of physical change would have been dramatically more than anything we currently now experience here in the West, and only if one compared this with some parts of modern-day China would the comparisons be similar.[27]

To complete our story of the telegraph, it quickly became apparent that here was a technology that could span the globe. Despite the problems with constructing undersea cables, the English Channel was breached in 1852 although it took another fifteen years for the Transatlantic cable to be laid.[28] Much of the science was acquired on the job, so to speak. Yet, progress was fast, largely because electric telegraphy was one of the simplest and robust technologies. Indeed, it is a technology that clearly demonstrates you need to know little or nothing about wave equations to be able to build a functioning telegraph. Maxwell himself did not publish his grand synthesis until well after the first telegraphs became functional. In fact, there was another technology that would come hard on the heels of the telegraph and this would be the telephone, which was first demonstrated by Alexander Graham Bell in 1880.[29] We will postpone any comment on the growth of these communications technologies until a later chapter because we first need to say something about the origins of computation. The computable city is a convergence of communications and computation as it originated in translating numbers, pictures, word, and sounds into binary representations. Much of this began too in the nineteenth century, and our story will continue in the next chapter sketching these origins. But lastly, let us try and draw out some key principles that tie information to cities and society in our quest to chart the key dimensions of the computable city.

THE DEATH OF DISTANCE

A central argument that we will develop throughout this book is that for thousands of years—in fact, from the time when we consider *Homo sapiens* to have emerged as a distinct species—technological progress has been pitifully slow. Only ten thousand years or more ago did man develop settled agriculture from which came cities as pockets of more intensive innovations in ideas and in the common utilization of resources. What characterizes the last 250 years, however, is that the rhythms of modern life have changed dramatically. Until the invention of steam and the revolution in mechanics, the pace of life was largely based on how far one could walk in an hour or perhaps a day, how heavy an object might be moved by a small group of individuals, and how much time it took to effect physical change such as that which still endures in the construction of massive buildings from pyramids to cathedrals. In fact, these limits on scale in terms of how far one could travel and how much one might move within fixed times had been unchanging from millennia to millennia. But from the Renaissance on, these thresholds began to be tested despite the fact that the Egyptians, the Greeks, the Romans, the Chinese, and every other civilization before were unable to break through these limits. It was the industrial revolution that we framed our argument around in chapter 1 that changed the game, so to speak, and as we have already seen, it was the railway and telegraph that broke the mold. Massive qualitative change has dominated the last two centuries in virtually every human dimension, for example concerning issues as basic as the human life span. For two hundred thousand years at least, man lived on average for about twenty-five years, but in the last 250 years, our life spans have trebled,[30] and prospects in this century alone suggest that even more dramatic change might originate from interventions in our basic physiology. Not only are our cities becoming computable, but we also stand at the dawn of a time when we ourselves will become so from advances in the fourth and subsequent industrial revolutions.

These technologies have left a strong physical trace on the city, but the transition has not been simply about our increasing control over energy. Clearly, the beginnings of the industrial revolution were based on controlling or rather enhancing instruments and tools that enabled us to move much heavier objects over much longer distances and within much shorter times, but as we have already noted, many more invisible technologies were in the making. The telegraph, for example, was such a simple and obvious technology that its impact on our control of distance was ultimately even more profound than the steam engine. But it would be at least a century before the information revolution would really make itself felt, and another fifty years before we could begin to see real evidence of this sea change in our focus of interest here on cities. Many have commented on this great transition from the premodern world to that which evolved in earnest from the mid-eighteenth century on, change characterized by the transition from a world where energy dominated to one of information.[31] In fact, the revolution in energy had to occur before this unlocked our abilities to use it, which in turn led to machines with which we could process more abstracted information. Nicholas Negroponte has characterized this as the transition from "atoms to bits," but rather than a continuous transition, it appears that the change from atoms to bits feeds on each, and they are not necessarily in sequence.[32] Physical imprints on cities and societies from mechanical devices and instruments run in parallel to informational imprints, and this complicates the picture even further. It does, however, suggest that technologies weave themselves in and out of one another in ever more complex ways, and thus that the history of the last 250 years is one of increasing complexity at all levels of life and society.

If so much has changed, then it is reasonable to ask how different cities are in the post Industrial Age from those in preindustrial times. Pretty different in most respects, one might argue, except perhaps when it comes to their form and function. One very obvious difference is that of size. If you can only reasonably walk four miles in one hour and if you judge, as many have done over thousands of years, that you do not want to travel more than a total of two hours a day to your home from your work and back, then it is clear that this puts severe limits on how big a city might grow. If all you have to help you traverse large distances is your own power or its extension using a horse and carriage, most work must be much more clustered than in the places where one might reside; cities will thus remain quite small. We noted in chapter 1 that cities before the modern era could not grow much beyond one million in population, and this is clearly demonstrable historically. Rome, for example, in its heyday, was never more than one million, around 5 percent of the size of its extended empire. A back-of-the-envelope calculation suggests that its density would then have been something like a hundred thousand people per square mile—considerably higher than most of the densest cities worldwide today. In ancient and preindustrial cities, congestion due to high densities was a recurrent condition. The speed of traffic in central areas was no more than walking pace, despite there being stagecoaches and related technologies as well as horse riders highly skilled at moving at much faster speeds of up to twelve miles an hour.[33] In fact, a frequent comparison is that the speed of traffic now in the densest world cities is about the same as in the medieval or ancient cities: not more than six miles an hour as in London in the first Elizabethan era compared to what it is now at the end of the second.

Before the nineteenth century, travel over much longer distances took what seemed to be a lifetime, with the number of really long trips across the globe constituting a significant proportion of an individual's working life. Columbus took some two months to reach the New World from Portugal, while by 1800, the time taken had barely been reduced, with it still taking at least six weeks. This was a significant proportion of one's working life, effective enough to cut off nations from one another, but nevertheless, it is remarkable how much long-distance travel took place in this preindustrial world. By the end of the nineteenth century, with steamships, the time taken to cross the Atlantic had reduced dramatically to about five days, but it took the technologies of the twentieth century—the automobile and the plane—to make a real impression on this kind of global travel. In the early nineteenth century, before railways were widely introduced, it took at least six hours to travel from London to Cambridge (compared to forty minutes by rail today). The fastest Royal Mail coaches to Glasgow undertook the trip in just less than two days, traveling at an average of ten miles an hour at their peak in 1837. It was almost impossible to cross North America in 1800, and it then took five weeks to reach Chicago and the Midwest. Only when the cross-continental railroads had been built was this time reduced considerably to about five days in comparison to present times, where it can take less than five hours by plane.

One of our themes here is that the smart city was just as evident in history as it is today. The technologies are different, for they are mainly based on information rather than energy, but an essential feature of our thesis is that the change from a world based on energy to one based on information is not a simple transition or sequence: formal computation is now involved. The industrial revolution, the catch-all phrase that dominates this history, is best seen as a sequence of revolutions that are overlapping, each involving different degrees of computability.[34] We have already noted that the first revolution introduced the internal combustion engine, culminating in the railways. The second revolution was about the development of electricity and was a prelude to the invention of the automobile, the airplane, and the age of mass production. The third revolution involved the invention of the digital computer, and the fourth, according to Klaus Schwab, is part of our current preoccupation with medical innovation, cyber-physical systems, global communications that build on the internet, and ways of automating production even further using new forms of AI that depend on big data. In short, one might almost think of the computable city as coming of age in this fourth revolution, and as we have hinted, a fifth based on widespread but weak AI is in prospect, almost upon us.

THE KEY THEMES

We will describe many of these changes in later chapters, but it is worth concluding with some of the key themes that will define our ideas about the computable city. A very obvious one involves the way cities are structured spatially with respect to how we move within them and how they are structured in terms of their form and function. It is very clear from our discussion here about the time taken to travel with respect to different technologies that the great transition has reduced this dramatically. In doing so, this has changed the world from one largely based on local actions and interactions to one based on how far our mechanical and related technologies enable us to realize similar contacts globally.[35] As we take less and less time to travel, and as our life spans expand, then what we are able to do changes significantly from what life was like prior to the industrial revolution. This prospect was first articulated at the end of the nineteenth century by one of the prophets of the future, H. G. Wells. In an essay published in 1902, he argued that transportation technologies were the key determinants of how we have lived and will continue to live in cities. He said, "the general distribution of population in a country must always be directly dependent on transport facilities"[36]—a simple enough prospect but one that would lead to much of Britain being covered in cities, all joining up with one another, a megalopolis "laced all together not only by railway and telegraph, but by novel roads . . . and by a dense network of telephones, parcels delivery tubes, and the like nervous and arterial connections." He wrote these words long before the automobile and the telephone and the airplane became the key instruments of travel and communications in the modern era, and in some respects, we will refer to this as "Wells' Proposition," which is one of our key principles around which the computable city is organized.[37] As we will show, the future is one in which distance will continue to be annihilated—the "death of distance" as Frances Cairncross has referred to it—which is another elaboration of this first principle.

Our second theme or principle relates to the growth of information as a key determinant of how we live in general and how we organize ourselves in cities. Our tentative definition of information, on which we will elaborate in the next chapter, involves representing our different realities in symbolic fashion. This traditionally has meant artistic and literary media—the written word—and during the nineteenth century this was largely associated with the way letters and books were communicated. In terms of our introduction here, this media was largely organized by the Post Office and its instruments in terms of the mail and the telegraph. To an extent, what has been happening since the industrial revolution began is that mechanical artifacts are being gradually replaced and complemented by devices that process information. The movement of physical objects has not become less, for we still live in a material world, but information has become ever more important in enabling us to communicate and to acquire and use material products. Some physical objects are being replaced by information, which embrace different kinds of behavior such as the way in which we communicate with each other, passing written messages using email and like technologies that are virtual.[38] But the principle that is key to understanding how energy and information determine how we function in the smart city involves the way these media interact with one another—through substitution and complementation. This we can define as our second principle: the interaction of energy and information.

Our third principle involves the extent to which we can measure human communications whether they be in physical-material terms such as in the movement of people and goods compared to information, which is symbolic. From the perspective of the city, much of what we might want to observe cannot be measured, even if it is represented in physical terms, largely because we do not have the machinery to be able to pin it down. We cannot easily observe our own human decision-making processes that underlie many behavioral responses, and as the communication of information using electronic media increases, it becomes increasingly difficult to measure such flows, despite there being instruments to do so. In fact, related to this third principle, which we will call "the invisibility of information," is the notion that cities are becoming ever more complex as we invent more and more technologies.[39] As the rate of change appears to speed up which is our current perception, cities are getting ever more complex and increasingly difficult to understand. This may well become the norm, and it is consistent with the fact that until the great industrial transition began, the level of complexity that human societies had been able to develop was governed by strong physical limits. Cities were smart in the old days, but the nature of that smartness has changed, and what constitutes the smartness of the cities of yesteryear is no longer quite the same as the smart city of today. In fact, we must grasp this idea and take to heart the fact that the smart city of tomorrow will always be different from the smart city of today.

This then is our fourth principle, which we will define as "the increasing complexity of cities." Indeed, the very notion that cities are getting more complex is deeply enshrined in the idea that the city is computable, with this computability itself evidence of this complexity.[40] As we introduce ever more formalized processes that govern our actions in the contemporary city, these are becoming embedded in the very fabric and structure of the city itself, adding to its computability, and it is in this sense that we consider the city as being computable. This is somewhat different from the notion that the city and its citizens are smart. For a city that is defined through the deployment of new technologies, there is no guarantee that it be smart, for much depends on how its technologies are deployed. Thus, we consider the idea that the city is computable to be a much deeper concept than that of the smart city. This is key to the argument we will elaborate on as our story unfolds, and to this end, we need to start at the dawn of the computer era, which is the subject of our next chapter.

3 TURING'S LEGACY

The idea behind digital computers may be explained by saying that these machines are intended to carry out any operations which could be done by a human computer.

—A. M. Turing (1950)

THE IDEA FOR A COMPUTER

Alan Turing was born in a rather elegant Regency house, a little northwest of Paddington station in Maida Vale. A familiar blue plaque marks the spot which is about three miles from where we ended our walk at another blue plaque celebrating James Clerk Maxwell's presence at King's College on the Strand. Turing's father was a commissioner in the Indian Civil Service, and thus he spent most of his childhood in boarding schools until he went to read mathematics at the University of Cambridge in 1931. Like many of the great intellects who we will come across in this book, he was an awkward child—a prodigy, of that there can be no doubt[1]—but he thrived in the rarefied atmosphere of Cambridge, gaining a double first-class honors degree and being elected a fellow of King's at the young age of twenty-two.

A generation before, one of the great mathematicians of the nineteenth century, David Hilbert, posed what he and others considered the key problems of mathematics.[2] Of the twenty-three problems that Hilbert defined as open questions to be resolved by the field, the proof of the second problem in mathematics was solvable from a consistent set of formal axioms, and this was something that caught the attention of the young Turing. Other mathematicians, from Leibniz writing in the seventeenth century to Turing's contemporaries, Alonzo Church and Kurt Gödel, were instrumental in posing and attempting to solve what they called the decidability problem, and it was rather natural that Turing, once he had formulated his own approach to Hilbert's challenge, would seek to work with his contemporaries, who were both living in Princeton. So, in 1936, Turing left Cambridge to spend two years in that august institution where he quickly polished off a PhD in mathematical logic and published the paper that would reserve him a place in all eternity as the intellectual father of computer science.

What Turing did was address the decidability problem in a rather obscure manner by formulating a symbolic sequence of instructions that would demonstrate that the procedures he initiated defined a means of solving any problem so formulated. In this way, he introduced an abstract method of computation, which was not a computer per se but an idea for a computer. Computers even then were something that were embodied in hardware, generally a mechanical device, but Turing set up his computer in an extremely abstract form. It consisted of a long paper tape of adjacent cells, each of which could be marked with a zero or a one using the binary code. This tape was a kind of storage medium, and sitting above each cell was a head that could read the cell's number, a zero or a one. The position of the head was defined as a state, and to get the machine to move on to another state, there was a rule book, like a computer program, that needed to be applied. The rules told the machine how to change its state by moving one cell to the left or right, depending on the number in the cell, and to change the number accordingly to a one or a zero, thus defining another state. In short, the state defined how the machine moved its location and how the location changed what was stored. This continued until it reached a state where the rules specified that the machine (or rather its head) would not move any longer, thus halting and hence generating a solution to the problem.[3]

How this solved Hilbert's decision problem was pretty obtuse. The Turing machine, as it subsequently came to be called, was almost a throwaway line in the mathematical proofs that were central to his famous paper that provided a solution. In fact, Turing's machine is almost trivial in its simplicity, although it is possible to see how another Turing machine might be fed to this as a set of rules—a program—and how this kind of recursion can demonstrate how different problems might be solved over and over again. Turing wrote his paper in 1936, but it was not published until 1937.[4] That date really marks the point where the idea for a computer was finally established. It was the year when both Konrad Zuse in Berlin and John Atanasoff and Cliff Berry in Ames, Iowa, began to construct hardware for what became, by the end of World War II, the digital computer: an origin story if ever there was one that we will return to a little later when we recount the ferocious battle to claim credit for the world's first computer.

At this point in our story, we have the idea for a computer from Turing, which is essentially an architecture for manipulating code that in its most elemental form can be conceived of as zeroes and ones, and from the nineteenth century, we have a very basic technology, which is a wire over which we can activate an electrical signal that is an obvious means of transmitting anything that can be coded. The wire became the basis of the telephone, and once we began to grapple with waves, then television came on stream. As we have implied, by the beginning of World War II, enough speculations and demonstrations of the transmission of binary code had been initiated to put all the components together that would turn Turing's dream of a universal machine into a reality. So many things came together and were accelerated by the war years that it was no surprise that the digital computer emerged, but this was still a painstakingly convoluted genesis, as we will see below. The prospect of the embedding of these machines into all aspects of social and economic life and into all technologies, including ourselves, was still a very distant prospect. If anyone had suggested that these arcane and highly esoteric ideas would ultimately be part of the very fabric of our lives and our social environments, they would have been laughed out of court. Yet, the seeds of a future built around lives informed by digital information and aided by this intelligence for how we might live in cities were about to herald a great transition, beginning the move to the smart city where much of what we now engage in would become computable in diverse ways.

In the meantime, Turing left Princeton in 1938, returning to his college in Cambridge, which was urgently preparing for war. Several of his colleagues were influential in government, especially in circles where their science would be clearly valued as part of the war effort. Turing himself, almost as a part-time hobby, was interested in codes, and although he kept this somewhat distinct from his interests in computation and the possibilities of artificial intelligence that he harbored, the link was not hard to fathom. His mentors must have recommended him for war work in this area, and in 1939, he joined the government's effort to break enemy codes as one of the first employees of Bletchley Park.[5] Many remarkable innovations emerged during World War II from this place but the level of secrecy was such that much of what was produced has only been finally declassified since the millennium. In fact, it was so secret that with classic British irony, the address of the Park was often given as Room 47, The Foreign Office.[6]

ROOM 47, THE FOREIGN OFFICE

Room 47, in fact, was the nucleus of the Government Code and Cypher School, now called GCHQ, which is similar, perhaps equivalent to the National Security Agency in the US. There are links here back to the Post Office, for traditionally it was there that the first security controls were mobilized by the government as far back as the eighteenth century when the GPO developed a series of services to intercept mail and to root out spies. In fact, as Bletchley Park grew throughout the war, its equipment was largely acquired from the Post Office, much of it being constructed at the PO Research Station at Dollis Hill in north London.[7] Turing joined Bletchley when there was hardly anyone employed there. He was part of an elite group of mathematicians and scientists, some recruited from the armed services, who set to work in what was euphemistically called "Hut 6" to crack the basic ciphers that were used at the beginning of the war to encode German signals traffic mainly but not exclusively to the submarines—U-boats—which had begun to devastate shipping in the North Atlantic.

The Germans used a coding machine called "Enigma," whose ciphers were changed every day and which required an arcane but complex set of routines that made the codes virtually unbreakable. In fact, through a judicious mix of accident and design, involving occasional capture of the machines themselves and observing the routine but systematic errors and lapses of security that plague any extensive encryption, Hut 6 was able to begin to break down codes and provide much-needed intelligence, first to shipping but then to the actual movement of the armies themselves. Turing himself devised one the key machines that was used to break Enigma in the form of a mechanical device called a "Bomba," following designs for an early version developed by the Poles just before war broke out. It is quite hard to portray the way this group worked because their style was a strange combination of reflection and insight ranging from wild intuition, dramatic guesswork, filling in the missing pieces of a jigsaw, and systematic and highly routinized search through many combinations of translations that ultimately converged on the messages in question. Turing himself often stood a little aloof from this group, but there was little doubt that as part of their practical work, there was a deep and lasting research element that Turing cultivated and which became central to how computation would ultimately come to dominate the world.[8]

In Hut 6, one of Turing's close colleagues, who effectively ran the group, was Gordon Welchman, another Cambridge mathematician with an eye for code.[9] If you examined any message encoded by Enigma, it would look like gobbledygook, with no obvious structure apart from certain standard references to logging the message itself in various ways that were consistent between particular classes of message. One of these clearly related to the location of the message, and this obviously pertained to where the message was sent from and to whom, that is, to the geographical locations implied by the message. This information by itself was enormously important in figuring out where different army groups and naval flotillas were gathering, and over time, Hut 6 became ever more adept at using this data as the key to determine future phases of the war: who was to be attacked as well as the strength of the opposing forces. It was Welchman who initiated what he called traffic analysis, which would enable maps of where the German forces were grouping and the direction of the next battles. In fact, breaking Enigma codes was essential to the effort in 1941 where the famous German battleship, the Bismarck, was pursued down the English Channel and finally sunk off Brest in Normandy. In fact, traffic analysis (which today we call "network analysis") was an essential element in the information that was generated from messages produced by Enigma. Although it was a by-product of Enigma messages, it required all the skills of those in Hut 6 to decode such information. In fact, there is one skill that often goes unmentioned, and that is a good knowledge of working German. Ironically, because those working at Bletchley Park were often recruited from the upper echelons of a class-ridden British society, many had some knowledge of the German language as well as a smattering of mathematics, and this blend of expertise was essential to the cracking of such codes.[10]

Although Turing spent his war years at Bletchley Park and was honored for his time there, there were many unsung heroes of that era and place who never lived to see the veil of secrecy lifted on the remarkable discoveries that were so instrumental in the development of the first computers as well as in winning the war. The Germans of course were continually modifying their encryption machines, and in 1942, it became clear that a new form of cipher traffic had begun to dominate the airwaves. The various listening stations reported much higher frequency encryptions, and the code breakers at Bletchley Park guessed that a new machine, probably reserved for the German High Command, had begun operation. This machine turned out to be called the "Lorenz" after the company that manufactured it, but it was nicknamed "Tunny."[11] This was considerably more sophisticated than Enigma, but in 1942, after a period of complete bafflement in breaking the code, a message was intercepted between Vienna and Berlin. This message had been sent twice due to noise in the initial transmission, and the differences between the two messages (due to the sender relaying the same message with the same machine settings but abbreviating certain words) were enough to provide a basis for beginning the extensive guesswork and intuition in breaking the code. A young recruit from Cambridge, Bill Tutte, was given this task, and for four months, he labored in making the best interpretations of the patterns in the code that he could, eventually guessing some key elements of how the machine worked. Remember of course that no one on the allied side had seen a Lorenz cipher before, and it was not until after the war ended that anyone did so.

Bill Tutte's inspiration led to an army of assistants rearranging code, spending long days translating the final messages, and almost immediately, Max Newman, who hailed from the same stable as Turing and Tutte (at Cambridge) and was running the group, decided an automated approach based on machine translation was required.[12] The group then sought the expertise of Tommy Flowers, who worked for the Post Office in telecommunications, an expert in electronic telephone exchanges, who after some trial attempts built what is now largely regarded as the first electronic programmable computer: the Colossus.[13] Colossus 1 was delivered to Bletchley Park from the Post Office Research Station in Dollis Hill in late 1943. Remarkably, the machine worked immediately, and it reduced the translation of a typical message from six days to six hours. An order then went out for the construction of nine more machines that were to be delivered to Bletchley over the next eighteen months, but the war ended in May 1945. This is a wonderful story in that the world's first digital computer was originated to help crack codes generated by another machine, another computer in fact, that nobody had ever seen. Many, including Winston Churchill, said that this breakthrough succeeded in shortening the war by at least two years. Some even said that the allies would have lost the war had it not been for the collective efforts of Bletchley Park and the individual efforts of Turing, Welchman, Newman, Tutte, Flowers, and many others.[14] Of course, with the usual British aplomb and the insidious desire for secrecy, these stories remained buried for many years. Only since the millennium has the Lorenz story come to light, and only in the much longer term will the argument about who did what, where, and when with respect to the first computers be placed in perspective. Indeed, there may never be an authoritative and definitive history of those times.

IN PHILADELPHIA AND PRINCETON

To provide some of this perspective, we must recross the Atlantic back to Princeton, where Turing had put together his PhD under the tutelage of Alonzo Church and the watchful eye of John von Neumann—Von Neumann who first entered our history in chapter 1.[15] Von Neumann was widely regarded as the great genius of the mid-twentieth century, making profound contributions to everything from quantum mechanics to game theory and being responsible for much of the theoretical development and practical impact of the atom and hydrogen bombs, while on the way, making major contributions to the theory of computing which drew him to Turing. In a sense, the idea for a computer goes back to prehistory, certainly to the Greeks in the ancient world and the Chinese before the Common Era. But it was only at the beginning of the seventeenth century that the notion of the binary digit as the essence of any such computation was first formalized. In his book *The Advancement of Learning*, Francis Bacon, an intellectual favorite of Elizabeth I, introduced the notion of a twofold difference in coding language, saying, "let all the Letters of the Alphabet, by transposition, be resolved into two Letters only."[16] Forty years later, Blaise Pascal constructed a genius of mechanical engineering in his adding machine, while sometime after, in 1689, Leibniz introduced the binary notation quite explicitly. But it was not until George Boole produced his magnificent *Laws of Thought* in 1847—at the time when the telegraph was first being developed—that the idea of operating on numbers using binary was introduced in the familiar form we know today, which involves the logical operators AND, OR, XOR, and variants thereof. It took nearly another hundred years for Claude Shannon to associate these operations unequivocally with switching circuits, which he did in his master's thesis in 1937.

At this point, it is worthwhile stepping back a bit and summarizing what we have unearthed about computers so far before we launch into the computer era itself. This we have rather boldly and precisely suggested begins in the year 1937. Of course, the date is roughly right, for by then all the elements were in place. The idea for a computer—its architecture in terms of its memory, program, and input—had been proposed by Turing and extended by von Neumann. Its media, based on the ultimate representation of all its input data in binary code, had been established quite slowly over many years, as we have just recounted.[17] And the hardware for computation, which was fast enough to solve reasonably big problems in a fraction of the time taken to do this by hand, had been established as being electronic rather than mechanical. Shannon had demonstrated how the media could be associated with this using electric circuitry (although this was implicit anyway by the early nineteenth century in the discovery of electricity), and a variety of early versions bringing all these ideas together had been established in the rudimentary computers first built by Atanasoff and Berry, by Zuse, and also by Howard Aitkin at Harvard, all just before the outbreak of World War II.

The early computers often did not meet Turing's requirement that they should be able to simulate any other computer, that is, themselves, and in this sense were not Turing complete, although they were digital and electronic. The first mechanical device that was complete in the Turing sense was due to Charles Babbage, whose Analytical Engine was sketched out but never properly constructed in the middle of the nineteenth century, containing all the elements of the modern digital computer other than being driven by electricity. Babbage was a polymath, almost a scientific dilettante, whose many projects were often never finished, but he managed to establish the necessary elements for the hardware of a computational device that was initially designed to generate tables of logarithms, which he then proposed for universal computation. His colleague and to an extent his protégé, Ada Lovelace, daughter of the mad poet Lord Byron, produced the algebra and the algorithms—in short, the software to drive such a machine. She also produced the clearest conceptual statements of the logic of the machine in her written papers, but because the machine was never completed—it was always being extended into another version—Babbage's ideas went with him to his grave.[18] Only in the 1930s were his ideas rediscovered and picked up again by Howard Aiken at Harvard who began to fashion his own version of the Analytical Engine, this time using digital technologies. Like many computer pioneers, Aitken's mission was to solve differential equations very quickly, and to that end, he conceived of a machine that was highly relevant to the war effort, receiving funding not only from IBM, which he actively sought out, but also from the US Navy that established a program at Harvard that lasted until well after the end of the war.

These various parallel developments, including those at Bletchley Park, were not unknown to one another, but the origins of digital computing and computers betrayed the early rivalries that beset all innovation. In 1940, John Mauchly from the Moore School of Electrical Engineering at the University of Pennsylvania met John Atanasoff at a conference where they began to talk about computing. Mauchly was working on analog computation machines at the time but proposed developing an electronic computer in 1941, then soliciting the help of his colleague Presper Eckert. In 1942, Atanasoff went to work for the Naval Ordnance Department, but he continued to meet with Mauchly many times, and only in 1944 did he learn that the computer they were developing, which they called the "ENIAC," had been virtually completed. Of course, Atanasoff had not taken out any patents on his early creation. The story now becomes more convoluted as von Neumann enters the picture. The US Army was always involved with the ENIAC, for the liaison between them and the Moore School was made through Herman Goldstine, who worked at the Aberdeen Proving ground where von Neumann was a frequent visitor in his dealing with missiles and bombs as part of the Manhattan project. It was in a chance meeting between them at Princeton Junction one day in 1944 that von Neumann learned of the ENIAC, and he quickly made contact and one might even say muscled into the project in his inimitable but charming way.

The scene was quickly set. By June 1945, Goldstine had put together a report for von Neumann that essentially was a statement of the essential logic of the machine. The report made no reference to Mauchly and Eckert but was widely circulated among the cognoscenti in the emergent science of computing. This was not a particularly unusual thing for von Neumann, who tended to move quickly from one project to the next, but in this case, he decided that computation and computers were so important that he needed to develop his own. His appointment for several years had been at the Institute for Advanced Study (IAS) in Princeton, where a select band of the world's great theoreticians resided—Einstein, Gödel, Morse, Weyl—but imagine their horror when von Neumann, despite his own credentials in the world of theory, proposed that a computer be built there. The notion of that peaceful place being disturbed by soldering irons, thermionic valves occasionally exploding, continual high-pitched noise, and so on was entirely inconsistent with the idea of the Institute that at that point in 1945 had only been in existence since 1930. Moreover, the notion that much of this computer work might be part of the effort to build a hydrogen bomb was something of an affront to the fellows of the Institute, despite the fact that the chief architect of the Manhattan Project, Robert Oppenheimer, himself an eminent physicist, had become director of the Institute at that same point in time.[19]

It was no surprise that Mauchly and Eckert did not join them. They had already embarked on the construction of a better machine—the EDVAC—with army funding, and in any case, they had been taken unawares by von Neumann's own ideas, which seemed not only to build on their own but also extend them into working on the bomb. In some respects, it was Atanasoff who got the raw deal. The suspicion that Mauchly drew many ideas from him and that von Neumann then took the same from him and Eckert might seem like a kind of poetic justice, but in the end, it was von Neumann who could command the respect and funding from the wider scientific and official communities that would support his computer project in the Institute. And thus, the IAS became home to the development of a machine that in the best traditions of the way all great innovations begin, especially in computing, was built from spare parts and considerable engineering ingenuity from the team that von Neumann assembled, run on a day-to-day basis by his military factotum, Herman Goldstine. The IAS machine implemented what has since been called the von Neumann architecture, which is much the same as that used in the ENIAC and several other machines in Britain just after the war.[20] And although it eventually became the inspiration for a series of IBM machines in the mid-1950s, von Neumann himself was preoccupied with other matters, particularly with the Atomic Energy Commission. His premature death, however, in 1958 more or less put an end to the project. Yet, much was happening elsewhere, and once again, we will cross the Atlantic to see what the team from Bletchley Park were up to in the aftermath of the war.

MAINFRAME MACHINES AND THE ROAD TO MINIATURIZATION

Mauchly and Eckert left Penn in 1947 and set up their own company to build the successors to the EDVAC in what emerged as the UNIVAC machines. These were ultimately acquired by the Sperry Rand company, which in turn led to the line of computers associated with takeovers by Burroughs, Unisys, Honeywell, and eventually the Lockheed Martin Corporation. Howard Aitken continued at Harvard, and his association with IBM continued to influence the design of their machines. Goldstine who had taken over von Neumann's IAS computer project in 1954 also moved to IBM to head up their mathematical sciences division at Yorktown Heights when the Princeton project ended in 1958. In some senses, in the 1950s, the influence of these early pioneers was closely woven together in the development of what came to be called mainframe computers, with much of their early funding coming from the defense industries and the armed services.[21] However, although these computers moved quickly into business applications, the machines were still largely run by the engineers who were involved in their construction, and users were usually kept well away from the hardware itself, knowing little or nothing of the operating systems around which these machines were organized.

At Bletchley Park when the war ended, the team quickly dispersed. Turing went to the National Physical Laboratory (NPL) in London where a project was started to work on the design of a new computer called the "ACE." Turing's original excitement about the project quickly disappeared as the NPL became bogged down in the red tape of the civil service, which had been almost entirely absent from Bletchley Park. This did not appeal to Turing's casual but firm view of what science was all about, and in any case, his thinking about AI, which had always been his interest, began to mature, and he needed more time and space to reflect on its implications. In 1948, he moved to the University of Manchester where one of his Cambridge and Bletchley colleagues, Max Newman, had decamped in 1945 to be professor of mathematics and to start a major computer project that led to one of the first mass-produced computers in the world, the Ferranti Mark 1. Newman assembled a small team whose core were the telecoms experts Fred Williams and Tom Kilburn who first assembled a version of the ENIAC/EDVAC machine with some cooperation from von Neumann. This machine, called the "Manchester Baby," was a prototype for the later machines in that it was the first to use a stored program, as well as embodying a variety of innovations in hardware involving mercury delay lines and such like arcana of these early computers.[22] The Manchester machines spawned others, all made by Ferranti in Oldham, which by the late 1950s had been built for Cambridge and the nuclear laboratory in Harwell. These machines were still being widely used in the late 1960s.

Newman and Turing were not involved in the detailed construction of these machines, but they provided intellectual support for the construction effort and for the use of these machines in mainstream science. Newman returned to his work on topology, while Turing began a much more sustained effort in thinking about how biological (and indeed social) systems evolved in terms of their form. The dynamic models that he developed are now widely known as part of the science of complexity, reflecting how patterns in various species evolve and finding applications in systems as diverse as cities and economies. This is something that we will return to as our ideas develop about how cities are becoming computable in later chapters. But in 1954, Turing committed suicide in the most tragic of circumstances that we can only surmise related to his conviction for homosexuality. This history is well known, and it is just possible that his death was an accident, but his career, to an extent, had also been at a watershed. Both he and John von Neumann (who incidentally had asked Turing to be his research assistant when he was at Princeton in 1937) had moved on from the theory of computing to AI and evolutionary modeling, Turing in pattern formation and von Neumann in cellular automata (CA), both generic systems that required fast computation.[23]

Von Neumann, as we have noted, died of cancer early in 1957, just three years after Turing. Both their lives were thus prematurely terminated, Turing being forty-one and von Neumann fifty-three. One wonders what they would have done if they had lived longer, for both were in the prime of their lives, and between them, they are now regarded as the most influential people of the twentieth century in computer science. Indeed, in his 2019 novel *Machines Like Me*, the novelist Ian McEwan paints a picture of an alternative future of Britain in the 1980s where Turing is still alive working at the cutting edge of AI and centrally involved in the manufacture of a race of intelligent computers: robots.[24] The focus in this book on these pioneers has not been simply to fill in the gaps in the history of how computers developed; it is as much about the way these men worked, the way they used their intuitions to almost guess how the best computer architectures might be constructed, and what the wider impacts of thinking in these terms could anticipate through the enormous impact the digital revolution would have on society at large. These men and others like them, whom we will introduce as we continue our story, provide us with an object lesson in how unpredictable the world is but how those closest to the cutting edge of new technologies can provide us with intriguing and insightful glimpses of the future.

The University of Cambridge produced many who went on to work at Bletchley Park, and it no surprise that a parallel effort in computing started there at the end of the war. Max Newman jump-started the effort at Manchester, but Maurice Wilkes, who had worked not in Bletchley Park but at the Telecommunications Research Establishment (TRE) in Malvern during the war, set up what came to be the Computer Laboratory in Cambridge. TRE was in some respects the twin of Bletchley Park, focusing on radar during the war, and many who worked there, such as Williams and Kilburn who designed the Manchester Baby, cut their teeth in that remarkable establishment. Wilkes, in fact, had attended Mauchly's and Eckert's lectures at the Moore School, and on reading von Neumann's summary of the EDVAC in 1946, he decided to build his own machine.[25] The machine, called "EDSAC," went through many versions and was replaced as the scientific computing service for Cambridge by a version of the Ferranti Mark 1 computer, the Titan, in 1962. The Computer Laboratory is now one of the powerhouses of research and development in British computing, with a key focus on networks, semantics, and AI at its forefront. Cutting-edge companies such as Deep Mind have their origins there.

The mainframe era began with these first computers, and they quickly found major applications in the general area of transactions processing, in particular in payroll, the population census, sales, revenues, and related areas where the routine generation, accounting, and processing of numerical data was dominant. Basic scientific research, the original inspiration for computers, dominated applications during this time, and for the next twenty years into the 1970s, computing was largely concerned with developing larger, faster, more robust, and reliable machines that could serve many different users. These, however, remained behind closed doors, so to speak, operating through batch processing, which meant that users developed their data and programs offline, as we say now. Computation was managed in batches, where the user would submit a job consisting of input data and a program of instructions to be loaded and processed by the machine under the supervision of the operator. In fact, the idea of the computer having its own operating system was not something that was of great concern to users then as it is today. As we noted in the Preface, the turnaround time for most users was measured in days. By the 1960s, in most university environments where users were working with many different types of problem, they would usually expect to see the outputs of their program about twenty-four hours after it had been submitted to the operators. Usually, the input media were punched tape, punched cards, and magnetic tape under the control of the user, but often intermediate outputs using the same media were generated by the computer itself to be used in later stages of programming the task in hand.

Forecasts about the future of computers and computing have always been wide of the mark. We noted the words of the founder and chairman of IBM, Thomas J. Watson, in chapter 1, who was reported as saying in the 1940s, "I think there is a world market for maybe five computers." In fact, in IBM's case, as soon as the war ended, the company went hell for leather for the manufacture of as many mainframe machines as possible, of ever-increasing speed and size. But in computing as in many other technologies, as we argued earlier, it is impossible to predict the future. Even the pioneers, especially the pioneers one might argue, were skeptical that their machines would take off in the way they began to in the early 1950s. This inability to wrestle with the ideas as to what the next stage in the industry would bring is one of the hallmarks of the last eighty years. When minicomputers followed mainframes, and the personal computer (PC) was just on the horizon, Ken Olsen, the founder of the Digital Equipment Corporation (DEC) who produced the iconic PDP and VAX minicomputers and workstations, said in 1977, "There is no reason for any individual to have a computer in his home,"[26] and this was despite the fact that one his colleagues, the MIT's Jay Forrester,[27] had an online terminal in his home as early as 1968. Ten years after Olsen made this statement, PCs were in one in every six American households and one in every two by the year 2000.

We need to do two things so that we can unravel the revolution that began with the construction of the first mainframes. First, we must anticipate and define the different kinds of computers that have come to dominate the contemporary scene because these relate to the increasing power and decreasing size of machines. This progress of miniaturization, which began almost as soon as the first machines were produced, has led and is still leading to extensive embedding of computational resources into the different environments all around us. This, in fact, is what is making the world and our cities computable. From a situation where some doubted that we needed more than a handful of computers at the beginning of this revolution, some now say that computing is now so all pervasive that very soon there will be one computer, or rather simply a skin of computing across the planet that we will all be able to access. Second, we need to explore the path to miniaturization, which began in a rather different way.

The quest for smaller machines in physical terms began almost immediately, but it was not until the early 1960s that what came to be called minicomputers became significant. These, as their name suggests, were scaled down versions of mainframes, and at first, they were highly individualized, often designed for particular niche sciences that were the prerogative of small specialist groups. Right from the beginning, there was some sense of interacting with the machine with respect to its users, and an early concern with user interfaces was a natural response to the fact that user and machine were becoming ever closer together. Meanwhile, mainframes were being increasingly accessed online using various forms of wired connection. Minicomputers were also grouped into clusters to break down the tasks often run on mainframes, and in this way, they began to act as alternatives to mainframe computing. From a different direction, very explicit scientific computing tasks led to a new generation of supercomputers that involved problems that required highly intensive computations and were often extensive in size. Supercomputing is often associated with parallelization, where arrays of computers are used to solve problems while minicomputing often merges into workstation technologies that have all the compute intensity of a minicomputer but all the functionality of the PC through its graphics interfaces.[28]

As computers have become smaller in size and as users have come closer to them, it almost appears that any one type of computer can be interfaced with any other. In fact, mainframes have not died out as the sizes and types of all computers have been massively elaborated. Mainframes are still being used for robust, large-scale applications, sometimes configured as servers but offering the possibilities of running very different processes on the same machine, with the mainframe being partitioned into separate virtual machines. In elaborating the way computers have become all pervasive over many tasks, they have been embedded into different physical infrastructures, which is the signature or hallmark of the smart city. We will return later to these differences in type because they are associated with different kinds of intelligence and information that pervade the smart city. There are other developments that we could recount that fill out this story even further. There is a whole history of automatons and analog computing devices from the abacus onward that were widely developed up to World War II, but these ended fairly dramatically with the development of these first digital computers. However, the first computers were monsters in terms of their size, and we must return to the US to chart the breakthrough that took place quite coincidentally in New York that set us on the path to miniaturization. As we implied in our first two chapters, without this, much of what we will say in the rest of this book would not have been possible, and the idea of a smart city in a digital society would have been stillborn. As we will see, the computable city is an essential resultant of this path to miniaturization.

Computers developed very quickly during World War II, moving from machines that were largely mechanical analogs of how we might break down elemental mathematical operations based on simple algebras and formal logics to electromechanical devices such as those that cracked the Enigma codes.[29] Fully-fledged programmable computers such as Colossus based on vacuum tubes—thermionic valves—then emerged, and this enabled the switching mechanisms of digital computation to be handled more efficiently and quickly. In 1948, however, John Bardeen, Walter Brattain, and William Shockley working at Bell Labs,[30] which had been built more than half a century from Alexander Graham Bell's telephone company as one of the world's powerhouses of electronics research, discovered that they could transmit electricity in such a way as to amplify the power of the switch using the materials of silicon and germanium. These semiconductors dramatically speeded up the power of a device that enabled this switching. Its much smaller size appeared to offer the digital computer a more efficient method of computation, and Bell Labs quickly began to manufacture the device for this purpose. A vote among the engineers working there led to the device being called the "transistor," and so the quest to miniaturize the key components of the computer began.

Like all such discoveries, there was some dispute between the three originators as to whose idea it really was, but they shared the Nobel Prize for this work in 1956. Shockley left Bell Labs in 1948 to establish Shockley Semiconductors in the place where he was born, in the apple orchards south of San Francisco, and through various rather rapid processes of mergers and acquisitions, his company, or rather its employees, set up the microchip firm Intel in 1955. Shockley's initiative as well as contracts from the US defense industries led to the establishment of a massive network of researchers, engineers, manufacturers, and entrepreneurs working on semiconductor technologies located in what came to be called Silicon Valley.[31] The rest, as they say, is history.

THE PATH TO THE SMART CITY

With a little reflection, it should already be clear to all of us that the smart city is a generic concept that is impossible to define. Although we will use the term extensively, this book is more about how the city is becoming computable, indeed how the city is beginning to turn into a web of computable technologies that are beginning to underpin urban life itself.[32] In this book, we will provide many oblique insights into what we mean by the smart city, and we can start by noting that the smart city must be composed of smart citizens. A city is only as smart as those who identify with it, use it in diverse ways, and are concerned for the quality of life it provides for its residents.

Smart cities are by no means the prerogative of the Digital Age. As we have noted, cities have always been smart, to a greater or lesser extent, for they have always contributed to innovation and social interaction quite independently of the technologies that dominate the historical context in which they exist. Here, our focus will not be exclusively historical with respect to past eras, for our emphasis is on the city of the recent past and of course the city of the future in terms of the emergence and evolution of digital technologies that are becoming all pervasive. But our motivation for making cities smart—which is shorthand for making citizens smart—involves many themes that emphasize how we can make our cities more sustainable using very long-standing ideas about how to make them more equitable, more efficient, more beautiful, easier to live within, and more inclusive places where the human condition can be better fulfilled.[33] In short, the smart city is a place where we aspire to various norms.

There is thus no one definition, but many different themes are woven into the various perspectives on the smart city developed in this book. To an extent, the smart city is not a new idea in that as new technologies have been developed, specifically from the beginning of the industrial revolution, cities have incorporated them as tools for better living, particularly in terms of changing the effects of space and time, annihilating distance in many respects, and changing the way we organize ourselves in the spaces that define the city. As we have noted, the first industrial revolution spawned the internal combustion engine that led to the railways, enabling cities to grow well beyond the bounds of all earlier cities for the first time. Mechanical technologies made this possible, and with the development of electricity, transportation yet again went through another revolution in terms of the transmission of information by telegraph, telephone, radio, and television as well as through materials and people. These technologies, however, were largely noninteractive. They have had a significant impact on the shape of the city, but they did not really change the way we interacted with one another, and thus they tended to support previous patterns of work and social life. Only when the computer arrived did the technologies of the city become truly interactive, and it is this interactivity that we take as our starting point in thinking about the city from the mid-twentieth century on. It is in this sense, then, that we say the contemporary city and all future cities are computable.

In a later chapter, when we come to define our standard model of the city from which many of our theories of how cities form and function emerges, we will be in a position to look at these past technologies on the structure and processes that define the city. But now, it is worth identifying two key forces that have enabled our cities to break out of their past strait-jackets as they have been struggling to do since the first computers were introduced seventy years or more ago. The first of these forces is *miniaturization*, which, as we have shown, dates from the dawn of the Digital Age through the invention of the transistor and, in its wake, to the microchip and microprocessor.[34] Without this, the second force would not have been realized, and this is the all-pervasive spreading out and networking of computers and universal machines into more and more places and people. This has led to the *embedding of computers* into the many environments around us—certainly the built environment, but also the natural and the physical as well as the physiological environment of ourselves and animal populations. This embedding is accompanied by the development of sensors, which are rather dumb devices recording physical and social processes, but usually controlled by computers, although computation and sensing frequently merge with one another.

When computers were first invented, they were used largely for scientific purposes, for exploring and predicting complex systems,[35] but quite quickly, they came to be used for much more routine purposes, for short-term predictions and the control of various scientific and business processes. In terms of cities, we can now see a very clear difference between using computers for predicting long-term change in cities over periods of years and even decades. In this context, we are using computers to inform what we might call the low-frequency city. When computers are embedded into the environment and used to generate very short-term data and control, over seconds, minutes, hours, days, and so on, we are dealing with the high-frequency city. We will return to this distinction later, but for the moment, it is an important signal on the way to fleshing out the way computers are being used to make the city smarter, to automate the city, to control it, and to plan it over a series of very different time horizons.[36]

This multifaceted usage is the hallmark of the smart city, but it has only been through miniaturization that computers have been scaled to the point where their costs are negligible and their power enormous, with their consequent embedding into any system one cares to define. Automation, of course, has been the hallmark of cities since the industrial revolution began, but it is only with the onset of the Digital Age that our cities are acquiring a digital skin that enables very extensive communications between every place and every other and between everyone and everything that interacts with the city. This physicality is one the defining characteristics of the smart city. Mechanical and electrical technologies are clearly key to the evolution of digital functions, but cities can also be smart with respect to organizational technologies. One of the significant features of the post Industrial Age is the development of new methods and tools for organizing the use and applications of these many technologies, and in this sense, good organization is key to the development and deployment of digital technologies.

To conclude this chapter, we must say a little more about miniaturization and where this is taking us, as well as something about the extent to which the first computers realized Turing's and von Neumann's early visions of being universal machines. Gordon Moore, one of the key people who broke away from William Shockley's semiconductor company to found Intel in 1957, observed in the mid-1960s a remarkable trend in the production of the silicon chips taken from data in his own company. In one of the most prescient papers of all time, which he called "Cramming more components onto integrated circuits" published in *Electronics* in 1965, he simply observed that over a period of about ten years, the number of components on a chip seemed to be doubling every year. He revised this a little in the mid-1970s, where it appeared that the doubling was occurring every eighteen months, and this, he argued, was the defining quality of the process of miniaturization. If you now examine what has happened to this exponential growth, then it has continued unerringly for the last fifty years. There are many consequences of this, but essentially this has meant that the cost of processing has fallen by half every eighteen months, while the speed of processing has been doubling over the same period. By the 1970s, this had come to be called "Moore's law," and many have speculated on how long it will last.[37] The limits on miniaturization involve fundamental limits pertaining to the speed of light, and there is speculation that the number of components being packed on a chip will slow down very rapidly in the near future, but this has not happened so far. The emergence of quantum computing is tied up with some of these limits, and there is some mystery surrounding what is possible in terms of how the geometry of the chip can be rearranged to keep miniaturization on course.

Let us be clear about what this means. As the number of components doubles every eighteen months and as time elapses, this number rises exponentially. At the start, there might be, say, ten components, eighteen months later twenty components, after thirty-six months forty, fifty-four months eighty, and so on, all packed into the same space. After fifty years, the number of components is nearly a hundred billion, and this somewhat mind-boggling number gives an idea of the very dramatic decreases in cost, increases in speed, and increases in the storage capacity of such chips. When computers change as dramatically as this, the world changes too, and we begin to do things very differently. What we will show in the next chapter is that by the 1970s, this scaling down in size had reached the point where computers moved to the desktop within the ambit of the user. This changed the nature of computation very dramatically. It led to the notion of software that was portable and could be accessed by very large numbers of individual users, and this in turn generated demand for very generic programs such as word processing.[38] One element of this revolution, which, in fact, began with the first computers, was the development of very different applications of computing, thus invoking the idea of universality in computation. Picture processing—computer graphics—came rapidly onto the agenda, and we will complete our chapter with an illustration of how the pioneers realized these possibilities from the very beginning.

We have already hinted that the first development of computers in the immediate aftermath of World War II was largely for military purposes. The pioneers were strongly divided about the use of such machines in peacetime, with von Neumann, for example, seeing his own IAS computer project as being an extension of his war work on the Manhattan project, which involved constructing and testing the atomic and thence hydrogen bomb. Before the end of the war, the US Navy commissioned a group at the Servomechanism Laboratory at the MIT to design a digital computer along the lines of the ENIAC primarily as basis for a missile defense network that was already in play when the war ended. Jay Forrester, who we will come across much later in a rather different context as one of the key scientists building city models, took charge of the project and, for three years, developed a powerful computational facility called "Whirlwind," which contained several innovations involving magnetic core memory and stored program facilities. The machine became the basis of the US SAGE missile system, which dominated US global defenses in the 1950s and was still largely operational in the 1960s.

Forrester, in fact, made many innovations, but the system required a basic kind of graphics, so that those using it to mobilize missiles and to position them globally could sense their geography. In the late 1940s, several pioneers used oscilloscopes to monitor the operation of their machines, but it soon became apparent that these could be used as rudimentary graphics displays. Such idiosyncratic uses were so appealing that in December 1951, Forrester appeared on Ed Morrow's show See It Now, a Saturday evening current affairs TV program on CBS, and demonstrated how one could show the launch of a rocket on such a scope.[39] He finished his part of the show by demonstrating how the same computer, the Whirlwind, could be programmed to play Santa Claus, thus showing how the machine could produce music: a portent for the future and one that would become central to the way we interact with computers now and perhaps forever. This, then, is Turing's legacy, which marks the path to a universal machine of which the computable city is an essential part.

4 THE PC REVOLUTION

I think it's fair to say that personal computers have become the most empowering
tool we've ever created. They're tools of communication, they're tools of creativ-
ity, and they can be shaped by their user.
—Bill Gates (2004)

THE COMPUTER USER

If as a user you were to be transported back to the high point of the mainframe computer era sometime in the mid-1960s, you would never see a
computer unless you were one of the priesthood who attended the operation of the machine. As mainframes got larger in both size and power, the
laboratories in which they were spawned became closed environments, and
once production of these beasts began in earnest by companies such as IBM,
increasingly they were housed in cloistered, air-conditioned rooms that no
one other than the high priests were allowed to enter. Indeed, there was a
distinct dress code. One often had to wear a protective hat, change outdoor
shoes into slippers, and don the white coat of a laboratory technician so
that any dust that might pervade the environment would be reduced to a
minimum. At least, that was the theory.

The hierarchical organization, in fact, went further. The high priests who
managed the system and interacted with the scientist-engineers involved
in the design of the machine itself were part of an army of gatekeepers
who approved access to the machine.[1] Gaining entry was no mean feat. You

needed to make a case for solving a problem that could only be addressed by the computer. Many routine operations such as processing the payroll took pride of place in commercial computing, although in less structured environments, particularly in universities and research centers, you could get access, but only if you were able to write your own programs and establish that you had enough expertise to run them or rather for the high priests to actually run them. If you were interested in processing graphics or text rather than numbers, and thus more focused on qualitative problems, there was no place for you yet in using computers, for this would only come as the machines became more personal and accessible. And if you had no expertise in being able to articulate your numerical problem in computable terms, it was most unlikely that you would ever find anyone who could let you come close to the machines or even the media used to construct the programs and the data.

Computation that took place in these machine environments was quite strictly separated from the places where users prepared their programs. These were pretty arcane settings too in that they involved not only writing the program but also then translating it into the appropriate media so it could be fed into the machine. Data was always separated from program, and this required punched paper tape or cards to be prepared for each, boxed up, taken physically to the computer center, and delivered to the reception staff, who would spirit the submitted job away to the machine. The job would then be placed in a queue to be run by the operators in the prescribed order. Additional media pertaining to data that might be stored, for example, on magnetic tape needed to be identified so that the operators could load such tapes. The program was first loaded, and once it began, it would call for data input from the cards and/or tapes, which all had to be assembled in the correct order for the program to execute. The output was usually in the form of reams of tele-printer paper, while sometimes if the user required intermediate data to be used in a subsequent job, this was generated as more cards, punched tape, or magnetic tape.[2]

Manchester was still one of the main centers of computing in Britain nearly twenty years after the first Manchester Baby was built by Fred Williams and Tom Kilburn under the inspiration of Max Newman. In 1967, some thirty years after Turing wrote his famous paper, the Manchester machine had evolved to the point where its manufacturers, Ferranti and Plessey, were producing versions for the commercial world.[3] The machine in the university was still a prototype. It was arguably the first supercomputer and reportedly the largest anywhere in the world in terms of memory and speed. Yet, it still involved data inputs that were produced quite separately from the computation itself. Back in those days, you would have found me producing jobs for this machine where my data and programs were punched out on seven-track (hole) paper tape using a device called a Flexowriter. The length of the tape was voluminous, for it contained many lines of code and data. If you made a mistake, then the tape would have to be spliced and reconnected using special sticky tape, carefully positioned so that it did not cover any holes in the adjacent portions of the tape. The tape was coiled, put in boxes, and then taken to the computer center for submission to the operators. The turnaround time when the job had been completed was usually one day. The discipline all this imposed on the user was sufficiently rigorous to ensure that considerable care was exercised at every stage of the preparation, lest something simple and obvious were to go wrong that led to the program failing, thus wasting a good twenty-four hours. If the spliced tape were joined, and one of the holes in the tape was even touched by the sticky tape, the job would fail with a parity error, only for this to be fixed in a few seconds and the job resubmitted to be run again. However, this was never, ever on a cycle (for me at least) of less than twenty-four hours.

I do not have any pictures of what it was like to prepare programs in this way, but I remember many times being surrounded by punched tape, with yards and yards of the stuff filling most of the room around me while searching for a line of code where a typing error had been made and a new piece of tape needed to be inserted.[4] Very quickly, one learned to read the code you had typed on the punched tape from the holes themselves, just as those close to many kinds of electronic coding from Morse onward acquired a facility to read binary or whatever the code might be. As the data and programming were separated almost entirely from the computer itself, early users had no experience whatsoever of operating systems.[5] Moreover, it was most unlikely that a user would ever run a job on anything other than a single dedicated machine, for most machines were tailored in some way to the local environment, the university, the commercial business, or whatever, and this meant that most computing was not portable in any sense. In fact, if the computer went down, as the Manchester Atlas did in early 1969 when part of the machine caught fire, the stream of jobs was diverted to the second Atlas at Harwell (one of the UK's Atomic Energy Laboratories), but there was little confidence that the Manchester jobs would run, and most of them did not. In fact, when the Manchester machine went down, it was said that half the computing power of the UK was lost, as indeed was the case in question but only for a matter of days.[6] It was, in some respects, not only the high point of the mainframe era but also the high point of the British contribution to computing. The zenith had been reached, but from then on, the industry in the UK rapidly went downhill, but that too is another story, too dismal, in fact, to recount here!

There is one feature of this early experience with mainframe computing that anticipates the idea that the entire city is fast becoming automated and hence computable. In some senses, these machines were far removed from most working and social environments. There was no sense of networking other than users physically visiting the machine to submit jobs, and there was no notion that different machines might be connected to one another to provide a much more extensible computing infrastructure. But the nature of the jobs themselves, to an extent, held the keys to the future and to the emergence of the smart city. You may well ask what I was doing working with computers back in the day. Well, almost as soon as computers emerged, a vast array of scientific problems that spanned the social as well as the physical and natural sciences appeared to be computable in some measure. In the early 1950s, architects, planners, and engineers began to explore how cities could be represented in mathematical terms, as flow systems involving traffic and land use, and very quickly rudimentary computer models followed. This was all fairly abstract, but there were strong policy implications in that such models were designed to produce conditional predictions forming the basis of a rudimentary science of cities.

These were hardly smart cities, for there was no direct embedding of the machines or their data into the urban environment, but the models enabled plans to be fashioned, which in turn could be implemented to make cities more efficient and more equitable—to make them smarter in the traditional sense of the word. In fact, such model applications, particularly those involved in predicting traffic, became widespread, with most larger municipalities using computers to make such predictions before they agreed any traffic scheme. As part of our continuing exploration of the smart city in the rest of this book, we will spend an entire chapter talking about such developments when we come to deal with the nature of computation in cities. But for the time being, we will simply note that right from the beginning, the idea that computers could be used to think about cities was on the agenda. Indeed, computer applications from the onset covered a much broader spectrum of uses from economic forecasting to business applications, from governance across many scales to welfare provision, suggesting a cornucopia of uses that we will at least hint at in this chapter and the next.[7]

It would be quite wrong to give the impression that the world was then composed of isolated large remote computers, for there was a hierarchy of such machines, ranging from electromechanical calculating devices to small specialist machines and thence to the largest mainframes. In fact, during this period, all kinds of experimentation was taking place in developments around the mainframe, while chip technology was advancing in leaps and bounds.[8] Mainframe computers were not always shielded from their users. Where computing was regarded as being more central to mainstream scientific research in physics and engineering, which had always been dominated by extensive calculation, faster and more immediate access was required. By the late 1960s, several new systems evolved, which were more personal in that their users could interact directly with the machine, usually through systems where they could enter their own data and programs directly. These ultimately came to be referred to as minicomputers, and they formed the vanguard of the march to miniaturization and the personal computer, the PC.

THE LONG MARCH TO THE PC

In the early 1960s, John Kemeny, who had been Einstein's research assistant at Princeton and had worked with von Neumann on the IAS machine, developed a system of computing and programming that was designed to enable mathematics students to learn how to use computers primarily to solve mathematical problems quickly and easily.[9] The language he and his colleagues at Dartmouth College invented was called "BASIC," which was a much more user-friendly version of the various high-level languages of those days, mainly FORTRAN and Algol.[10] The real innovation, however, was in the interactive system that he introduced, which was called the "Dartmouth Time-Sharing System,"[11] where users could run their jobs from terminals—consoles or teletypes—that enabled them to interact with the computer as if it were their sole machine. Many simple computations that, in fact, were well beyond the ability of any manual calculation could thus be run simultaneously, and the first computer classrooms thus emerged. Not only was Kemeny's BASIC language innovative in and of itself and many versions of it are still being used, but his time-sharing system was also one of the first to introduce the notion that a network of users could be tied electronically and virtually into a machine while sharing its use. The arrays of terminals and the network of users thus constituted the first kind of embedding of computer infrastructure into the wider environment. Users still had to find a terminal and log on, and this was usually done in the confines of an organized computer facility, but the notion of computation spreading out—becoming more pervasive—was very much on the cards by the end of the 1960s, with Kemeny's system spurring on the development of ever-smaller and more accessible machines.

It is hard to paint a picture of the robust innovation that has dominated the computer industry ever since its inception, but even as early as the late 1950s, smaller machines were being developed. In particular, the Lincoln Laboratory at the MIT, where, in the last chapter, we came across Jay Forrester's Whirlwind computer for the SAGE missile system, was a hotbed of development. Out of this same milieu came the development of the PDP computer, which was arguably the first minicomputer.[12] The PDP was, in fact, a pretty much scaled-down version of the first computers dating from the ENIAC, the UNIVAC, and the EDVAC, but it was not really regarded as a fully-fledged computer, as reflected in its acronym PDP, which meant "Programmed Data Processor." These machines were largely used by physical and mathematical scientists for mainstream computing problems, and their great advantage was their accessibility to the user, who still had to have some basic expertise in their operation but only enough to enable the machine to be programmed.

Kemeny's BASIC and his time-sharing system, although by no means the only such system to have emerged by the early 1970s, dovetailed nicely with the development of these new kinds of machines called "minicomputers." A long line of them came into production, not only the PDP, which was manufactured by the Digital Equipment Corporation, but also machines built by Data General, Hewlett-Packard, and others.[13] The focus from then on was downscaling the mainframe, which was made possible by the remarkable miniaturization that was taking place due to Moore's law and the gradual realization that computation was also beginning to merge with ideas about disseminating processing power to networks of users. Let us stick with the mainframe for now, for the user's experience of computing was continually changing during this period. In particular, as soon as minicomputers were introduced, the user had to learn something about how programs needed to be run and manipulated on the machine itself, and this invoked the need to know much more about how operating systems worked.

To explore this, our next stop is Canada in the mid-1970s at the University of Waterloo, where there are several stories and events defining this headlong progression into the domain of the personal computer that by the end of that decade was about to launch itself on an unsuspecting world. The University of Waterloo is best known for the Blackberry, the personal device that became the darling of the executive for secure email before it was knocked out of the game by the smartphone that took all its functions and more. The company that manufactured the Blackberry, Research in Motion,[14] was a spin-off from the university in the early 1980s, and it was associated with a remarkable engineering program informed by state of the art mainframe computing, a large part originally funded by IBM Canada. Waterloo was a new university set up in 1957, with its computer facility one of its key assets. This was as much to do with the mission of the university to make its engineering students the best in the world as well as its pioneering of more basic research. In 1967, it sported an IBM System/360 Model 75, the largest computer in Canada, which was located in a fairly brutal concrete modernist structure at whose heart the machine could be viewed in operation. The building was often referred to as a "cathedral to the computer," and around the priesthood who attended the machine, there were many smaller but related initiatives.

In one of the engineering buildings, a local branch of the main computer center was clustered around a cafeteria system. Here, undergraduates (and anyone else for that matter) would assemble their programs and data on punched cards. They would then join the queue to load their own cards into the hoppers, proceed to collect these once they had been read into the machine, and then, at walking pace, progress to the area where they would tear their own outputs from the printer, the job having been run and finished in a matter of minutes from the time they first joined the queue. Twenty-four hours at Manchester was reduced to two minutes at Waterloo, but of course, the computation itself was limited, and the cafeteria system was mainly useful for quite modest jobs, relevant to the teaching of computing. Nevertheless, research computing was interwoven with these other more educational uses, for the fact that one had to run a job within very tight limits set by time and size if one wanted almost instant output imposed great discipline on the task in hand.

Waterloo did much more to make computing accessible to its users. A version of FORTRAN was introduced called "WATFOR" in the 1960s. The Waterloo terminal emulator system WIDJET was being developed alongside the cafeteria system so that users could run their jobs directly on the mainframe. And there were several PDP machines scattered among various research groups. If you were to visit the engineering department in those days, then this would really look like a building whose infrastructure was built around computing. By the early 1980s, a lot of this had been replaced by PCs, while at the same time, office functions were migrating onto those same machines, and a degree of invisibility was creeping back into the infrastructure, but in 1975, the environment was still dominated by highly visible small and large machines.[15] In fact, as time went on and the PC era gathered pace, much of this overt computing spread onto the desktop, and there was less evidence of large rooms given over to the machines. Of course, many university and research environments had similar installations, notwithstanding the eminence of Waterloo. And with hindsight, although he had little to do with this kind of computing per se apart from passing the "cathedral to the computer" every day on his way to the office, one of the university's most famous sons, Bill Tutte, whom we noted in chapter 3 for his massive contributions at Bletchley Park (which were still classified, hence unknown), was making important contributions to graph theory and eventually to computer graphics. As we will see if it is not apparent already, in the development of new technologies, it has always been a small world.

Computers began to scale down as soon as they were invented, but this was largely a silent process, particularly because mainframes and the ways in which we interacted with them showed little sign of changing. We continued to punch cards and write programs offline, take our jobs to the computer center, and wait painfully for the output. But since the end of World War II and during the long economic boom that followed it but which was coming to an end in the late 1960s, miniaturization of computer components was proceeding apace. The integrated processor was invented in 1957, while the idea that chips could be programmable—you could use a chip to execute a program as well as store data in memory—was the breakthrough that came in 1971 at Intel where the microprocessor or computer on a chip was first developed.[16] Minicomputers were also making their presence felt, while local time-sharing systems and rudimentary networks based largely on the hub (computers) and spoke (users) model were slowly abolishing the archaic practice of the user having to take programs and data physically to the machine itself. It was into this context that the microcomputer—which was quickly relabeled the "personal computer" or "PC" as much by IBM as anyone else—arrived with a vengeance in the late 1970s.[17]

It was the experience of the PC rather than any other computational technology that impressed on all of us Turing's message that the computer was a universal machine. The mainframe was widely regarded by the public as a large and expensive piece of office equipment that appeared to have the dubious value of existing purely to indulge a rather elitist big science. It was somehow linked to typewriters and calculators, but the notion that it might do everything was widely regarded as far-fetched, especially as "everything" was only defined if you could express your problem in numbers. As most people assumed, this was science at its most obtuse, and as soon as one began to explain that many things could actually be reduced to zeroes and ones, the whole project appeared fanciful. Many of us had firsthand experience of this conundrum. In the early 1980s, from my own experience of building computer models of cities, we were replacing our mainframe machine with a series of VAXs. These were minicomputers originating from the first PDP machines manufactured by the fast-growing Digital Equipment Corporation.[18] They gave us much closer access to the machine and of their operating systems to which we were linked through visual display units—screens that enabled us to type program commands, input numerical data, store these as files, and access simple libraries of routines that could be combined into quite powerful programs.

In parallel, the PC was being rapidly introduced as a desktop machine for data and word processing. Those who used computers in our group were entranced by both developments, but this created a massive dilemma for our office staff. The suggestion that these PCs could also be used for word processing, thus replacing electric typewriters, which were the main tools used by those in administrative positions, was greeted with horror. This led to a clash of science with bureaucracy in that the administrative staff simply did not believe that one could create letters and print them directly from machines that they observed the academics using purely for science. This gap in comprehension was taken to an extreme when PCs arrived in the office that had been blocked for doing anything other than word processing, their disk drives being covered up from adding new software to the preinstalled word programming packages. Imagine the scene, which happened quite literally, when the nerds prized off the blocks across the drives to run their own scientific software. This, of course, was done after hours, once the administrators had left the office, again impressing on all involved the great gap between science and its administration,[19] reinforcing, of course, the universality of the same machines for executing dramatically different tasks. The lack of understanding about what computers could and could not do reached a critical point at this juncture, as computers began to spread out into all walks of life, and this was tantamount to the story we noted in chapter 2 of the woman who wished to send money to her son using the telegraph, worrying about whether (and where) the loose change would get lost en route! It was early evidence that computers were being fast embedded into the environment around us for uses that were far away from the science that had spawned them, thus demonstrating yet another classic manifestation of the universal machine.

SCALING DOWN: COMPUTERS BECOME PERSONAL

Most innovations in technology cannot be attributed to a single person or organization. We have already seen this in the development of the telegraph, which was invented simultaneously by Samuel Morse in the US and William Cook and Charles Wheatstone in the UK, but there were unnamed others at the time who were working on similar kinds of device. In terms of the digital computer itself, Zuse in Germany, Atanasoff and Bell in Iowa, Aikten at Harvard, Mauchly and Eckardt at Penn in the US, and Flowers at Bletchley Park in Britain invented the machine more or less at the same time without much knowledge of one another's work. The personal computer was not really much different. There are almost too many original inventions to recount that led to the PC, and as this history is continually revisited, more and more original and idiosyncratic contributions are being found, even now. Moreover, those already in the game never appear to be those who make these innovations and extensions. Most of the companies that were designing, manufacturing, and marketing the mainframe did not consider there was any market for a personal computer, to the point where many went on record as stating that such a diffusion of this technology would never happen. As we noted in the last chapter, this was true even of those developing minicomputers such as the Digital Equipment Corporation, despite the fact that their own line of PDP minicomputers had almost reached the point in the early 1970s where they were portable enough to be considered as personal computers.[20] In short, new technology often mostly disrupts those who already own the same kinds of technology, with the existing guardians of that technology seldom nimble enough to move on to the next wave of innovation, despite them being only too aware that this is possible and often necessary. The origins of the personal computer represent a classic case study in such disruption. It is evidence of the great Austrian economist Joseph Schumpeter's idea that existing technologies that have not yet reached their sell-by date are creatively destroyed to make way for new ones.[21] In fact, to some extent, it is evidence that new technologies do not subsume old technologies, but rather that the owners of the old simply disappear and are swamped by the development of the new.

For the first time, the PC also represented a very clear and obvious statement that computers could be used both for science and for controlling the very media that was embodied in that science, for putting computers into situations where control was at the forefront. This was a quite close to the idea that one could use a computer to simulate a computer, which as we pointed out in the previous chapter was and continues to be Turing's legacy. Indeed, although the microprocessor was a computer on a chip that could be programmed in its own right, many of the initial software contributions to getting the PC up and running were designed by simulating the architecture of the PC on a bigger computer and developing programs for the simulated version rather than the PC itself. The earliest history of personal computing between 1970 and 1975 provides remarkable evidence that computers are both simulators of themselves as well as mechanisms to make sense and control the world in which they are embedded.[22]

The first commercially available PC was the Altair, which reached the market in January 1975 as a kit that needed software. Two supersmart American whiz kids wrote a version of the BASIC language to enable users to store their programs on the machine and run them without resorting to the arcane practice of manipulating multiple switches. The version that was packaged with the machine was only written after the machine was announced, when the two guys who wrote the software became aware that such a machine needed software, making it more than simply a hobbyist venture: they had never seen it, nor did they until their program was finished. They produced the software on a minicomputer, simulating the instruction set that was used to control the Intel chip that was used as the microprocessor in the personal machine. It is quite remarkable that reportedly the program ran first time: little wonder that the two went on to found one of the world's biggest companies, Microsoft, which has dominated software for the PC ever since.[23] Indeed, one of them is one of the richest people in the world. We will say more about Bill Gates and his colleague the late Paul Allen as we continue to fill in the backstory to the emergence of the smart city.

As noted, the first commercially available PC was the Altair, which was the product of a challenge by the magazine *Personal Electronics* to the hobbyist community in the US to build such a machine, given the latent demand for such a device in the early 1970s. When the Altair came to the market in January 1975 through the magazine, the demand for software was immediate and unassailable.[24] Although a dedicated hobbyist could program the Intel 8080 chip directly by manipulating the instruction set for the microprocessor, to extend the reach of this very basic computer to the wider circle of hobbyists and to embrace even a fringe of business professionals who had hobbyist sympathies, many peripherals were needed. These pertained to input and outputs as well as graphics displays, keyboards, and, of course, basic software. Indeed, the very term "software" did not become widespread until the computer became personal, in that professional users of mainframes previously relied on specialist computer center staff to enable them to use software of various sorts, without those users knowing very much about the software in question. The PC changed all this, and a new vocabulary based on software quickly evolved.

The first PCs were thus introduced by many tiny, cash-poor companies, largely in and around the San Francisco Bay area, more specifically in Silicon Valley, with the entire movement very strongly influenced by the somewhat laid-back style of the counterculture from the 1960s. The Altair, in fact, was produced further afield in New Mexico, but most of the machines built by a succession of start-ups such as North Star, IMSAI, Crememco, Sphere, Processor Technology, and so on grew so fast that they quickly bit the dust, but not before a vibrant momentum for the PC had become established. Many innovations spun off from these initiatives, ranging from the introduction of retail stores selling computers to magazines proclaiming their applications as well to an even larger industry of peripheral manufacturers. But out of this great morass of devices in the mid-1970s came one computer that stood above all of them, and that was the one literally constructed in the apocryphal garage by Steve Jobs and Stephen Wozniak. The Apple II achieved instant fame through the brilliance of its design due to Wozniak and the persistence, charisma, and vision of Jobs.[25] By 1980, Apple was on the way to becoming a household name, but it, like many of its competitors, was still regarded as a company producing toys by the more serious peddlers of mainframes and minis. Essentially though, by this time, such computers came with floppy disk drives for storage, input and output, rudimentary dot-matrix printers, and visual display units essential for graphics associated initially with games that drove the industry forward before more business-orientated applications software developed.

THE SOFTWARE REVOLUTION AND BEYOND

What moved these first computers into the mainstream was very definitely software applications. The key differences established between remote mainframes, somewhat smaller minis but still remote from the general user, and these new personal machines involved the convergence of software with the hardware. Not only were individual users immediately exposed to hands-on contact with the hardware, but also, for the first time, the distinction between the operating system of the computer (usually shortened to some acronym with "OS" in the title) and individual applications, be they games, word processing, spreadsheets, or drawing packages, was thrust into the hands of the user. In fact, what quickly emerged was a hierarchy of skills from those closest to the machine themselves in terms of what the hardware could do to those who became expert in the OS to those focused solely on applications in some higher-level language such as BASIC. The OS depended to a very large extent on the chip that was used in these machines, with Apple using a Motorola microprocessor, and others using chips from the Intel family, and this tended to divide personal computers into two groups. A further distinction also emerged between machines that were built around an open architecture where third-party producers could manufacture add-ons rather easily and those machines such as Apple that closed their architecture from letting others develop add-ons in terms of their hardware. Software too was essential, and quite quickly some killer applications that were not available for mainframe or minicomputers emerged. We will discuss games later, but it was spreadsheets, particularly the product Visicalc, a flexible tabular processing sheet with movable windows useful for any kind of numerical calculation with applications to business as well as science, that drove the revolution forward.[26] For the first time, in the early 1980s, the PC looked as though it might become a serious machine for business, perhaps taking over some of the functions of larger, more expensive, more centralized, and, of course, more remote main frame machines.

This evolution was dramatically accelerated by the biggest computer company in the world, IBM, deciding to manufacture its own personal computer, which was first introduced in 1981. The emergent industry lived in fear of "Big Blue," as it was known, creeping into their terrain, for IBM had been so all powerful and dominant in computing from the beginning that many thought that personal computing may have a short shelf life if it entered the field. The company had so much money and clout that both those who manufactured software and hardware for PCs felt that sheer financial muscle would blow them out of the way. In fact, this did not happen. An ingenious mixture of Bill Gates's Microsoft bidding to build the operating system for the machine, a playoff against Gary Kindall's CP/M, the only other operating system that had any chance with IBM, and the fact that IBM, unlike its past decisions, wanted to outsource many functions to third parties led to a rather open system that did in some senses take the world by storm in 1981 when it was introduced. There was nothing special about the IBM PC, but it was well tested and robust, and it provided what users wanted for a low-cost home and business machine with a brand name.[27]

The scheming and counter-scheming that went on in those early days may not be so different from most other situations of rapid growth. There was excitement at the new challenges of bringing computers to the masses. Moreover, the prospect of earning obscene amounts of money by selling what were still regarded as complex and little-understood devices fast turning into consumer products quickly dominated the industry. The IBM PC became the ultimate seal of authority on an industry that was quickly organizing itself into much more professional activities and serious applications. IBM, in fact, departed from its entire previous practice in developing computers by allowing an open architecture that others could exploit in the form of hardware add-ons and bespoke software. Even cloning the entire machine was possible and this led to many new manufacturers entering the market, building PCs that emulated all that the IBM PC could do. Bill Gates managed in his inimitable way to persuade IBM to let his company, Microsoft, develop the operating system (OS) for the PC. He called this "MSDOS," which was referred to by IBM as "PC DOS" but which quickly assumed the generic name DOS. This effectively saw both the end of the alternate OS in the PC market, which was CP/M, and the rise of Microsoft, which was to become the world's dominant software vendor by the late 1990s.[28]

Software also propelled the industry to shed its hobbyist mantle and to move quickly to more serious applications. The PC led to new inventions in computation itself in the form of software in business, science, games, and recreation. Various mainframe and minicomputer companies only realized when IBM released their machine that an enormous new market was opening up, and machines that various vendors such as DEC, Hewlett Packard, Xerox, and similar companies were slowly working on arrived too late to save the day. IBM did, in fact, succeed in driving out many small and often idiosyncratic manufacturers and vendors, and by the end of the 1980s, they, together with Apple and Microsoft, dominated the market. In fact, the environment continued to be volatile. Apple had been in deep trouble in the mid-1980s, only to revive itself through its Macintosh machines, while IBM itself lost its luster in the PC market as the attention turned to workstations and other forms of computing. One of these was a further scaling down to laptops. The Japanese made important innovations in this domain just as they were continuing to do in consumer electronics where they dominated the market in personal handheld music players, cameras, small TVs, and so on. The scene was ripe for the arrival of highly personalized devices, but this was still some years in the future, and before we explore how the most miniaturized machines arrived, we must say a little about the revolution in inputs and particularly outputs that the PC occasioned.

Interacting with a PC was very different from engaging with a mainframe. Numerical data was usually entered using a keyboard or from data stored on a floppy disk. Outputs were on dot-matrix, laser, or other forms of printer, but the predominant focus was on graphic operations, which were displayed using some form of screen. Prior to the invention of the PC, screens had been in use for showing code and for rudimentary graphics on what were called "visual display units" (VDUs), but these were restricted to very specialized workstations that displayed pictures by tracing an electron beam over a display tube. In the next section, we will say a little about this approach to graphics, but very quickly it was recognized that the PC was an ideal medium for drawing and playing games—hence the development of an area of memory in the machine that could store any pictures that were necessary for the activities involved. In short, if pictures could be stored in memory, they could be retrieved and displayed on a screen, and as soon as the early computers arrived such as the Altair, the Apple II, and then the IBM PC, a new focus on display became inevitable.[29]

In fact, there was a personal computer of sorts as early as 1972, but it was developed behind closed doors and was never considered a marketable product until Apple began to promote the visual desktop in its Macintosh machine in the early 1980s. The Xerox Corporation, known for its copiers, was never really serious about computers, but it developed a research laboratory in Silicon Valley on the back of a buyout from Scientific Data Systems in 1969. This was its Palo Alto Research Centre (abbreviated to Xerox PARC), and it defined its mission as developing new computer technologies without any obvious market in mind, with an exception perhaps being niche governmental agencies. In this sense, it was much closer to a university research laboratory. To say it lacked direction would be wrong, but it departed radically from the goals of its parent corporation and indeed the goals of most of the rapidly developing PC industry. Xerox PARC made many remarkable innovations, which have found their way into mainstream computing today.[30] It is no exaggeration to say that the look and feel of contemporary desktops and smartphones can be traced back to the research groups at PARC, which, from about 1971 to 1976, developed the original "windows" system, accessible through the mouse, as well as the local area network (LAN) that tied individual computers together through servers containing common facilities, as well as the use of printers and remote disks for storage and printing. Bob Metcalfe who invented the Ethernet at Xerox PARC—one of the first LANs—also coined Metcalfe's law that suggests that the power of a network in terms of shared resources increases as the square of the number of devices connected to it. We will discuss this again a little later, for our next chapter will focus on networks, but besides windows, the mouse, and networking coming out of PARC, new computer languages closer to real English and AI also came such as that called Small-Talk developed by Alan Kay.

Xerox, of course, decided not to market any of this technology. Intellectually, much of it belonged to Douglas Engelbart, who was a visionary computer scientist at the Stanford Research Institute in the 1960s, and Ted Nelson, who argued vehemently that computers should be designed in such a way that user motivations should condition the nature of software, not the other way around. Engelbart had invented the first mouse and demonstrated how windows could open up the user experience, while at the same time illustrating how information could be linked using hypertext. Nelson put a spin on all of this thinking laterally about the user experience, but none of these visionaries really developed these ideas to the point where they might even have been considered marketable and generalizable.[31] However, the Xerox PARC facility was hardly a secret, for many in the computer industry knew about it. Yet, it was not until a contemporary introduced Steve Jobs to the facility and its innovations that the prospects for these kinds of tools being developed for a personal computer really blossomed. Jobs was blown away, overwhelmed, vowing to use these technologies in his newly envisaged Macintosh computer. Indeed, he did, although the gestation still took more than five years since his first glimpse of these ideas at Xerox PARC for them to be absorbed into the products of his company Apple.

There is one last development from the 1980s that influenced the infrastructure for personal computing, and that was the development of the workstation. If you think of computation as evolving both from the top down and the bottom up in parallel and simultaneously, then the minicomputer was becoming smaller but more powerful from the top down, while the PC was doing the same from the bottom up. The Macintosh-PARC windows and mouse system was rapidly ported to compute-intensive PCs and ever-smaller minis. By the mid-1980s, these workstations had emerged as powerful stand-alone machines, exploiting new network technologies and enabling technically proficient users, not the public at large, to generate highly creative science with a strong graphical orientation. In some senses, workstations, particularly those developed by SUN (Stanford University Network) and Silicon Graphics, represented the state of the art in all computing by 1990.[32] The particular focus, however, was that the predominant operating system for these computers was UNIX, the system developed in the 1970s at Bell Labs and widely regarded as a computer scientist's ideal way of interacting with computers. UNIX to some extent is unforgiving. It is powerful and closer to the architecture of the machine than DOS, and thus the workstation divided serious scientific users from those with more personal interests, although the PC was fast catching up. In fact, although it is a little beyond our present history, PCs would eventually reach the point where workstations simply disappeared, converging into more powerful clients and their servers by the year 2000. Moreover, by then, all of these were networked. Indeed, the message of our next chapter is reflected in SUN's slogan that "the computer is the network." This cliché reflected the convergence of computers and telecommunications, as well as blurring the lines between different types of machine. Mainframes were merging into minis, minis into micros, and the idea of a distributed client–server architecture was on the horizon. But before we reach this point in our next chapter, we should take one last look at what the PC revolution brought to computing, and that was one of the most significant. More than anything else, it encapsulated Turing's idea of the universal machine, and it is critical to one of the major perspectives we use in thinking about cities. This was based on computable pictures not numbers—on graphics—and it is central to the idea of the smart city, for we still conceive of cities as pictures, maps, models, albeit computable media that traditionally has always been soft rather than hard.

MORE UNIVERSALITY: COMPUTING PICTURES

Most of us still tend to think of computation in its narrowest sense as manipulating numbers, although this is now fading somewhat. But it would have been quite clear if you had been present at the construction of the earliest machines of any vintage, be they mainframes, minis, or PCs, that computation was being reflected in other ways. You would have been struck by the flashing lights from the way electricity was deployed in computation. This, in itself, can be graphically entrancing, and it is not surprising that as soon as these machines emerged, their inventors speculated that they could be programmed to produce pictures. In short, the various lights that could be switched on and off at will—programmed, that is—could be designed and patterned to produce intriguing graphics. Forrester's demonstration in 1951 of the launch of a rocket on Ed Morrow's show *See It Now*, which we noted in the last chapter, is an early example.[33] Earlier still, Maurice Wilkes, the originator of the EDSAC at Cambridge, demonstrated at his laboratory's Christmas party in 1949 how his machine could be made to play Jingle Bells as well as simulating the flashing lights on a skeletal Christmas tree etched out by the electron beam acting on the machine's oscilloscope. Fast-forward thirty-five years to William Gibson's images of cyberspace in his book *Neuromancer*, where he defined an emergent computerized future as "Unthinkable complexity. Lines of light ranged in the nonspace of the mind, clusters and constellations of data. Like city lights, receding."[34] The movie *The Matrix* and similar films say it all. However, we rarely see the constellations of lights associated with processing today, as most of our computer's electronics are hidden from view, encapsulated in beautiful boxes that give the impression of order and simplicity where, in fact, there may be far less or even none.

To compute anything other than numbers, the data ultimately has to be converted into numbers and then back again to the media of choice. If the input involves pictures, these have to be represented either geometrically by numbers, scanned in some way, digitized into a set of vectors, or represented as a grid of picture elements (pixels), which is usually called a raster. Outputs can then be displayed on a screen or printed to a line or laser printer. Usually, the picture in numeric form has to be stored somewhere in the machine or at least output in such a way that it is easily input again. What the PC brought to every user was the ability to store pictures in its memory. A special area of the machine called a "buffer," sometimes a "screen buffer" if directly associated with the display, was reserved for such pictures, and although buffers were in existence before the PC for the mini and even occasionally the mainframe, they were too cumbersome to even consider for anything other than an occasional demonstration of computer graphics.

In fact, prior to the PC, most graphics output was either plotted by automated instruments akin to traditional drawing tools instructed to draw all the lines computed and stored as points and vectors, often offline, or plotted on a line printer where the eighty-character set dominated how the density and size of the picture output could be achieved. In fact, the line printer was often deployed to overprint characters thereby simulating the density of development in a thematic map. One of the earliest computer programs to achieve this was produced in Russia by Nikolay Konstantinov in 1968 who generated Kittie, an animated cat.[35] The line printer was used to print a series of images of the cat, thus forming an animation that was assembled as frames (sheets of computer printout) and then videoed. At about the same time, in 1967, the first computer animation of a town was simulated by Alan Schmidt using the same technique: printing a sheet from the line printer of the built-up area of the town of East Lansing, MI, where each sheet represented the built-up area of the town at five-yearly intervals.[36] When strung in temporal sequence, these provided key frames for the video that was subsequently made. In fact, the technique was then widely popularized by the Harvard Laboratory for Computer Graphics, where it was used it as the basis for their symbolic/synergistic mapping package by the name of SYMAP, which was very widely used from the early 1970s until PCs stole the show a decade or so later.[37]

As visual display units became faster and less turgid to operate, line printer–style maps were printed on screen at the command line. In fact, storage tubes were also used from the mid-1950s to print traffic-flow maps, the first known example being traffic flow centered on downtown Chicago developed for the Chicago Area Transportation Study Commission[38] in 1959. But PCs displaced many of these types of graphic, for as memory got cheaper during the 1980s and machine memories and ancillary storage such as disk drives got bigger, the PC screen became ever more visual. In fact, by 1990, most PCs were converging with workstations whose traditional frame buffers reserved for pictures began to merge with the screen memory itself, which was increasingly being used for graphical interaction. Thus, two types of graphics emerged: screens for pictures of increasing resolution (size and detail) and screens dedicated to graphical user interfaces (GUIs) based on the PARC-Macintosh model. Switching between the two thus became commonplace, but this practice only lasted until the windows interface to all PCs and workstations became the norm. Then, one could overlap as many pictures as one liked as each picture had its own individual window, its own virtual screen.

Maps first began to be automated in terms of their production in the 1960s but it was not until the workstation became more widely used for rudimentary computer graphics that the distinction was made between maps as geometric objects, sometimes called "vector maps" composed of points, lines, and polygons, and the much simpler raster or grid-based maps, where each distinct area of the map was the same based on regular geometric tessellations such as a grid, nested hexagons, and so on. Essentially, it was satellite remote sensing that drove the map toward its raster representation, which was a more neutral way of representing geographic information. Vector and raster graphics, in fact, formed the core of the first representations of geographic space, encoded in geographic information systems (GIS), which had become popular on workstations by the late 1980s, embodying special-purpose software that enabled professional and scientific users to construct all manner of maps and evolve various functional analytics associated with their manipulation. These developments and their subsequent translation to web-based representations led to the widespread development of maps for navigation, and these have become essential today for location-based services. In this sense, as we will see, they are a key component in the evolution of the smart city.

Representing the third dimension of geographical space from buildings to landscapes also emerged during the PC era. In fact, in the late 1970s, the first wire-frame representations of buildings were developed, the most well-known being the Skidmore, Owings & Merrill visualization of downtown Chicago and other US cities.[39] Although these kinds of visualizations were one-off, developed on workstations of various kinds, as soon as the PC emerged, 3D representations of buildings became one of the first applications through packages such as AutoCAD created by the parent company Autodesk,[40] which burst onto the scene in 1981. The 1980s saw massive development in creating digital pictures and extended scenes in 3D, with animation following closely on the tail of such representations. We will say more about this below, but it is worth noting that workstations then began to incorporate the basic functions of creating 2D and 3D pictures within their chipset, with companies such as SUN and Silicon Graphics leading the way with their development of what they called "geometry engines." As we will see a little later, these workstation companies also bit the dust as the PC became ever more powerful and graphics moved from the desktop to the network, hence to the web.

Of course, by the millennium, the PC too had almost been replaced by the laptop and early handheld devices. So, from a situation in 1950 when the user took their data to the computer so that information could be generated from it, in 2000, the computer had come to the user. In fact, computation would not have evolved in this way if there had not been some switch in the relationship between the user and the machine. Originally, the user came to the machine, without anything other than the physical network linking the place where programs and data were prepared to where the computation took place. By the end of the century, this network had been replaced by wires and wireless networks, which piggybacked on earlier developments of the telegraph, the telephone, the radio, and the television. Networks were originally implicit, but alongside the PC and into the 1990s, they became quite explicit. It is time to chart this progress and to describe how computer and communications merged and converged. We must complete the jigsaw, but before we do so, let us impress the idea that as computation has become all pervasive, the network has become both essential and central. As we will see in part II, after we explore networks in the next chapter, the contemporary model of the city is, in fact, the network and the correspondence of networks, for moving materials and people with transmitting information lies at the heart of the computable city, which we will elaborate in the rest of this book.

5 NETWORKS: THE FINAL PIECE OF THE JIGSAW

The Internet is the first thing that humanity has built that humanity doesn't understand, the largest experiment in anarchy that we have ever had.
—Eric Schmidt (1999)

ARPANET AND THE NETWORK LAYERS

We never quite extended our walk in chapter 2 to the house where Alan Turing was born in Maida Vale,[1] for our information mile ended at King's College on the Strand. But from King's, if we head northwest toward the east end of Oxford Street, we encounter Soho, London's dominant entertainment quarter adjacent to theater land and the red-light district insofar as London has one. Various inventions featuring in the history of information technologies that underpin its science were made nearby. Michael Faraday began his experiments in electricity toward the end of the Napoleonic Wars and continued these in 1820s and 1830s at the Royal Institution in Albemarle Street.[2] William Smith produced his geological map of Britain in 1815, which was a forerunner of our ideas about evolution through the fossil record, which embraces the evolution of the landscape; it hangs in the Geological Society of London at Burlington House in Piccadilly.[3] And in 1854, John Snow discovered how cholera was spread from his analysis of water from the various pumps dotted around Carnaby Street, arguably producing the first geographic information system.[4]

As we head north to Bloomsbury, we reach Darwin's house, where he first lived when he came back from his voyage on the HMS *Beagle*.[5] This was on the site of a wing of University College, where we are destined. The link from Smith to Darwin and thence much later to the genetic code and its computation, some of it developed in both King's College and University College, cannot be dismissed.[6] But on our way is the house in Soho where John Logie Baird, maverick inventor, first demonstrated a live broadcast of moving images with grayscale tones in 1926. Baird did not invent television alone, for like many before and thereafter, this technology was developed simultaneously by several inventors.[7] As we head up Gower Street, passing the place where Darwin lived, we encounter our destination, which is in the northeast corner of the quadrangle of University College—in fact, very close to the place where much of statistics and genetics were developed by Galton, Pearson, Haldane, and Huxley in the late nineteenth and early twentieth centuries.

In the basement of the Katherine Lonsdale building,[8] there is the room where the first internet connection outside the US was established in 1973. In 1958, the Pentagon set up the Defense Advanced Research Projects Agency (DARPA) to push the US to develop military and space technology to meet the Soviet threat after the launch of Sputnik. In 1966, its Information Processing Techniques Office (IPTO) decided to initiate a small network consisting of funded scientists working on defense-related projects primarily in US universities so that they could design a system to expand their computational facilities using time-sharing on one another's machines. The first link was made in 1969 between UCLA and the Stanford Research Institute, and it was extended immediately after to UC Santa Barbara and the University of Utah. From then on, nodes in different research centers were added at a rate of about one a month. By 1973, it was felt that the ARPA-NET, as it was then being called, was ready for global expansion, and thus using a microwave satellite connection with a relay positioned off Norway, a link was made to UCL.[9] At this point in time, there were some twenty-five machines connected to what we now call the "net."

It would be a mistake to think that the ARPANET represents the only origin of the internet. The network was primarily for time-sharing, that is, to increase computational resources for those who were networked. But this was based on pooling computation in different locations, unlike most time-sharing to that date, which was based on enabling different users to engage in computation simultaneously on the same machine. What was not quite anticipated at the time was how difficult it was to share resources using different machines, as their operating systems were quite different. For a long time throughout the 1970s and into the 1980s, considerable efforts were made to devise standards for operating systems that would enable this kind of sharing to be possible, and it was only in the mid-1980s that such sharing was made possible using the Transmission Control Protocol/Internet Protocol (TCP/IP) attributed to Vint Cerf and Bob Kahn. Yet, the original network was largely used as a test bed for digital communications, and it is not surprising that it began to be used for messaging.[10]

In fact, sending messages to other users of the same machine had been done for a long time. Even the Dartmouth Time-Sharing System used by students had this facility, but sending messages to someone on another machine was quite a different matter. First developed by Ray Tomlinson who worked for the main contractor for the ARPANET, Bolt, Beranek, and Newman (BBN, now Raytheon), a protocol was devised for sending such messages as files that would attach to the remote user's machine and convert themselves into a form of mailbox—hence the origin of email, where Tomlinson even used the @ symbol to indicate the name of the recipient's machine or site where the machine was located, prefaced by the name of the user.[11] To an extent, although the practice of emailing did not really enter the mainstream other than in universities and research institutes before about 1990, this date in 1972 more or less coincided with the high noon of the General Post Office in the UK, when its employment reached a record high, when traditional snail mail reached its zenith, and when the PC first made its entry in the US—yet another significant threshold date in the evolution of the networked society.

Back, then, to University College. The idea of transmitting data—one of the main uses foreseen for the ARPANET by its early Pentagon advocates, Bob Taylor, Larry Roberts, and John Licklider—was to use packet switching. This was the practice of dividing data into discrete packets and switching them from computer to computer—or rather relay to relay This, in fact, had been proposed in 1965 in the UK by Donald Davies working at the National Physical Laboratory, who built a demonstration network. A similar idea was developed simultaneously in the US by Paul Baran at RAND and embodied in the new ARPANET by Leonard Kleinrock, but the UK contribution was quite well known, and these ideas were all collated at a meeting in 1967 in

the US. Donald Davies was judged by the British Government to be working on material that was too sensitive for him to attend, and thus it fell to
his colleague, Peter Kirstein, to represent the UK. Peter was a professor at
UCL, and thus the first link outside the US for the ARPANET was established
from this source.[12] In fact, the funding for this first link did not come from
the military but from the GPO on the British side, which had always been
in the vanguard of developments in telecommunications, as we detailed
in chapter 2. The ARPANET and its nonmilitary successors such as NSFNET,
which evolved as the system grew, were mainly in universities and funded
by government in one form or another. The number of computers connected
grew from the initial four in 1969 to nine in 1970, eighteen in 1971, twenty-
nine in 1972, and thence to forty in 1973, fifty-seven in 1975, 213 in 1981,
and exponentially from then on, geographically diffusing and rapidly connecting with other networks that were being developed in the private and
public sectors. During this period, a basic form of organization for the internet was being set up. The domain name system was established in 1983,
which introduced the familiar .com, .org, .gov, .edu, and such like extensions
now writ large across the net and across the world.

The key question that is often asked is whether the ARPANET can be
considered the progenitor of the Internet. To an extent, this was indeed the
case, largely because the ARPANET was a global network. Indeed, the very
technology of networks was spawned from academia, as many innovations
in computation have been. Software technologies such as file transfer protocols and TCP/IP were developed in close association with the ARPANET,
while developments in more local networking were closely linked. Bob Taylor, who in many ways was key to DARPA in the 1960s, became the director
of Xerox PARC in the 1970s and oversaw the development of Bob Metcalfe's
Ethernet technology, which rapidly formed the basis for LANs everywhere.
Frequently, the internet is defined as a network of networks, but most definitely, the ARPANET was not this, for it was primarily a means where computers, not networks per se, could communicate.

The role of the telecoms companies is key to this, for there is little doubt
that Bell Labs, the research arm of AT&T—the Bell telephone network company—was heavily involved in funding the hardware for the ARPANET and
was itself very closely tied in with the IPTO,[13] DARPA's key link with nonmilitary contractors that were involved in the effort. To an extent, all that
can be said is that without the ARPANET, there would be no internet. The

story, however, is still incomplete, for there are a host of other developments that are closely woven into the history of mainframes, minis, PCs, and something not yet considered here—supercomputers of various shapes and sizes, but all of them big in terms of speed, storage, and compute power. To obtain access to such resources, it was necessary to communicate with such machines remotely, and this was essential, of course, in big science such as in nuclear research. Thus, there was little doubt that the ARPANET represented the first construction of a global infrastructure that is key to a wired society, to the post-industrial or fourth industrial revolution, to the smart city, and to the kind of digital infrastructure that is making the city ever more computable.

The ARPANET was not a network of networks but rather a network of computers that existed alongside the development of other networks, which in principle could be linked to one another, thus raising the prospect that network infrastructure exists in and across many layers. We can, in fact, think of computation itself as existing in many layers all the way from the outputs that are displayed on the devices, which we directly use to reap the benefit of computation, down to the assembly languages used in basic computation. Ever more abstract variants are buried deeper and used to translate data and programs into a form that can be manipulated by the chips that drive the binary switching that ultimately delivers the fruits of that computation.[14] This layer cake of instructions that determine computation is embedded in layers of networks that enable anyone using a device to communicate with others. No one on earth understands how this incredible array of networks and computers functions in such a seemingly coordinated way. Few now understand, other than in somewhat abstract terms, how such global computation takes place, and when we come to examine the computable city, the dominant view is that all these layers simply coalesce into a system that works.

The internet functions because systems that are built from the bottom up contain enough checks and balances to ensure that, as individuals, we act rationally within our own limits, enabling such collectivities to hang together in such a way that the whole apparently functions independently from the parts. Two hundred and fifty years ago, Adam Smith in his classic book *The Wealth of Nations* made the same point: the economy holds together as if there were a "hidden hand" guiding it. But of course, there is no such hand.[15] The mechanisms that make this possible relate to human self-organization from the bottom up. The same is true of computation and communications, which we do well to remember in the search for the essence of the computable city. In fact, as will become apparent if it has not already, there is no such essence. When computation spreads out in the way it already had by the time the net was a significant force, the idea that there is one cause, one paradigm, one motivating force is as far from the mark as it can be. This will become clearer and clearer as we continue to elaborate our argument; it relates rather strongly to the unpredictability of the technologies that are being generated in very rapid succession and which ultimately appear to be converging on one another.

THE NET EXPANDS: LANS AND WANS

In contrast to the ARPANET, which was the quintessential global network, LANs represent the tightest of bindings linking computers, printers, and servers together for well-defined common purposes. In the last chapter, we introduced the origins of such networks at Xerox PARC, where, in the early 1970s, Bob Metcalfe established a LAN linking the various Alto computers together. He developed what came to be called the "Ethernet," which allowed common resources such as servers with memory and printers as well as scanners and digitizers and a whole range of peripherals associated with inputs and outputs to be linked together and thus shared. LANs are essentially based on networks that are constructed in-house by those who need to be relatively close—in the same building or even the same room—when it comes to sharing resources. Printers need to be loaded with paper, servers rebooted from time to time, and thus manual servicing is always required with such LANs that makes them local. LANs are thus networks that can be constructed locally and do not require any more global networking per se until they get linked to wider area networks (WANs). The major difference between WANs and LANs is one of scale, but usually WANs are based on leased telecommunications lines and, in this sense, are provided by independent companies intent on giving a much wider service to those involved in communicating data and computation. In short, WANs enable LANs to connect to many other networks of a similar kind—other LANs for example, as well as the whole range of more global networks that now form the internet itself.[16]

In fact, there exists a much more elaborate hierarchy of networks that we rarely see and which were built up dramatically after the ARPANET was started and the Ethernet invented. This multilayer composition has not been stable throughout its life, and it is still continually changing, being enriched by new networks on all levels. During the 1980s, the topology of networks appeared significant, only for this to pass as newer forms of network and ever greater capacity were continuously added to the global infrastructure. As the decade wore on, commercial network providers began to fashion various kinds of email and related consumer services for those who began to see their PCs as key links to a world of information.[17] By 1990, these providers were beginning to merge, split, and coalesce as the internet itself took shape, and from the time when the basic protocols were adopted, the emergent internet grew ever faster as academia, the national labs in the US and Europe, and the cutting edge of the telecommunications industry sought to build a system that from the beginning was open to all.

When the ARPANET was first put together in the 1970s, it had more than enough spare capacity to enable many indulgences relating to manipulating and communicating different data. A close group of computer scientists drove the project forward as a test bed to develop methods whereby different machines could communicate so that, ultimately, machines could be shared with respect to single scientific tasks that required computer time not available locally. In this sense, the motivations were similar to those of today, where much of our computational resources are moving or have moved to the cloud, which is essentially a network of remote machines and servers whose location is essentially unimportant, at least to applications users.[18] In fact, the growing need for networking was established quite early among the close group of ARPANET pioneers—Roberts, Taylor, Cerf, Kahn, Metcalfe, and others we noted earlier—who worked in and around the IPTO, BBN, and the initial installations at Stanford, UCLA, UCSB, and the MIT. SATNET was set up in 1977, USENET for news and groups in 1979, closely followed by CSNET for computer scientists.[19] BITNET, which was based on sending packets—files attached to electronic mail—was set up in 1981 and in 1983, the Department of Defense separated the ARPANET from its military applications, which were wrapped up into the new network MILNET.

At this point, there was little thought about where the net was heading. In the computer-science fraternity, there was the desire to share resources— data and compute power—and there were some killer applications involving visualization on the horizon, but there was no generally accepted idea that everyone on the planet would be using the net within a generation or two, connected through layer upon layer of networking that would become the new infrastructure of the post-industrial society. Adding networks was the name of the game in the 1980s, with a scramble for how the ARPANET might actually become the network of networks, then the current somewhat casual definition of the internet. Regional networks were established while the National Science Foundation founded NSFNET, which began the transition from ARPA to the internet. In 1990, the ARPANET was retired at about the same time it became clear that to organize the resources on the internet, much more powerful and user-friendly software was required.

Most of what we have reported here originated in the US, but as in the original development of computers, some of the significant actions lay elsewhere in Europe. Networking proceeded in parallel from the links made back in the late 1960s with Donald Davies at the NPL, who, as we have noted, developed a variant of packet switching, and Peter Kirstein at UCL, who pioneered control protocols in internetworking, taking the lead. A reasonably vibrant community emerged in Europe promoted by the UK's Joint Academic Network (JANET). Because big science in Europe required much more coordination between countries with different traditions and languages, considerable effort was put into joint laboratories of which the preeminent one was CERN (Conseil Européen pour la Recherche Nucléaire). This was to pay enormous dividends just at the end of the Cold War, for it was here that the World Wide Web was invented. Its resonance with the very development of the first computers at Bletchley Park for code breaking and in the Manhattan project that produced the atomic bomb could not have been greater,[20] although the scale of the effort in developing the web rested largely with one person: Tim Berners-Lee. Indeed, there are stark contrasts between Turing and Berners-Lee in terms of the relatively low key and somewhat idiosyncratic way in which their remarkable inventions were generated.

By that time, many resources devoted to searching, transferring, and communicating on these various networks had become established. Services with archaic names such as Gopher, FTP (file transfer protocol), Archie, and Usenet and various email Listservers, all built around the TCP/IP protocols, mushroomed at this time only to be absorbed quickly and relatively painlessly into the web and its various browsers that developed in the early 1990s. In 1992, when the web finally became established, the number of hosts connected to NSFNET reached one million, signaling that the internet had at last arrived. The growth from the first two nodes at UCLA and

SRI, which were activated in late October 1969, had been explosive. The number of computers in total had been equally dramatic, with some one billion estimated by this date. Networking would see this total expand even more dramatically. Like 1937 and 1972 before, 1992 represents a key transition to a world where the web would penetrate all our lives and change many of the practices and activities that society had taken for granted since the Renaissance, since the industrial revolution, certainly since classical times, and probably since the Stone Age. This marks the transition from the material to the digital, the physical to the ethereal, and from atoms to bits as Negroponte so seductively referred to it.[21]

There is little doubt that the growth of computation and software during the 1980s was both dramatic and chaotic. No grand plan can ever exist for a complex system that grows from the bottom up. The computable city and the wired society evolved from the many pieces that were put in place by a deluge of innovations that competed fiercely with one another, a few triumphing, many falling by the wayside, but all of them contributing to the enormous growth and diffusion of information technologies that, by 1990, the date of the first web page, saw nearly a million hosts connected to the internet in a world where for the last time we could hazard a guess at the number of computers. From about 250 machines in 1955, ten years after the first, there were something in the order of fifty-seven million in 1990 in the US (of which some 95 percent were PCs). Something was urgently needed at that point to draw together many of these disparate strands, and this something was the World Wide Web. Some may say it came just in time; others will say that it was inevitable anyway.

THE WORLD WIDE WEB

As we have just noted, it was the physicist Tim Berners-Lee working at CERN in 1989 who first developed the idea of a web page with links across the internet to other pages, where these pages represented the access points to various resources required by the internet's users: libraries, data files, news, videos—the list of possible digital resources viewed in these terms is endless. Berners-Lee argued that some way of organizing a one-stop shop for these resources was needed, and he proposed to his managers at CERN that such a system was both necessary and desirable.[22] His bosses disagreed, but at least they gave him time to pursue the project, and to this end, he developed three key components that would establish the structure of the web as it is today. Berners-Lee had explored hypertext back in the early 1980s. So, his system was first based on the idea that you could mark up any item that you wished to use as a pointer to other resources. HyperText Markup Language (HTML)[23] emerged. Second, each web page or site would have a locator called a Uniform Resource Identifier (URI), more often now called a Uniform Resource Locator (URL), an address that uniquely located the resources. Finally, he defined a protocol for retrieving resources, which he called a "Hypertext Transfer Protocol" (HTTP), which enabled the retrieval of the resources requested.

Berners-Lee demonstrated his system on a NeXT computer, which had a highly graphic interface and sophisticated operating system (which was to become the basis of the revived Apple Macintosh platform once Steve Jobs returned to the company, Apple having acquired NeXT in 1996). It took three years for CERN to agree that Berners-Lee's software should be distributed freely, and it is probably this that did most for the open-access environment that exists today with respect to much of the internet. To an extent, although Berners-Lee fought for open access, it was the generic notion that widespread access to something such as the internet was only possible if good enough graphical user interfaces could be established. This prospect, in fact, went back many years to the very origins of computing. We mentioned Vannevar Bush in chapter 1 and his prescient and highly influential article in the magazine *The Atlantic Monthly* entitled "As We May Think." There he sketched an all-purpose personal computing workstation called a Memex with a graphical user interface that would enable any user to retrieve information from a world populated by such devices that were networked together. Bush assumed such devices would be everywhere. In short, he assumed the internet, and in his imagination, his vision was as much if not more than the World Wide Web itself.[24]

The Memex also assumed something such as hypertext, but this was to remain dormant until the seeds of the internet were sown at ARPA in the early 1960s. John Licklider, the first director of IPTO in 1963, wrote a memo envisaging what the emergent ARPANET might be like. His somewhat tantalizing, sobering, yet serious memo was, in fact, addressed to "Members and Affiliates of the *Intergalactic* Computer Network" (my italics), which, in good faith, mirrored the optimism of those times that computers would come to populate a world that was entirely networked. Leonard Kleinrock, who was on the ground floor at UCLA when the first internet connection was made in 1969, had also written a highly influential paper entitled "Information Flow in Large Communication Nets" as part of his thesis in 1961 at the MIT. This also mirrored the excitement that scientists had for the idea that the world could be connected digitally.[25] We have already noted the contribution of Douglas Engelbart at the Stanford Research Institute, who, with Ted Nelson, evolved and demonstrated hypertext in the 1960s, influencing the remarkable effort at Xerox PARC, which was central to both personal computers and LANs and, ultimately, the internet. In fact, in 1967, Engelbart presented what came to be called the "Mother of All Demos," where he illustrated a system of personal computing that was based on video conferencing, hypertext, word processing, and online communications all controlled by a mouse, overlapping windows, and real-time editing. The Memex, what else?

Berners-Lee himself had created ENQUIRE in 1980, which was a rudimentary hypertext processing system, and he was well aware of all these developments as ways of interacting with computers. Although his system did not quite go the whole hog and produce the ultimate graphical user interface,[26] it became very well known in the early 1990s, and it was picked up by researchers at the US National Center for Supercomputing Applications (NCSA) at the University of Illinois at Urbana-Champaign. There Marc Andreessen and Eric Bina developed a browser based on Berners-Lee's open standards, which they called "Mosaic."[27] It was not the first graphical browser to the web, but it was the one that took off, illustrating once again that technologies are developed simultaneously in different places with what appears to be random chance in establishing which technology becomes dominant. In fact, Mosaic was available on my SUN workstation sometime in 1994, and I must confess now to not knowing that there were any other graphical browsers. What was remarkable was that after the painful experiences with text-based browsers without any of the protocols that Berners-Lee had made available, the Mosaic browser acted seamlessly in accessing other web pages from other users, of which there were very few in those early days but enough to catch the imagination of all those who toyed with this new technology.[28] In Arthur C. Clarke's phraseology, at the time, this technology seemed like magic.

By the year 2000, stitching together the various tools and web resources that one had to do in the early 1990s was long gone. Indeed, from the desktop or the laptop, the new search engines—Google for example—made this process painless. Of course, every time a new and often smaller device was introduced, the process had to occur again, but each time, it was less and less painful. Slowly and surely as the last decade before the millennium wore on, various fragments of software were integrated as the data and computational infrastructure of the wired society came together. The first thing to happen after MOSIAC was introduced was its commercialization. Some say that Larry Smarr, the Director of the NCSA, told Jim Clark, one of the founders of the graphics workstation company Silicon Graphics, about Mosaic, and this led to Clark initiating funding for a new company—Netscape—to promote the browser with Andreessen. This all happened rather quickly in the wake of Mosaic in 1994. A streamlined product emerged only to be challenged by Microsoft, who, in the wake of these developments, introduced their own browser—Explorer—which was eventually bundled in with their Windows operating system.[29] Explorer more or less forced the demise of Netscape, although the focus on search continued, with various other browsers being tested, of which Yahoo was the only one to survive the introduction of Google in the early 2000s.

We will return to Google in a moment, for this really does mark another stage in the development of the internet, but in the interim, the focus on computing through the internet became dominant. Much software began to move to the internet from the desktop, and a variety of internet engines emerged. We have noted Yahoo, which was founded in 1994, but internet access was finally introduced into the more commercial platforms such as Compuserve, America Online, and Prodigy while entertainment and shopping sites began to be formed. Amazon, Craigslist, and eBay were founded followed by Netflix. In 1998, Google was started in the apocryphal garage, but it quickly moved to more luxurious quarters, and in the couple of years before the dot-com bubble burst, it established a new era in methods of search.[30] The 1990s had been frenetic in terms of investment in infrastructure, particularly in networks, but the mania for investing in anything that looked as though it could access a massive audience of connected buyers and sellers, increasingly the public at large, led to a bubble that finally burst, with many embryonic as well as older established technology companies failing. The disruption caused to the new economy was considerably less than to the old, but the first era of the internet was ending. From now on, the world would be dominated by an infrastructure that was deeply rooted in contemporary society, an infrastructure that was still growing and transforming at an alarming rate, and one that would be destined to change many past practices, organizations, ways of thinking about the world, and ways of making money.

THE INFORMATION EXPLOSION: STORAGE, SEARCH, AND THE ONLINE WORLD

There is a somewhat crude but rather effective model that explains the way computation has developed since digital computers were first invented in the middle of the last century. This involves an evolution of functions, beginning with a very strong emphasis on hardware, which constituted the most significant costs in making computation possible until about the time when the PC was invented. By 1980, hardware costs had fallen to the point where the focus changed to the cost of software, which began to dominate the picture. To an extent, the cost of software engineers and others, particularly computer administrators and even users, represented another way of chart-ing this changing dominance during these years, but as software became more automated, the focus changed again to the cost of data. So, hardware followed by software followed by dataware is one image of this evolution over the last eighty years, but this now is changing too as orgware, which of course involves peopleware, becomes ever more significant. Paralleling this evolution is the merging of computers and communications—the con-vergence of computer hardware and networks into digital platforms—but as this continues to evolve, the focus now is on everyone having access to generic forms of computation through information devices such as smart-phones, externally embodied sensors, and web services of many different kinds.[31]

To an extent, this transition from hardware to software to data and thence to organization mirrors our approach to the computable city, reflected to some extent in the logic of this book. But at this point, we must really grasp the nettle that in the 1990s, we crossed the Rubicon from individual idiosyncratic uses of computers for fairly specialist and often esoteric tasks to much more widespread access to data and information. To a very large extent, this transition was made possible by the invention of global net-works and the way we were able then to access them—largely pictorially or visually—through the World Wide Web and from small highly mobile

devices such as smartphones. It is hard not to underestimate the importance of the web for the world we have now reached. This is entirely dependent on information infrastructure that is largely based on free software
and the ability of millions of people to provide their own information of
many kinds, also free of charge. To get some sense of what this means, we
only need to examine the number of web pages that we have access to. The
key issue is that when we obtain access to information over a network, we
are able to communicate with anyone who is attached to it. If we have a
network of, let us say, ten users who are able to share their individual information with one another, there are a hundred possible links that might be
used to transfer data. If we assume there is little point in sharing information
with ourselves, then there are, in fact, only ninety such communications,
but if the network were to increase to twenty users, there would be twenty
less than four hundred possible links. As the network gets bigger, the information that can be transmitted between users increases exponentially with
the square of the number of users. With fifty users, there are approximately
2,500 possible swaps of information, with a hundred users, ten thousand,
with a thousand users, one million, and so on.

   In fact, it was back in Xerox PARC in the early 1970s that Bob Metcalfe
invented the Ethernet and, in so doing, coined what has come to be called
"Metcalfe's law." There are many so-called laws of computing, and we will
look at some others toward the end of this chapter, but what Metcalfe effectively said was: "the effect of a telecommunications network is proportional
to the square of the number of connected users of the system." In fact, this
definition from Wikipedia is George Gilder's translation of what Metcalfe
said about the language of networks, where Metcalfe himself spoke of
devices attached to the network rather than network users, but the effect is
the same. In this sense, the law suggested that the value of the network in
terms of the information that it generates increases exponentially with the
number of users.[32] To illustrate what was happening in the 1990s, once the
web had been developed, we need a simple but effective measure of data.
So, let us take web servers or, in fact, websites. The explosion has been massive. In August 1991, there was one website—that developed by Berners-Lee
at CERN. There were ten by 1992, a million by 1997, a hundred million by
2006, and a billion by 2016, and in mid-2022, there are just short of two
billion.[33] The number of users per web page varies considerably, but on
average, there are about 3.7. Every site could, of course, be linked to every

other, in principle at least, and with two billion × two billion possible links, then the information explosion is complete, with connectivity almost as much as it can ever be. There is little point in speculating what these superlatives mean, for once computers began to be used in this way after 1992, the entire infrastructure of what the information society means began to change. But before we look briefly at some of the implications of this, it is worth explaining the sheer scale of databases and organizations that are in the process of generating information in this manner.

What the web began to deliver immediately was access to conventional information resources, which we progressively digitized and transformed from their traditional print and related non-digital media. Encyclopedias and dictionaries were the first candidates, and Wikipedia, which began in 2001, is classically representative of such digital data that can be both searched and continually improved. In fact, the first generation of web technologies prior to the introduction of Wikipedia involved the delivery of passive information that users could simply browse, but in the early 2000s, the web became interactive.[34] These later technologies, referred to as Web 2.0 in comparison with the passive technologies of Web 1.0, heralded an era when the web became active and interactive and where users could not only browse and absorb but also explore, generate, and create new information.[35] Many of the social media sites such as Facebook, Flickr, and Twitter that were started from 2004 to 2006 provided new forms of information that is continually being added to, transforming as users create and destroy data in real time. There are now many millions of such sites based on video, photographic data, social network links, and so on that are continually adding to this enormous store of data that the internet has made possible for us to develop and access.

As soon as the web developed, it was extended to deal with commerce, with Amazon being set up in 1995 and PayPal in 1999. We need a framework to show how computation is evolving to deal with these many functions, which are being fast ported to the online world, and as this is so much a part of the computable city, we will leave this until later chapters. There we will examine the development of online retailing, online education, health, and the many other functions that take place in cities and society that are being transformed by the new information infrastructures whose origins we have examined in this chapter. Much of this is about a new era of software whose presence is in the form of small programs called "apps," accessible on any device from the very smallest up.[36] This software is very much about access to information and to the online society, where market functions are key to new forms of online living that are fast emerging and which, as we will see, are beginning to underpin the computable city. But before we explore some of these possibilities by way of conclusion to this exploration of the role of networks in computation, let us take a look at what the web has made possible in terms of searching for structure and meaning in this great cornucopia of data and information that is now on offer.

When you generate massive quantities of data, either in unstructured raw form or in processed form as information, it is essential to structure this for the purpose at hand. This is the role of search, which essentially is based on giving structure to large quantities of data, for example typically that contained in web pages. Such structure of course is essential to add meaning to the data with respect to the purpose of the search in the first place. It is not surprising then that as soon as software became available on the PC for situations where numerical and textual data became the subject of processing of any substantial size, search tools were added to the software. In comparison with database technologies and word processing, which reflect modest search but nevertheless require sophisticated software to enable good search to be accomplished, when it comes to the web, then search takes on a different dimension. As soon as the web emerged, search engines followed. To an extent, Mosaic was such an engine and so was Netscape Navigator. Microsoft's Explorer, which dominated search in the late 1990s, won the so-called browser wars, but it was left to Google which began in 1998, to really focus exclusively on the task of search. Microsoft inevitably introduced their own engine, Bing, but this was ten years after Google. In fact, what Google did was to establish a simple but powerful business model that linked advertising to the results of search.[37] By making search very powerful and free to most domestic customers at least, then advertisers flocked to the site. The revenues involved have enabled Google to finance many improvements not only to web-based computing but also to developing software and hardware for everything ranging from autonomous cars and phones to new web programming interfaces, as well as providing an extensive range of free web-based services, ranging from Google Maps to Google Calendar to OpenOffice-style functions.

The number of Google searches per year was 1.2 trillion, but this was in 2012. There is some sense that this number has now reached more than two trillion, but the actual number is hard to estimate in any definitive sense.[38] It appears now that about half the world's population (some 3.8 billion people) use the internet each day, but whatever the precise figures, the scale is daunting, and the implications for everyday life, education, retailing, and so on are enormous. Since 1992, the year when we consider that the sea change in all-pervasive computing began, our ability to store data has increased at the same kinds of exponential rate associated with network connections (Metcalfe's law), web pages, and the costs, size, and speed of computation (Moore's law). Simply at the level of external storage devices for PCs, these have fallen in cost and increased in memory so that most of one's personal data can be stored on a small (in physical terms) flash drive. There are important implications here for our ability to make sense of the data that is captured and stored; data is now being generated at such rapid rates that although we can store it, we do not have the facilities to process it, even search a lot of it, and much of it is simply dumped and hence lost.[39]

We are getting ahead of ourselves. Since the millennium, there have been dramatic developments in search that lie at the core of the way the internet is accessed and used in the online economy and for social media. We will chart these trends in the last part of the book, since we will now turn to cities and their computation, which will establish the range of functions that the computable city makes possible. Once we have completed this, we will return full circle to these developments that are part of what James Martin many years ago called the "wired society"[40] of which the computable city is one manifestation. But before this, we will conclude by pulling together the major themes that define the computer revolution and the digital transformation. We call these the "laws of computing," of which we have already anticipated some, but here they provide a fitting closure to the history of how the digital world got going and continues to develop.

THE LAWS OF COMPUTING

It is just over eighty years since that auspicious period in 1937–1938 when the first significant stirrings of the digital world beckoned. Alan Turing's famous paper published in that year established the idea that the computer was a universal machine, and this built on the contributions of the logicians Gödel and Church at Princeton. In parallel and simultaneous fashion, as we noted in chapter 1, the first digital computers were built in Germany and the US, with these engineering efforts best articulated by John von Neumann, who unofficially took charge of these efforts toward the end of World War II. These were key pointers to the dramatic developments that heralded the era we have recounted in this and previous chapters. When we reflect on this remarkable period, which is as close as we have ever got in the history of the world to a time when science fiction turned into science fact, we see many inexorable trends that make sense of these developments and also signal ways in which we might speculate about what will happen in the next eighty years. By then, we will have reached the end of the twenty-first century, the population of the planet will no longer be growing exponentially, as it is already appearing to stabilize,[41] and, with luck, we will have learned how to deal with the current crises involving aging, inequality, global warming, and the host of difficulties that continue to beset us, as well as new unpredictable grand challenges such as the COVID-19 pandemic we are still living through in the early 2020s. One of the objectives of this book is to provide the context to speculating on what this world will be like with respect to cities and information technology by the end of this current century.

To this end and at this point, it is worth standing back and saying something about the way computers and communications have developed and will continue to do so. In this manner, we will describe in somewhat crude terms what we and others sometimes refer to as the "laws of computing" or the "laws of computation." These are hardly laws in the sense of the physical sciences: they are not hard and fast or falsifiable in the scientific sense, but they are contingent on the past while being uncertain with regard to the future. They do, however, give us some sense that there are regularities that can be exploited with respect to how we might see the shape of this future. We have already anticipated two that are well-known: first, Moore's law, initially articulated by Gordon Moore, one of the founders of Intel, which charts the path of miniaturization, the decrease in cost, the expansion of speed and memory, and the overall value of the microprocessor, which has continued to grow exponentially; and second, Metcalfe's law, which we have recounted with respect to the value of a network as more and more devices or people get connected to it. Both laws imply exponential growth, as indeed do the others we will sketch here.[42]

Two other features of this history of the way computers and communications have evolved is, first, in terms of the convergence of what have always appeared to be disparate developments and, second, the way computers in terms of their hardware, software, and dataware have spread out and diffused both geographically and in terms of their take up within the population, permeating all aspects of contemporary society. Most of what we have recounted has been focused on the US, occasionally including other parts of the English-speaking world where the majority of developments have taken place during the last eighty years. But once the World Wide Web was established, this began to change. We have barely recounted the role of China in current developments, but the fact that three or four of the top eight social media sites are Chinese suggests that the next eighty years will be somewhat different in terms of the contributions geographically to the future of the wired society and the computable city.[43] We have only touched the surface of the extent to which computation has established itself globally in the last fifteen years or so, and in one sense, current trends and events appear more confused, less stable, and less regular than ever. This is being made ever more complex by the development of computation and communications globally, by the end of the American empire, by the twists and turns in the world geopolitics in places like Ukraine and Taiwan, and by the fact that the focus on new forms of computation is on developments at a much more individualistic level. This, we hope, will become clearer once we begin to develop our ideas about the computable city and its smart counterpart in the next part of this book.

We noted Moore's law in chapter 3, where we described how he penned his famous law in an article in *Electronics* in 1965, predicting that the number of transistors that you could cram onto an integrated circuit has been more or less doubling every year since 1959. In fact, there is some memory that Douglas Engelbart, the enfant terrible of everything from windows to internet computing, made the same sorts of observations in a paper in 1959 of which Moore was aware. This may well be an urban myth, but Moore himself never considered that his paper in 1965 would attract the attention it did because he never assumed that his observation would stand the test of time. In some respects, this is an amazing admission because those closest to the technology could not see where it was heading, even though these innovators knew the history of computing since the invention of the transistor nearly twenty years before inside out. In 1975, Moore, in fact, revised his estimate of doubling every two years to, on average, around eighteen months, where it has been ever since.[44]

The rate of change is beginning to slow now. It appears the doubling time is more like three than two years, and there are estimates that Moore's law will come to an end in 2036 or even earlier as we approach various physical limits. This future is uncertain because there are so many features of the period doubling to be taken into account. Moore phrased his law in terms of transistors, but it has been translated in simple proportional terms into the exponentially decreasing cost of memory, increasing speed of processing, and increasing amount of memory. But even the way the fabrication of silicon chips takes place and the economics of the factories in which to achieve this make a difference. Although Moore's law in its strict sense is ending, there are various developments around the corner that are likely to lead to ever-increasing economies of scale, in the manner portrayed in his law. Indeed, Ray Kurzweil, known for his ideas about future exponential growth of many technologies in contemporary society, suggests that Moore's law goes back long before 1959, back to vacuum tube technology and related electromechanical devices, and in this long time frame, the doubling rate might even be less than eighteen months.[45] The precise number does not matter because this is merely a set of observations, hardly a law in the sense in which we articulate them in classical physics. The key issue is that we can expect continued advances in miniaturization that are as much to do with how we organize computing in the future as with increases in clock speed and memory.

We have described Metcalfe's law in this chapter, and to summarize, we might define it as the value of a network increases according to the square of the number of devices connected to it. We qualified this a little earlier by saying that the number of messages that might be transmitted related to the size of the network measured by its links rather than the nodes that define it, nodes being places where messages originate or are destined for. Clearly, as more and more people connect, then the possible number of links grows exponentially, and this illustrates the power of the network in transmission. If a message strikes a chord among its participants, it is easy to see how the network effect can make it viral. There are many variants of Metcalfe's law, and it was George Gilder[46] who helped generalize this law in his book *Telecosm: The World After Bandwidth Abundance*. However, his own particular law—Gilder's law—relates to the total bandwidth of broadcast network communication systems that he argued, in 2000, had been tripling every twelve months since the early 1980s. This implies a rate that is even faster than Moore's law, but it is particularly volatile due to the fact that the economics of network provision are dependent much more on economic demand and geopolitical influences than on technological constraints and possibilities.

There are several other laws that pertain to the growth of computing, but most relate to changes in hardware. The growth of software has not been charted in the same way, apart from throwaway comments from luminaries such as Fred Brooks, who, in 1975, said, "adding human resources to a late software project makes it later."[47] A more recent law coined by or after Mark Zuckerberg pertains to data, and he has said, "the amount of information that people share doubles each year."[48] This may well be the case as it relates to Metcalfe's law, but individually, it does not take account of information limits, overload, and capacity. If you search the web, then you will find many speculations about how computing has and will continue to change and evolve. What is strangely absent is any real focus on the geography of computing, on the shape of cyberspace, and on the diffusion of computation and its technologies. This is apart from the very obvious spread of computing through its innovations into different types of expert communities. In fact, this is exactly what the smart city is all about, and although we cannot take this any further now in terms of this short history of computing, in the rest of this book, we will explore how digital information and its processing technologies have come to underpin the organization, form, and function of the city. This will herald possibilities and opportunities that have the potential to make our cities smarter, more efficient, and more equitable through our ability to understand them better in terms of computation. But it is a potential that is, by and large, yet unrealized. It will depend on what kinds of computation we engage in with respect to our personal and collective aspirations for more efficient, equitable, and sustainable cities. The future in this sense lies in our hands, as it always has done.

II CITIES AND URBANIZATION

At one level, cities are rather simple structures. Throughout history, they have been organized physically around some center, which traditionally has been the focus for exchange—the marketplace where citizens gather to trade with one another. The extent to which different populations can control the space around a city's center determines who lives where and how much space they consume. In the industrial city, the rich, who can afford better transport, tend to live further away from the center, live at lower densities, and use more space than the poor. This we refer to as the "standard model," which we introduce in chapter 6.

This model is now being radically transformed by new information technologies. The role of distance is dramatically changing as we substitute and complement our own physical movements with information. This is the "death of distance," which we outline in chapter 7. The way distance is being annihilated by changes in space and time reflects the ways that the bonds that tie different parts of the city together around its traditional center are coming unstuck. In fact, the city is increasingly dominated by networks of computers and sensors across many levels, and to provide a sense of how our use of the city is changing, in chapter 8, we recount ways in which new kinds of networked spaces are being constructed. We refer to this as "cyberspace."

In chapter 9, our focus on information generated in real time gives rise to what we call the "high-frequency city," where the emphasis is on the very short term—the 24-hour city. This is in contrast to what we call the "low-frequency city," where our interest is in how cities evolve over years, decades, and centuries. In fact, many of our established theories about how the city functions pertain to how they change over long periods of time, but the juxtaposition of low- and high-frequency events is adding to their unpredictability and increasing their complexity, richness, and diversity at every level.

6 THE STANDARD MODEL

Modern geographers claim Thünen. That is their right. But economists like me, who are not all that taken with location theory, hail Thünen as more than a location theorist. His theory is a theory of general equilibrium.
—Paul A. Samuelson (1983)

CITIES AND URBAN GROWTH

Throughout the nineteenth and most of the twentieth centuries, our dominant view of cities was that they were a necessary evil. Cities, the argument went, were necessary because they enabled people to share their labor in one place, thus generating ever more wealth. They were essentially machines for increasing our collective wealth through a division of labor that worked because we were all clustered together enabling us to exploit the fruits of our labors more easily. But the environment they created was of much lower quality than the agricultural countryside from which their incipient urban populations originally hailed. The cities that emerged during the industrial revolution were congested, polluted, and rife with disease, and the widespread reaction was to call for a return to the countryside.[1] Cities were seen as monstrous, ugly, and unnatural with respect to an agricultural past that had dominated society for at least the last ten thousand years.

The conventional wisdom about the evolution of cities is still based on the notion that sometime around the end of the last Ice Age, ten millennia or more ago, humankind moved from a nomadic, almost tribal existence to settled agriculture. This transition slowly increased prosperity. Around 4000 BCE, the first cities appeared, probably in the Middle East but arguably too in China and perhaps the Americas. These cities were essentially places that served the agricultural hinterland, and as various empires came to be established, they also acted as focal points for administration as well as the production of small, specialized crafts that served their wider populations. There is some dissent about the simple sequence based on nomadic cultures giving way to settled agriculture, which in turn spawned cities to distribute the surplus. Some urbanists such as Jane Jacobs have argued that pockets of urbanization existed even in preagricultural times and that the mixing of nomadic and agricultural existence with urban pursuits lasted almost until classical times, to the time of the Egyptians and the Greeks.[2] But what is very clear is that the world population and the population of distinct cities grew very slowly, and that for most of human history, the role of cities in the wider order of things remained modest, small even.

A best estimate of the proportion of the world's population living in cities is around 4 percent by the time of the Renaissance, reaching some 6 percent by the start of the industrial revolution (about 1780 CE) in Europe. Then, urbanization really took off as cities grew to exploit the new technologies further that were being invented as well as to implement them in daily living so that a more prosperous existence could be achieved. The percentage of people living in cities reached 11 percent in 1850, 16 percent in 1900, 21 percent in 1950, and some 47 percent in the year 2000.[3] The proportion crossed the magic threshold of 50 percent in 2008, and in 2023, it now stands just a little less than 60 percent. By the end of this century, to all intents and purposes, we will all be living in cities, but by then, we may well be off the planet; and in any case, when we all live in cities, the very notion of what constitutes a city will be questionable—or at least different from what it has been hitherto.[4] It is hard to unravel all this from the development of digital society that we sketched in the first part of this book, but we will attempt to so in what follows as we explore how what we know about cities intersects with their technologies, electromechanical as well as digital.

Cities evoke many contradictory emotions, ranging from love to hate, but when we go back to the nineteenth century, the industrial city always generated negative responses with respect to its growth. Urban growth, many thought, should be strictly controlled, limited physically by direct intervention with respect to where growth might take place, often by strangling urban growth in the first place. Indeed, many literary commentators were disparaging about the very idea of the big city. William Morris, the nineteenth-century reformer, said in 1892, "Even London . . . is sordidly vulgar in its rich quarters, noisome and squalid beyond word in its poor quarters."[5] Inhibiting urban growth became the watchword of those times, and the notion that this might wreck the economy was nowhere to be seen. In this milieu, institutionalized planning systems focusing on green belts to limit growth and new towns to channel that growth to small towns in the countryside, all set against the concept that such a world should be low-density, green, and suburban composed of garden cities and small towns in the countryside, became the dominant ideology. This decentralized world was substantially reinforced by the first technologies to emerge during the industrial revolution. The internal combustion engine that led to the passenger railway gave urban populations the ability to commute much longer distances to work in the central city, and as the nineteenth century progressed, rail systems and other forms of public transport such as the electric tram began to shape the city into one with a dominant central core and ever-lower densities of housing around it. When the automobile was invented at the beginning of the twentieth century, the idea of the city spreading out into the countryside where there was less pollution and congestion became the dominant form.[6] The telegraph and the telephone, perhaps even the radio and television, all reinforced these patterns so that by the middle of the twentieth century, the idea of a city was one in which a dominant core provided the focus for lower-density living associated with sets of radial routes providing conduits for the population to live at ever greater distances away from the city center, the central business district (CBD).

Cities have thus always been the focus of our interest since they first emerged, and it is not surprising that attempts to explain and understand their form and function go back almost to prehistory, certainly to classical times. Because cities evoke our aesthetic senses in that we can manipulate their geometry across many scales, the simplest explanations of their form are the most obvious. They grow from a core that then evolves as the place where people come together to trade or to rule or to worship or all of these things. The city center through history has thus been this focal point where palaces, marketplaces, cathedrals, and government exercise their control over the city population, which clusters around such functions. In the classical world before the modern era, which we date from the Renaissance in Europe, cities were tightly constrained and usually bounded by defensive walls to keep the enemy at bay. The juxtaposition of functions within the walls did not lack order, for certain activities took pride of place, often for historical reasons, but in general, the shape of the city and even its architecture were relatively simple to explain.

In classical times, most cities were small, seldom more than a few thousand inhabitants, with well-defined cores and highly compact, dense residential districts. Some were laid out on a grid, which was often the most convenient and obvious way to organize things, especially when cities were being established quickly as in the case of colonial settlements, army camps, and so on. Although the writings of the Greek philosophers, particularly Plato,[7] dwelt on the ideal size and layout of towns,[8] the logic was quite simpleminded in that it did not even attempt to explain why cities might be laid out in different ways and what the advantages of particular layouts might be. There was no pretense at any social and/or economic logic to such descriptions and prescriptions: explanations were largely absent. It was not until the early modern era, the time of Leonardo da Vinci, that there was any semblance of economic logic to what cities were all about, and even then, this was still dominated by a concern for geometry and architecture: proportion, perspective, and motion[9] were the dominant forces in explaining the rationale for cities in the Renaissance and beyond until the industrial revolution began.

There do not appear to be any detailed explanations of how cities form and function prior to da Vinci. There were hints, of course, wherever scholars and commentators sought to describe the form of cities and explain it in terms of physical patterns. Da Vinci's focus on organic analogies, however, led him in many directions, but one was toward the creation of ideal cities whose internal circuitry related to geometrical layouts that functioned to enable cities to "breathe" properly. His designs for Milan, for example, produced during the late fifteenth century, suggested strong segregation between various incompatible functions that could generate disease, while he split Milan into ten new towns at least with respect to the design that he fashioned.[10] This is reminiscent of the later nineteenth century and the way the garden cities movement gathered pace around the idea of enabling urban growth in small clusters such as new towns. His designs had enough circulation space to reduce the transmission of disease to a minimum, to transport waste products so that contamination was reduced, and to organize the whole to ensure that the parts functioned in self-organizing fashion[11]—a lesson perhaps for the post-pandemic world in which we now live. Da Vinci almost reached the point in his speculations about cities for developing explanations that reflected the way local economies and polities functioned, but not quite. Although the Renaissance, as its name suggests, involves a rebirth of knowledge from the classical era, the conditions were still not quite right for the systematic development of a theory of cities, and this had to wait for more than another three hundred years before such a theory came to be invented.

THE ISOLATED STATE

The year 1776 is another of those milestone years that divides the old world from the new. Three events, quite unconnected, mark this transition, the first being the American War of Independence. It was here that the colonists who settled the original British colonies on the eastern seaboard of North America finally threw off the yolk of the British monarchy, establishing a republic[12] that went on to become the dominant player in successive technological revolutions that were to emerge over the next 150 years. On July 4, 1776, Thomas Jefferson finally presented the First Continental Congress with the articles for independence, and the deed was sown. In some respects, this act initiated a series of events at the end of the eighteenth century that brought new freedoms to the modern world, the French Revolution and the Napoleonic Wars representing the evolution of democracies that set the political context for the first and subsequent industrial revolutions. In fact, our second event is just this. In 1776, James Watt's steam engine, which he had perfected over some twenty years, finally led to applications in various commercial enterprises.[13] The path was thus set for the development of machines that would form the basis for the transportation of people and goods throughout the first and subsequent industrial revolutions beginning some two hundred years ago. This was another mark in the sand between the old world and the new. But the last event in our trilogy was of quite a different nature, perhaps more abstract that either of the first two, and this was the publication in 1776 of Adam Smith's *The Wealth of Nations*.[14]

Smith's book established the logic for economic life that has more or less prevailed as the basis for our understanding of contemporary society ever since. In many respects, his seminal treatise laid the groundwork for most of economic theory that has developed ever since. In fact, Smith himself spoke of cities many times in terms of the economic system that he articulated in his book. His focus on the role of transport costs anticipated the first coherent statement of how cities or rather space in cities might be organized, which was due to von Thünen, whom we accredit here with what we have called the "standard model." Von Thünen's work will indeed be our starting point,[15] but as we will see, it is unlikely to stand the test of time, for the role of distance and space, which is so central to cities, is being fast eroded by the new digital world. In the next chapter, we will firmly establish the way in which the city is being changed in a world entirely dominated by digital communications. But first, let us outline the model that was introduced in the early nineteenth century.

Johann Heinrich von Thünen was born in 1783 to a farming family in the region of Germany known as East Friesland, part of Lower Saxony, which borders the northern part of the Netherlands. He received an education dominated by the study of agriculture, with a strong bias toward the science of crop rotation, although this did not stop him from learning a fair amount of algebra, particularly calculus, which was essential for his subsequent contributions to economic theory. Prior to his own farming career, he was able to visit many potential estates in northern Germany, particularly in Mecklenburg, where, in 1810, using his father's legacy, he was able to purchase an estate of more than a thousand acres near the village of Tellow, some twenty-five miles southeast of Rostock. Here, he set about improving what was generally poor-quality land using the methods that he had learned from his studies and many that he was to develop himself to improve the quality of the yields on his estate.

Von Thünen was, it seemed, a relatively mild-mannered man, with very strong organizational skills and a real mission in life to use science for the benefit of those around him—his family, the workers on his estate, and the population in his community.[16] This, of course, was an optimistic age, and all of Europe was in the ferment brought by the French Revolution and the revolt in the Americas. Modern science was just beginning, and it was not surprising that von Thünen was imbued with this spirit with respect to his chosen profession. For the best part of his working life, not only did he embark on successive improvements to his estate, but he also painstakingly compiled a complete set of scientific observations as to the practice of his business, producing what some have called the first data for an econometric analysis of a local economy—his estate. Von Thünen did not really publicize his contributions, although he was intensely aware that his analysis was generally not recognized by agricultural practitioners or by those working in economics, who were few and far between in the first half of the nineteenth century. Adam Smith had set the tone in 1776, and then economists such as Malthus, Ricardo, and Cournot followed, but von Thünen, as far as we are aware, only knew of their work from afar. Nevertheless, by the time he was in early middle age, he had amassed enough evidence to publish the first volume of his speculations on the nature of a local economy built around a market for agricultural produce, which appeared in 1826 as *The Isolated State* [*Der Isolierte Staat*]. As we will see, this was not only the first formal contribution to the way economic activities were organized in space, which lies at the essence of how cities are structured, but also the foundation of what has subsequently come to be called location theory[17] and thence urban economics,[18] which contains much of what we know about the amount we pay for transport and land in different positions on the Earth's surface.

This seems a far cry from the form and function of cities, but if you bear with me, we will get there. Von Thünen was in effect both a practitioner and a theoretician, and the power of his analysis lay in his ability to provide a theory of how an ideal land market worked based on a mass of facts—"stylized facts" we might call them today—which he used to demonstrate the mechanisms that he exposed with respect to how such markets actually functioned and the spatial form that they took. Remarkably, von Thünen's contributions have largely gone unrecognized in economics. Alfred Marshall, the great nineteenth-century economist who coined the terms "agglomeration" and "economies of scale,"[19] called von Thünen "the great unrecognized," for it was he, von Thünen, who first demonstrated equilibrium theory and introduced marginalism into economics, some of which is demonstrated by the model he produced for his agricultural market. In fact, some argued that von Thünen's real work is not in the first volume of his book published in 1826 but rather in his second volume and the appended papers on the determination of the natural wage he produced that were published posthumously in 1863, only to have to await a century before being translated into English. Indeed, Fernand Braudel said of von Thünen that he "ranks alongside Marx as the greatest German economist of the nineteenth century."[20] Be that as it may, his contribution to the study of cities only really took off in the 1960s when the obvious translation of his ideas into a model for cities provided the foundations for urban economics.[21]

What did von Thünen actually say? His basic idea now seems so obvious that as soon as one describes it, one is prone to critique it. He essentially said that if you generate an agricultural product that you wish to sell, the cost of transporting it to the market must be added to the price. As you locate the source of your production further and further away from the market, the greater the transport cost and the less the profit you will generate. Now, at a certain (far) distance away, the transport cost will wipe out all the profit you make, and therefore you would no longer be able to produce the good. This would then mark the boundary of cultivation. There is, however, a twist in the tail. To produce the product, you need to pay for the land on which the product is grown, and this payment, which is effectively the economic rent, must be added to the cost of production. However, in a normal market, where the profit one gains is just enough to keep you in business, the rent will vary inversely with the transport cost. This means that as you get closer to the market, you generate more profit, and the landlord will then levy a rent that is proportional to this profit. Thus, what von Thünen was actually saying was that the rent or price of land would decline with increasing distance from the market center. This prediction was also implicit in Adam Smith's work, while David Ricardo, some ten years before von Thünen, also noted that transport costs were central to the location of industry and agriculture. But Ricardo's message was much more focused on the relationships between the fertility of land and economic rent, and thus it was left to von Thünen to make the relationship to transport cost explicit.[22]

His model, of course, was only applicable to local economies at the scale of his farmstead, but he argued that the principles would be the same for much larger estates and even for small towns that existed within an agricultural hinterland. His example was based on a hypothetical market at the center of a state, which was shielded from any outside influences that would distort the picture. Such an isolated state would be encapsulated within a wilderness; its landscape would be quite flat, with no rivers or other transport routes, and farmers would locate so that they would maximize their profits. In such a system, each product would generate different profits at different locations. Those that needed to be closer to the market with higher yields and the need to get to market quickly than those further away would incur rents that were higher nearer the market. What would happen is that these more profitable activities would outbid those less profitable from occupying nearer locations. This process would order the production so that a circular belt of the most profitable would occupy the ring of land use nearer to the center, the next profitable the next nearest band, and so on until the margin of production was reached, with the least profitable land with the lowest yields defining the boundary of cultivation. The process of ordering was akin to solving the equilibrium set of locations that enabled the local economy to function efficiently in generating normal profits. Von Thünen's great contribution to our understanding of how markets were organized spatially was thus the fact that rents for activities would decline with increasing cost or distance from the market center, and the location of different land use associated with these rents would reflect this.[23]

Von Thünen, in fact, speculated that the concentric rings of different land use devoted to different agricultural products would follow those that he considered represented ideal types: highly perishable, intensive farming and dairying would occupy land nearest the market center, timber and firewood production needed for heating would occupy the next ring, then there would be extensive field crops such as corn, and finally at the furthest distance from the center, there would be ranching. Wilderness would lie beyond. Each activity would pay rent according to its needs to be near the center, and this related to what was being produced, the nature of the demand, and the way it might be transported to the market. He also indicated how the pattern of concentric zones would be distorted if the state were not so isolated—if, for example, a particular transport route such as a river were to be part of the landscape, or even a railway, although his speculations predated passenger railways, which only began in 1825 in England and did not go anywhere near his estate until the late nineteenth century, long after he had passed away.

THE INDUSTRIAL CITY

Translating these rather basic ideas to cities from agricultural markets might seem quite straightforward, but it took more than a hundred years before the jump was finally made. It came in the 1920s from observations in Chicago, where the city appeared to be composed of concentric zones of different activities—this time populations of different ethnicities and income levels and industries of various types—that varied with respect to the cost of location at different distances from the CBD. Not only did population densities decline with distance, but there was also a succession of land use that displaced one another as the city grew in size and the population became richer. The model that was proposed was rooted in social segregation, but this, in turn, was the product of different income levels and of the fact that the cost of transport increased as one moved to the edge of the city, while the cost of land (rent) fell. Somewhat ironically, although the cost of land fell as one moved away from the CBD, poorer people were trapped in the inner areas of the city where they paid high rents for small units of space, while the rich could afford to travel to the edge of the city, thereby paying substantially less for land, enabling them to purchase ever bigger properties on the edge of town. This model was first developed by the Chicago school of social ecologists,[24] in particular by Robert Park and Ernest Burgess whose book *The City* introduced and demonstrated this notion of concentric rings differentiated socially as well as economically. Indeed, its publication rang a chord with many commentators on the city, who saw the same sorts of pattern based on concentric rings of like land use repeat themselves time and again in many cities globally.

Another thirty or more years elapsed before von Thünen's economic logic came to be linked to these observations of city structure. In the late 1950s, another German economist, Martin Beckmann,[25] suggested that a model of the city in which rent declined inversely with transport cost from the center (CBD) could be generated from basic assumptions involving the way an individual wishing to purchase housing could maximize their utility as reflected in the trade-off between the price or rent of land and transport costs. Beckmann pointed out that this had only been shown with any finality in an agricultural context by von Thünen, but in 1957, his speculation that this applied to cities as well as agricultural markets was the first time a formal association was made. It was, however, left for William Alonso, a graduate student at the University of Pennsylvania, to work up the full derivation of von Thünen's model from microeconomic theory in the early 1960s and to apply it, somewhat hypothetically, to a contemporary urban system that mirrored the typical industrial city that had emerged over the previous 150 years.[26] A flurry of activity followed, and a small group of influential economists, some of Nobel Prize–winning stature such as Robert Solow,[27] fashioned the model into the core of what came to be called the "new urban economics," perhaps the most elegant statement that we have had to date of how the form of cities related to their function in economic terms.

What Alonso demonstrated was not only that rents fell off inversely with transport cost from the center of the city, but also that population densities declined in the same way, which generalized nicely to a whole range of different population and industry types. Elegant theory, of course, demands consistent observation, and although there was some rather general correspondence between what cities looked like and what these models predicted, the match was not that good. In short, just as most theory in the social sciences falls short of the kind of match that will convince us to use that theory to predict even the short-term future, so it was with urban economics. As we now know, unpredictability is the hallmark of this kind of theory. As well as the various objections relating to the lack of a good correspondence between theory and reality, there were some pretty sweeping assumptions made that go back to the origins of these ideas themselves. First and foremost, the city was, is not, and can never be an isolated state. Even von Thünen in theorizing about an agricultural economy discounted any outside influences that would distort the regular landscapes that his model could produce. When it came to cities, it took an even greater act of faith to see the city in this way. Moreover, cities are so much bigger than agricultural estates. In 1810, when von Thünen began his work, the population of Berlin was only 170,000, while New York was a mere 96,000. London was at its preindustrial limit of 1.3 million, and even though all these cities would explode in population during the nineteenth century, the stretch involved in scaling von Thünen's model from the agricultural estate to the city was a stretch too far. This issue of scaling up is important because once one scales to the level of world cities, you begin to see that the city no longer has one center—the monocenter—it can have several—"edge cities" as they have been called.[28] The city is more likely to be polycentric than monocentric,[29] and one of the problems with the von Thünen–Alonso model is that it is very hard to adapt and apply to a city with more than one center. To an extent, this is the social sciences many-body problem.[30] The theory is worked up for one body, it might even be capable of handling two, but any more than this and it breaks down, or rather the complications begin to overwhelm the form that eventually results. Predictions thus become ever more arbitrary.

The model might just have been applicable at the end of the nineteenth century, but when urban growth really took off after the invention of the automobile, this sowed the seeds of its demise. Cities too began to change with the relative decline of industry in terms of the number employed due largely to automation and the massive expansion into services. At the end of the nineteenth century, although the CBD was largely given over to market functions—to services and commerce—vast swathes of the inner areas of cities were dominated by industry. Beyond this and even mixed in with it was working-class housing, which then gave way to middle-income groups in the next ring outward from the center. Further out still was housing for the rich, with the dominant transportation patterns being public transit nearer the center and the automobile further away in the suburbs. However, as cities grew, some began to fuse with one another, complicating even further the notion of the isolated state. Some cities also grew to the point where the relative balance between their center and periphery generated edge cities in peripheral locations whose advantages in terms of accessibility gradually wiped out the initial advantages associated with the CBD. Indeed, the future of the largest cities would now appear to be that of cities that fuse together, composed of many centers that are increasingly specialized and diverse, polycentric rather than monocentric in terms of the way they grow outward, inward, and upward.

Von Thünen's model and all the subsequent contributions to the theory of how the land market resolves the location of competing activities is based on the notion that distance in terms of transport costs determines how the city is configured. It is perhaps strange that the notion of a city on a featureless plain still held sway when von Thünen's theory was first adapted to the city by the urban economists, for by then, it was quite clear that the concentric rings of like land use around the center are massively distorted by the presence of transport. Radial routes cut through the urban landscape primarily to funnel commuters to the periphery, but concentric routes such as ring roads and beltways also emerged once the automobile began to dominate transport in the early years of the last century. To an extent, these radial routes, which we can picture as branches originating at the center, represent the most efficient way in which people and goods travel to the edge of the city, and the tree-like forms that evolve are those that seek to minimize the energy that is used to enable activities in the city to function through movement.[31] We have come across this before but at a much larger scale. In chapter 2, we noted that as the postal system evolved in Britain, the system of post roads radiating out of London probably represented the most efficient way of linking the capital to its provinces, and the dendritic form that emerged represented a pretty characteristic feature not only of the way towns and cities grew from early modern times through the first into the second industrial revolution, but also of the way entire nations developed their internal connections. In Britain, post roads turned into trunk roads and thence into freeways or motorways, and as cities grew, the skeletal structure of these primary highways came to be filled in with routes of lesser capacity but nevertheless preserving the continuing hierarchical efficiency of such structures.

However, our knowledge of how systems as complex as the city function remains quite limited. Since the work of the urban economists, beginning in the 1960s, our theorizing about cities in economic terms has hit a brick wall. The problem is not only that modern cities are more complex in terms of the clusters that make up their urban structure than the model can accommodate, but also that there are many nonspatial factors that intervene in their evolution and growth that the patterns we observe cannot be pinned down to a limited number of causes. Distance (or travel costs) still represent the main structuring device, but its influence with respect to many other forces is hard to unravel. You might also think that because we have some sense now that cities process energy in quite distinct ways, with energy flowing through a hierarchy of networks, then we are able to understand how these flows determine the way land markets work. But this is not the case. Even going back to a time when information technologies were in their infancy, back to our magical starting date of 1937, cities were then remarkably complex affairs composed of a multitude of networks that clearly worked together in diverse ways but were impossible to unravel with respect to a clear and unambiguous theory of how the city functioned.

Although we will not yet complicate matters by adding information technologies to this picture, for this is our quest in subsequent chapters, we will bring to the table the basic notion that the continuing industrial revolution marks a dramatic transition from a world based on energy to one based on information. We will develop this toward the end of this chapter, but what this transition means in the most obvious way is that cities are increasingly dominated by networks through which many kinds of information flow. There are complex substitution effects between flows of energy

and information, while the relative invisibility of many energy networks—
which has been a major problem in the past in understanding cities—is now
compounded by an even greater invisibility with respect to digital information. Because our understanding of cities is still mainly in terms of what we
see, this invisibility is complicating our world in ways that were never anticipated. Fifty years ago, it was thought that as we learned more about cities,
we would be in an ever-better position to understand their functioning and
then, in doing something about them with respect to making them more
sustainable, increasing the quality of life of their populations. But quite the
opposite has happened. Cities are getting more and more complex and less
and less manageable as more and more technologies are invented.[32]

With respect to theory, then, we are running to stand still. The transition
from energy to information is not one of mere replacement but rather one
of developing complementary flow networks that are increasingly entan-
gled with one another and which we find increasingly hard to chart and
observe. In fact, theoretical developments using the systematic paradigm
have not just stopped but are still being informed by new ways of thinking
about cities that are more geared to agglomeration, segregation, migration,
and innovation. A new urban science is in the making, and although much
of this has not yet become directly applicable to ways in which we might
generate better urban plans, new directions that we will hint at toward the
end of this chapter are emerging. Part of this is, in fact, the notion that this
science depends on flows and forces articulated through networks, which
we will turn to in the next section.

CITIES AS NETWORKS OF FLOWS AND FORCES

There is another way of approaching the standard model, which reflects the
various forces and tensions in modern society in terms of where we locate
and interact with others to engage in social life. Ever since the first cities emerged, the diversity and the prospects for human advancement that
they offered made them a sufficiently attractive force to lead to successive
waves of migration from their agricultural hinterlands, only limited by the
resources available with the prevailing technology. Right from the beginning, two related forces—the first, centralizing activity in the city, and the
second, its concentration—became evident. Centralization is not the same as
concentration as we will see, but in pre- and early industrial society, these

forces acted in much the same way.[33] Cities from classical times until the early modern period (the Renaissance in Europe) were built around their central market areas—the Greek and Roman forum and the various ceremonial and religious complexes that defined the source of economic and political power. The typical urban form embodied these centralizing forces, as well as compacting them into structures that enabled rapid and almost immediate social interaction. Cities seldom grew to more than a couple of miles in extent, and where several cities existed to serve a wider agricultural hinterland, they were spaced according to how far one could walk within about an hour, which was little more than four miles. Centralization thus tended to concentrate resources in cities, enabling the urban population to serve their hinterland. But even in past times, it was entirely possible to see some cities as central places whose form was not particularly concentrated—capital cities were often of this form—while other cities were concentrated clusters of activity but without exhibiting the centralizing forces necessary in the exercise of political power.

Centralization and concentration, however, are best seen when we examine their opposites—decentralization and deconcentration—which are the countervailing forces that first seemed to appear in cities at the beginning of the industrial revolution. As new mechanical technologies for moving people, largely in the form of railways and tramways, were invented, this gave the richer populations in cities the opportunity to move further out from the urban core where most work was still concentrated. This outward movement must be contrasted with inward movements exerted by the city's attractive force for generating more prosperity by working together. The tension between these opposing forces is effectively the same as the trade-off between what one might pay for living near the center and what it might cost to live further away.[34] It is analogous to the balance between the rent paid for living and working closer to the center and the transport cost of living and working further away: in short, this is the von Thünen model. New mechanical technologies drove these forces to push people not only to remain part of the city but also to locate far enough away so that as cities grew during the nineteenth century, they began to sprawl. Decentralization provided more space at decreasing costs (per unit of space), but initially, it was only tramways and railways that enabled such decentralization to happen. The growth that took place was very strongly determined by the railways that penetrated out from the core of the city into its hinterland. Because the number of stations along the various radial routes from the city was quite small relative to the distance traveled, suburbs clustered along these radials like beads on a string. These suburbs were quite compact, and the dominant modification to the standard model in the mid-to-late nineteenth century was one in which the von Thünen landscape was peppered with compact clusters that related to movement by rail to the central city.

To develop truly decentralized and deconcentrated development around the city, a much more ubiquitous form of travel was needed, and this came with the invention of the automobile in the late nineteenth century. Suddenly, city dwellers could locate anywhere within the city's hinterland, conditioned, of course, by the cost of travel and land but free of the strict radial routes determined by the railways. This kind of development rapidly became the norm everywhere in the industrialized West. In North America, where car ownership was almost universal by the mid-twentieth century, the industrial city began to unravel. Large-scale mechanical transport, largely publicly provided, began to decline as more and more people began to travel using the automobile. The city center became less attractive, and many CBDs went into rapid decline. New centers were created on the edge of town, called "edge cities" by Garreau in his book of the same name,[35] in fact at locations whose overall accessibility became massively improved by new urban road systems designed to cater for the car. The typical US variant of the standard model was polycentric, highly decentralized, essentially deconcentrated with very low-density sprawl.[36] You can see such variants of this model everywhere, particularly in the south and west of the US, but also in France and other parts of Western Europe, while it is being increasingly elaborated in yet other variants in places like China.

In less wealthy but developed economies that have a longer historical heritage that determines what is possible physically as cities grow, their forms are still quite decentralized. Activities that were once in the city center moved out to the periphery in the same manner as in North America, and edge cities devoted to retailing and commerce began to dominate the overall urban landscape but in a more muted way. Development is more compact, for there is an emphasis on rail and local bus transport, as well as more people walking and biking. These kinds of city still reflect a balance of inward and outward forces, but the balance is on the side of decentralized but compact urban form, not unlike what has been called "transit-oriented development." Expanding upward primarily in the city center has also become fashionable too, particularly with the construction of skyscrapers or "cathedrals of commerce," which tend to be prestige projects. They are not driven by the need for ever more space but rather by the desire to be close together, which is captured by building upward.

Now, imagine that these forces begin to change in their intensity, which is when the distance between one place and another falls in significance or more directly when the transport costs that covary with distance begin to fall. If transport costs fall to zero, locations in the city will be purely determined by the rent that they are able to pay, which now has nothing to do with transport costs. Activities thus become entirely footloose with respect to transport costs, and populations can continue to decentralize, creating ever-wider swathes of sprawl. In short, the city explodes, with many places appearing to have followed this line of development in the last fifty years.[37] In fact, transport costs cannot fall to zero, as there will always be some friction of distance, but as these costs have continued to decline, cities have spread out even further. When most people are traveling by car, then the urban form that results is low density, rather scattered, and in no way compact. If transport costs increase massively, heading ever upward, then everyone in the city will want to live at the center. Densities will rise, and the city will become ever more compact. In short, the city will begin to implode. These characterizations are extremes, unlikely to happen, but they do provide a series of thought experiments to explore what future cities might be like if distance became no longer significant.

BEYOND THE STANDARD MODEL: A NEW URBAN SCIENCE

When we come to scrutinize the role of distance in cities in the next chapter, we will explore a cornucopia of relationships that all relate to spatial impedance. These involve social networks, transportation networks, information networks, and so on, all of which are likely to have an enormous impact on the geometry, shape, and size of the city. These forces continue to complement and substitute traditional means of movement and travel with the flow of information, email, and web traffic, all of which are now much more voluminous than traditional modes of interaction. To an extent, of course, this is one of the great challenges that we describe in the histories we recount in this book: what the smart city of the future look will like when so much of our social and working life is replaced with information flows. The industrial city was largely composed of flows of energy, materials, people, and a limited amount of information. Where you get many networks crisscrossing and linking to one another, which is somewhat casually called "multiplexing," the pattern of flow becomes exceedingly complicated, and we do not have a good theory to deal with such networks. As well as the limits on our abilities to observe networks that are often physical but hidden from our superficial view, we do not have a theory as to how networks substitute and complement each other, nor do we have coherent methods for describing such overlapping networks and how they function in concert. This remains an important challenge.

The standard model is not simply an ideal type but a flexible structure that can be adapted at least intuitively rather than formally to deal with much more complex spatial systems. To apply it to cities composed of many centers, one can simply take the von Thünen model and apply it to every place, assuming that every place has a degree of monocentricity. This generates polycentric structures that are much closer to what we see throughout the world's cities than the monocentric focus that is our ideal type. It is even possible in an ad hoc manner to apply this model to every place, imposing equilibrium conditions on the final form and adjusting its parameters to secure a spatial equilibrium. This might be done in a simple iterative manner, but its limits relate to the fact that it is not possible so far to devise a procedure where the final equilibrium can be generated formally. In short, unlike von Thünen's model, it is not possible to generalize the model to ensure a closed analytic solution, which, to an extent, is still the long-term aim of this kind of economic theory.

In fact, in the various land-use, transport, and development models that we review in chapter 10, several of these models are consistent with von Thünen's ideas in that these models can be viewed as a generalization of the monocentric models to all locations. Sixty years ago, when Alonso began his exploration of the von Thünen model, the field anticipated that great strides would be made in its generalization to many locations and to many sectors and agents involving the spatial demand for activities in different locations and the process in which they might be supplied. But progress was slow, as much because this field was driven hard by a range of urban problems that continued to evolve, particularly those associated with questions of equity rather than efficiency. These social physics-economic styles of model were unable to address these questions in a very focused way, and the field began to shift toward different types and styles of systems model.

Social physics and urban economics were the main drivers of the early models and constituted an early form of urban science, where space and location were the key determinants of how we considered cities to be structured. However, a new urban science that takes a very different approach to articulating the functions of cities has emerged, and this depends on theorizing about how size, scale, and, to an extent, time, rather than space, are important concepts in this new world.[38] As cities grow and change in size, they change qualitatively, and as they grow, they invoke economies of scale, whose signatures are referred to as scaling. As cities get bigger, their attributes change more or less than proportionately, superlinearly or sublinearly rather than linearly, and although this has been widely established for whole cities and for their networks,[39] there are still some major questions to resolve as to how these relations change over space within the city, how they change as the hierarchy of functions within cities evolves, and how they relate to questions of poverty as well as wealth, which embrace a range of problems pertaining to spatial equity and segregation.

We could elaborate different approaches to urban science across a wide range of perspectives, but many of these, although providing essential background to any appreciation of the computable city, have not yet been translated into digital form and, to an extent, are limited in their representation using the kinds of science that we discuss here. In later chapters, we will see how elements of this new urban science are being developed for the high-frequency city, for citizen participation using digital technologies, for new forms of mobility that are being understood from big data, and for a host of problems that traditionally we have not considered as being part of urban science. In this sense, our treatment of the computable city is complete in that it brings us up to the present from which the future of this urban science, of the city itself, and yet undiscovered new technologies are uncertain, opaque, and, of course, unpredictable.

IMPACTS OF THE EARLIEST INFORMATION TECHNOLOGIES

In chapter 3, we introduced the idea popularized by Nicholas Negroponte at the MIT's Media Lab, who coined one of the most evocative clichés for the transition from industrial to post-industrial society. This, he argued, is the transformation from atoms to bits, from manufactured physical machines based on atoms to organisms built from the bottom up based on bits—that is, from hardware to software.[40] This is the transition from a world built around the use of energy to one built around information. But this transformation is much more than energy substituting for information. These concepts are not pure substitutes, they are complementary if anything, and flows of energy in cities are often paralleled by flows of information. Information is added to energy flow and vice versa, and this implies that the world is becoming ever more complex as cities become instruments not only to process energy but also to process information in parallel and simultaneously. Of course, we have known this for a long time because information did not simply emerge in the Modern Age. In chapter 2, we sketched the growth of printing, and we could go much further back to prehistoric times when humankind first began to communicate using pictures and simple words.[41] Our point is thus that this transition has been underway from long before the first industrial revolution began, except that it is now proceeding apace and it threatens to change everything in terms of how we represent cities and their technologies: from physical to digital, from hard to soft, even from real to ethereal, physical to virtual.

As we recounted in part I, information technologies based on the representation of many types of media were first invented at the very start of the industrial revolution. It was Samuel Morse (with Charles Wheatstone) who is accredited for inventing the telegraph, and it is to him we owe the apocryphal phrase that he telegraphed back his invention from Baltimore to the US Congress in May 1844, which said, "What Hath God Wrought?" To transport oneself back to those days when electricity was barely understood, it must have been truly amazing to see for the first time a wireless signal being repeated over a long distance—far longer than any other technology could communicate—so that someone who was not necessarily expecting it would suddenly be in instant communication with someone very many miles away. Of course, the telegraph's real impact was on connecting systems of cities together, connecting up the nation, although the devices were rarely considered to be portable, despite some effort devoted to developing a personal telegraph.[42] In this sense, cities did not change that much with the advent of the telegraph, but the telephone invented in the late nineteenth century by Alexander Graham Bell was a different matter completely.[43] In his typical unassuming manner, once he had made contact with his assistant, he said, "Mr. Watson—come here—I want to see you." The year was 1876, and it would be another century before such transmission became visual. Only now is it routinely so with smartphones, using social-media support with software such as FaceTime and in conferencing systems such as Zoom.

You would have thought that the telephone being largely portable but requiring fixed wires would have unleashed a communications frenzy in the US, but this was by no means the case. Telephones, as telegraphs before, were largely seen as business devices: the idea of mobile phones did not come onto the agenda for another ninety years or more. I first saw a mobile phone—it was a car telephone that could be taken from the car but on a long wire—in 1974 in Southern Ontario, but it was not until the early 1990s that mobile phones became small enough to become truly portable. What is somewhat strange is that very few people have mapped the diffusion of the telegraph and the telephone, but from what we do know, the telephone spread out locally in places where the richer populations lived—in the suburbs for example. The real story of the smart city, of course, should chart these early information technologies in much more detail than we have done so far, but as we have been at pains to point out, telephone traffic, like so much networked information, is largely invisible. A lot of the early development of such networks and their traffic has been lost.

Following hard on the heels of the telephone was cable, radio, radar, and then television, but all these innovations were essentially passive devices—to communicate information without letting the user manipulate it. Indeed, the very idea of the computer broke with the idea of one-way traffic that dominated radio, television, and all the rest of the technologies that were invented before the computer. It was programming, of course, that elevated the computer beyond all these other information devices as we made very clear in the first part of the book. Electricity was obviously one of the keys to the Computer Age, but it was programming that essentially divides early information technologies from the later ones. This is another important reason for charting the computer revolution from the year 1937, when the binary digit came into prominence as the basis of modern computation. In some senses, we still need to explore these early technologies a little more in the next chapter because they are key to the death of distance, which is one of the central constructs in our development of the smart city. In fact, there is a considerable way to go in terms of the development of the smart city, for we have more or less concluded our history at the point where the internet took on its basic form. But this was twenty-five years ago, and much has happened since then. It will take us the rest of part II to develop the background as to how computers are being embedded into cities, and in part III, we will describe the ways in which computers can be used to understand cities rather than control them. This contrast between understanding, which leads to simulation and prediction, is very different from the use of computation to control the functioning of the city, but by the end of part II, we will have a clear view of the conundrums and paradoxes involved in making cities smart and computable.

7 THE DEATH OF DISTANCE

This is just the beginning, the beginning of understanding that cyberspace has no limits, no boundaries.

—Nicholas Negroponte (1999)

THE BRITISH MUSEUM READING ROOM

The British Museum opened its Reading Room in 1857, and for the next twenty-five years, its most famous son toiled away constructing a monumental edifice that was to rock the twentieth century as no great work had done before. It is thus tempting to think of Karl Marx first claiming, then sitting in his favorite seat, G7, when the room opened, speculating about the death of distance as a prelude to thinking about the way in which capitalism was tearing down the barriers imposed by geographical space and time in the mid-nineteenth century. Here, Marx wrote much of his work,[1] particularly *Das Kapital* but also the *Grundrisse der Kritik der Politischen Öko-nomie*, his voluminous unfinished notebooks that he first started to put together in the library in 1857–1858. The museum where he worked was,[2] in many respects, another of the world's great focal points, at the heart of an empire, as was the General Post Office where we started our quest to unravel the computable city in our second chapter. Globalization was all around him, the museum had been fast-acquiring artifacts from around the empire such as the Elgin Marbles, and London had become home for many radical groups that saw the country as a haven of tolerance. Although we have been at pains so far to impress upon the reader that it is digital information technologies that have led us to a truly global world where everyone can be in touch with everyone else, the seeds of this revolution go back to the time of Marx and before—in fact, to the origins of the industrial revolution itself and perhaps even to the emergence of the modern world in the Renaissance.

During the nineteenth century, the idea that space could be annihilated by time caught the public imagination in a way that just occasionally happens when science and philosophy accidentally enter the mainstream. It is hard to know when the phrase was first used, but in 1844, the philosopher Ralph Waldo Emerson articulated it when he said of the development of steam power and the railroads, "Not only is distance annihilated, but when, as now, the locomotive and the steamboat, like enormous shuttles, shoot every day across the thousand various threads of national descent and employment, and bind them fast in one web, an hourly assimilation goes forward and there is no danger that local peculiarities and hostilities should be preserved."[3] Emerson, among others, argued that such technologies that transformed the speed at which we could move across the landscape and the sea were instrumental in changing the balance between the local and the global. It was no longer necessary to dwell on the local and the immediate present when there was the possibility of traveling long distances across the globe, rather rapidly in comparison with travel hitherto. For the first time, the local could truly merge into and embrace the global.

The term "globalization" was first introduced casually nearly a hundred years ago after World War I, but it only took on its wider usage with respect to the globalization of the world economy in the 1980s when it was popularized by the economist Theo Levitt.[4] Yet, Marx did more to establish the idea than most nineteenth-century social commentators, revolutionaries, and philosophers. Indeed, there is a deep sense that the famous pamphlet *The Communist Manifesto*, written in 1848 with Friedrich Engels,[5] rang a chord not only with globalization but also with the technological revolution that was transforming the way in which production and consumption were changing class structure. In fact, in the *Grundisse*, Marx articulated this transformation with great clarity when he said, "Capital by its nature drives beyond every spatial barrier. Thus the creation of the physical conditions of exchange—of the means of communication and transport—the annihilation of space by time—becomes an extraordinary necessity for it."[6]

THE ANNIHILATION OF SPACE BY TIME

To explore how technologies change space and time, we need briefly to recount what we know about the long-term history of how humankind evolved from a nomadic existence to create the kinds of settlements that today define the world's cities. From the time when man last walked out of Africa some sixty thousand years ago, and once the last Ice Age had finally begun to pass, human existence has been dominated by the ceaseless quest to control space and time. Space provides the container for resources that man requires to survive. It is clear that the invention of technologies from the most rudimentary to those we are currently evolving in the Information Age have been developed so that we can control our access and use of resources ever more effectively, hence engendering continually increasing levels of overall prosperity. This does not mean that our long-term survival is ensured, but it does mean that we can chart the progress of conquering space with time—of annihilating space by time—in terms of the extent to which we are able to control resources that are spatially distributed.

In fact, our nomadic existence in the form of our origins as hunter-gatherers dominated all our histories until the last of the ice began to thaw about ten thousand years ago and man moved from this existence to one of settled agriculture. Prior to such settlement, man usually spent three or four hours a day hunting and foraging for resources so that families might survive, but once agriculture emerged, this dropped dramatically to about one hour a day, which was about how far one could walk to support a more settled existence. The very first cities then emerged probably about seven thousand years ago, and the agricultural revolution began its transformation to an urban revolution,[7] but still technologies of movement were such that average travel of one hour a day remained intact. Despite the development of wheeled transport, which tended to be for the rich, it was not until the start of the industrial revolution that technologies really did begin to change how far we could travel and thus our control over space. The notion of space being annihilated, which gained such currency in the popular media in the first half of the nineteenth century, was dictated primarily by the development of the railways—by the Age of Steam.[8] This marked the beginnings of a revolution in energy, which dramatically changed how far we could travel. Its continued transition into the information revolution and the ways we can now communicate without physical movement is, of course, the focus of this book.

In fact, this average time budget of one hour's travel a day, perhaps somewhat more in the largest cities, continues to dominate contemporary life, and we will have more to say about this when we discuss Frances Cairncross's idea of the death of distance in the next section.[9] But it is worth introducing the various stages through which the industrial revolution has passed, for each of these serves to define the evolution of distinctly different machine technologies. The most recent classification into four industrial revolutions begins with the invention of steam power, which led to the mechanization of production. This lasted for a hundred years or so from the middle of the eighteenth to the middle of the nineteenth century, and then began the second revolution, which was based on the invention of electricity. This lasted some fifty years, and while the first revolution saw the invention of the railway, the second saw the invention of the automobile. It is really the third and fourth revolutions that preoccupy us here, for the third began with the invention of the digital computer, and to an extent, it has continued from the middle of the last century until now. The fourth, in fact, represents a fusion of technologies, with the mechanical, digital, and biological converging. Sometimes, this is the called the age of cyber-physical systems, but Klaus Schwab has popularized it as the "fourth industrial revolution," although it remains to be seen how the transition to further revolutions evolve.[10] A fifth is clearly in the making, unfolding almost as you read these lines and this is AI. There is a sense in which each technology develops and diffuses faster than the last, with some sense that the future will be dominated by continuous inventions, continuous revolutions.

There are many other characterizations of this sweep of technological history over the last 250 years, each with its own dating of when new technologies originated and developed, but broadly they all accord to Schwab's classification. Kondratieff, for example, defined cycles or waves of about fifty years that coincided with the development and subsequent dissemination of major technologies, which were marked by periods of rapid growth and then comparative decline that in turn gave way to more technological inventions that accelerated the upswing into another cycle.[11] The current wisdom about these cycles—and there is considerable ambiguity about their form and length—is that the first Kondratieff began with the development of steam power from 1780 to 1830.[12] The second marked the development of the railways from 1830 to 1880. Then, the third was based on the automobile and electricity until 1930, the fourth on the digital computer until about 1980 when the invention of the PC initiated the passage into the fifth. This ended in about 2010 with the Great Recession.

Elsewhere, I have called this the sixth wave[13]—the age of the smart city—but all this really accords to Schwab's fourth industrial revolution, which is a fusion of the mechanical, digital, and biological. It is also the age of machine learning, the beginning of autonomous vehicles, the second machine age, and the era of weak AI. In fact, some say that the cycles are overlapping, and certainly in terms of information technologies, it is convenient to think of initial inventions in the 1930s and 1940s forging the development of mainframes into the 1970s, their scaling down to minicomputers and the transition to PCs and workstations in the 1980s, which had run their course by the early 1990s. The internet was then popularized as the World Wide Web, and this then developed in parallel to the continuing era of extensive miniaturization where the computer at last became identified with individuals rather than larger organizations. These are the distinct periods that define the way computers, software, and networks have evolved, increasingly overlapping with one another, which we described in our early chapters.[14]

At each stage in this history, the technologies that enabled us to move have increased in their spatial extent over which they can carry people and materials, and it is the speed at which such technologies can propel us that defines this range. It is in this sense, then, that space is annihilated by time. Space is continually compressed in terms of the distance one can travel in a unit of time, and although the term space-time compression was not coined until David Harvey resurrected Karl Marx's writings on these matters some thirty years ago,[15] the ideas go back to the development of steam power in the early nineteenth century, as we noted at the start of this chapter. This has also been called "space-time distanciation," with an implicit play on the word "instantiation." Essentially, one can measure the degree of compression simply as the range of travel associated with the use of different technologies that enable us to move so far in one hour. Or if one prefers an actual numerical measure, we could use the inverse of this. Prior to the first industrial revolution, the range was about 3 km, the railways brought the range up to 40 km, the automobile to 50 km, and the airplane to 500 km, and evolving technologies such as trains based on magnetic levitation are bringing it toward 1,000 km per hour. Prior to the Age of Steam, ships moved almost as slowly as walking, but then they began to approach the average speed of railways, and this has remained pretty constant at about 35 km per hour ever since. Of course, these measures break down when we enter the age of digital communications[16]; for example, using email, modern phones, web pages, and so on to communicate, we have no equivalents. It might be argued that Harvey's discussion of these matters marked the end of an era, before the internet became widespread, and that, in some sense, the idea of space-time compression takes on an entirely different meaning when communication effectively becomes instantaneous.

In terms of the information technologies used to communicate written or visual information, it is still quite unclear if there is a distinct average that is conserved regardless of the technology. This is made all the more difficult by the fact that different communications technologies substitute and replace one another, and there is still no sense as to the long-term cost of these technologies and whether these will be constant. Many are still developing; hence, their costs of use are falling. Now, the US adult population spends about eleven hours a day using various kinds of media, and this is up from about nine-and-a-half hours some ten years ago. In terms of time spent on email, the average is about ninety minutes a day, while for social media, the time used is more than two hours a day and still rising. TV accounts for about five hours a day. As devices devoted to different functions of communications converge, these time windows are likely to become consolidated, but nevertheless, the notion that almost half the twenty-four-hour day is devoted to such media use is astonishing, not to say somewhat unnerving.[17] We will return to these issues later in this book, fleshing out the question of space with respect to our routine access in the next section, but for the moment, we will return to Marx, who has more to say on the matter.

Marx argued that societies (and cities as an integral function in society) could be articulated as a set of physical or "material" relations such as those embodied in the use of labor and capital, which necessitate production and consumption. His thesis was that as any society developed with respect to industrial technologies, these relations would be transformed, one consequence of which would be the people (the proletariat) rising up over the bourgeoisie, thus taking control of capital, the means of production. The essential implications of this transformation for cities, he argued, was the compression of space by time, which marked the continual transformation of production and consumption. His key point was that the processes of transformation were getting ever faster, changing at a quickening pace, in just the same way that new technologies were increasing time-space compression and altering the means of production out of all recognition. He developed this thesis over many years, but it is writ large in the manifesto he wrote with Fredrick Engels in 1848, now called *The Communist Manifesto*. There, Marx and Engels (1848, 1888 edition) said, "Constant revolutionizing of production, uninterrupted disturbance of all social conditions, everlasting uncertainty and agitation distinguish the bourgeois epoch from all earlier ones. All fixed, fast-frozen relations, with their train of ancient and venerable prejudices and opinions, are swept away, all new-formed ones become antiquated before they can ossify. All that is solid melts into air."[18] This sounds very much as though it was written by singularity theorists such as Ray Kurzweil,[19] but it predated such contemporary concerns for the future by at least 150 years, thus establishing that we are always rapidly heading toward a future that is almost entirely unknown.

CAIRNCROSS'S THESIS

In 1995, *The Economist* published a series of articles edited by Frances Cairncross[20] entitled "The Death of Distance." This was only five years from the time when Tim Berners-Lee produced his software for the World Wide Web, which we presented in chapters 4 and 5. The idea that we might link across such a web so that one could view pages of information developed remotely by someone else was a revolutionary concept. It took a couple more years for mainstream universities, largely in Western countries, to develop software to connect between different computers remotely using the web and for the wider public to begin to use email, which had been developed since the 1970s for specialist use. The notion that one could communicate instantaneously across such networks raised the prospect that people would no longer go to the office but work from home, that education and health care could be delivered remotely, and that the whole production–consumption cycle would be turned upside down and inside out by organizations that suddenly could operate virtually using various forms of information technology. After all, information by this time was outstripping all other forms of production, and the new means to manipulate and communicate it at a distance was all too clear. This was the prospect that Cairncross wrote about in her *Economist* article.

In fact, Cairncross actually appears to have coined the phrase "the death of distance," although there are many suggestive precursors. Heidegger, among others, talked of the "abolition of distance," and of course it is a short step to the annihilation of space by time if you consider distance to be the main arbiter.[21] As you might imagine with such an evocative term, it has had plenty of critics, all playing on Mark Twain's famous phrase that "reports of my death have been greatly exaggerated."[22] All these commentaries imply that distance is as important in thinking about cities and information as it always has been but that digital information technologies make it different. Cairncross had such an overwhelming response to her article that she produced a book with the same name that outlined her thesis in considerably more detail. In fact, she extended her thesis to many aspects of the economy that in principle might be affected by the development of communications technologies, which could substitute as well as complement existing physical connections and flows.

Before we introduce the key features of her argument, which demonstrate how economic and social functions might be transformed by the internet, we will define the actual impact of changes in communications technology in more graphic detail. We were at pains earlier to emphasize the notion that humans appear to travel no more than an average of one hour a day, despite the many kinds of transport technology that they have at their disposal. Without invoking any esoteric technical apparatus, we can figure out the area an adult human can traverse if they restrict their travel to one hour a day using a given transport technology, where we define distance in kilometers traveled for one hour. We know that, on average, any adult can travel 3.1 km if they walk at this speed. Assuming that they have freedom to travel in any direction, the areal range of their movements can be no larger than the circle around their origin, which has a radius of 3.1 km. We can easily work out that the total area around their origin point where they walk in one hour is 7.5 km^2. If they walk twice as far in two hours, then their range is 30 km^2, three times as far then it is 68 km^2, and so on.[23] Now, if the transport technology changes to rail, where the person can travel at the much higher speed of 40 km per hour, the area in which they can travel in an hour is 1,256 km^2, which is something like a square grid around their origin of some 35 km × 35 km.

In this way, we can figure out if the transport technology changes from walking to rail, the amount of space over which the individual has control in terms of access to resources in one hour's travel is more than 150 times greater than the resources they would control if they were to walk. This is an order of magnitude change, and it is easy to see how the invention of rail changed the limits on the size to which cities could grow. Before the industrial revolution, most cities did not grow to much more than 3 km away from their centers, and the very biggest cities pushed hard against these limits of the then technology that restricted their maximum size to fewer than a million people (at much higher densities, of course, than today). The city wall was thus defined by the envelope in which one could walk comfortably within one hour. The invention of steam power and the railways, of course, changed all of that.

The hourly limit that has been observed since ancient times would appear to be almost an incontrovertible threshold were it not for the fact that transport technologies are getting cheaper, faster, and more comfortable and adapting to other functions associated with work and leisure. Furthermore, although most cities in the preindustrial world never grew beyond a population of one million, the empires in which they were a focus were much bigger and depended upon technologies such as fast horse riders, who were able to command a much greater range of distances, up to fourteen days travel in some instances, with relay riders very occasionally covering as much as 500 km per day. In short, the range of travel for very different technologies has always pressed against these limits. In really big cities, usually with a population of more than about five million, this average hourly limit appears to be on the rise, and in world cities such as London, Tokyo, and New York, the average time spent is more like ninety minutes or more. In fact, we will speculate here that this limit is by no means a cast-iron threshold and that the development of communications technology introduces an instancy and immediacy into communications that could well see other limits being reached that trade off physical with virtual distance.[24] Indeed, in a future millennium, we may look back and see everything from the agricultural to the industrial revolution as being based on an average of one hour's travel time, and everything thereafter implying a very different physical constant that we do not yet know. The weightless world that new information technologies have foisted upon us will be full of surprises with respect to how we deal with time and space in the future.

The constraint of one hour a day also appears to be paralleled by an average fixed proportion of a family's individual budget that is spent on travel. This has been estimated to be around 14 percent of average total income, and this appears to have been stable for more than half a century in automobile-based societies. This implies that wealthier populations will spend more on travel absolutely, and this accords with their use of more expensive and faster technologies, which in turn implies greater travel distances and speeds. This limit too seems more precarious than the daily travel time limit in that we have little idea of how virtual information technologies will interact with these constraints, and there is little evidence so far for how human behaviors will adjust to these new technologies. Moreover, our perceptions of distance and cost of overcoming space are also likely to change as the set of technologies available to us to initiate communication also changes. Our cognitive abilities may well be changed as new technologies crowd one another in and out, fragmenting but also enriching the ways in which we might communicate with each other in the future.

When a transport technology increases the range over which we might move physically, the range of people and places that any individual can access increases exponentially with respect to the speed of travel. This has deep implications for accessibility to the entire space of opportunities and resources that an individual is able to access. The number of people that one is able to reach increases exponentially as do the resources available, and although there are limits on the number of people one can get to meet in a fixed time, the diversity and richness of individual opportunities increases faster than the number of acquaintances with whom one is able to interact.[25] That is, as the pool of opportunities grows larger exponentially, one's access to those who are most relevant to a particular individual in question grows similarly, and an individual can thus realize the best. Some of these ideas are now being explored in the parallel field of city science, which is beginning to merge with the smart-cities movement.

There is no doubt that the development of new forms of transport and communications are changing the physical configuration of space at every scale from the neighborhood to the nation and to the distribution of cities globally. In particular, barriers and borders are being redefined in particular with respect to the nature of global companies and Big Tech, such as Google, Apple, Facebook, and Amazon (GAFA).[26] The conflict between the nation-state and these companies, which are almost without territory, is changing the nature of community and citizenship. Moreover, every kind of work, physical and virtual, is rapidly being infused with new information technologies that change the nature of trade and exchange as well as processes of production and consumption. The nature of communications is thus being turned upside down. Face-to-face contacts appear as important as ever, and thus the focal points of cities in their cores, particularly big cities and world cities, seem to remain intact, but resource-based cities, which pioneered the first and second industrial revolutions, are slowly disappearing. We will explore the nature of work in this global urban world below, for this is being changed by the convergence of many different information devices and platforms. The notion of cities being organized around a few urban centers and large swathes of suburban residential areas is changing, as form no longer follows function. Doing everything that was once done in a fixed place but now on the move changes the nature of the demand for space and the way it is utilized. Since prehistory, certainly since the urban revolution, cities have been largely defined in terms of fixed places, but this glue is now coming unstuck, and the prospect of a very different world of cities is almost upon us.[27]

There is another important consequence of contemporary information technologies in terms of the death of distance. Our focus on the fact that new technologies of transport invariably increase the speed of movement, hence the range of spatial control, is paralleled by consequent falls in the cost of overcoming distance. Usually new transport technologies are more expensive at first than those that they are deemed to replace, but as they grow and become established, costs of access and operation fall, and this changes their control over distance. The richer groups in society, although starting with an advantage of use, soon lose this as technologies become essential for the ordinary citizen. This can be nowhere more evident than in the costs of telecommunications. For nearly a hundred years from when Alexander Graham Bell first demonstrated making a telephone call from the Lyceum in Salem to his assistant Watson at the Boston Globe, the device remained expensive to use, its cost increasing formidably for those wishing to make long-distance calls, largely due to restrictions on bandwidth available.[28]

In fact, once the convergence between computers and telecommunications really took off with the development of the internet and World Wide Web, telecommunications costs fell through the floor. The introduction of fiber optics and the vast provisioning of internet lines during the so-called dot-com boom in the late 1990s increased capacity massively, and costs dramatically fell to the point where capacity was no longer a constraint. Costs are now determined by issues such as monopolistic practice, regulatory powers, security, and geopolitics. In general, we can call anyone in any place at any time for almost negligible cost, and this more than anything else is distorting the role of distance in where we live and work. In short, another measure of space-time compression must be the unit cost of using a network, perhaps weighted by the accessible population connected by the network. To summarize then, space-time compression is thus a function of speed and cost of transmission where cost is strongly linked to bandwidth, that is, to the capacity to transmit a unit of data.

In the rest of this chapter, we will further elaborate Cairncross's thesis, but before we do so, we need to spell out a final notion that the death of distance is likely to be accompanied by a transformation of time. We have already pointed out that all technologies that have defined the industrial revolution so far follow the classic S-shaped diffusion curve, where there is rapid take-off of an invention or innovation in the early stages of its diffusion. This accelerates exponentially until a point of inflection after which the technology continues to mature and grow but at a decreasing rate until it works itself out, usually as its total market reaches saturation. In the meantime, another technology has begun, and it is ingested to the point where another diffusion takes off. Throughout the industrial revolution, these diffusions have become ever closer, and there is a sense in which they are converging. This is a somewhat frightening prospect in that no sooner has a technology become established, then another takes its place. Some evolve in parallel, and some are linked to one another. But the rate of change is now so great that we have little idea of what the collapse of these temporal trajectories means. It sounds sufficiently similar to the kind of singularity articulated by Ray Kurzweil,[29] but with respect to the death of distance and space-time compression, there are implications for both local and global structures, with temporal as well as spatial distortions. Marx and Engels said it all, as we noted earlier, when they referred to "constant revolutionizing of production, uninterrupted disturbance of all social conditions, everlasting uncertainty and agitation,"[30] which were the key characteristics of the age in which they lived. More than 150 years later, this message chimes down the ages as being more relevant than ever.

RECONFIGURING THE STANDARD MODEL

In the last chapter, we outlined an idealization of the city based on the notion that everybody involved in economic exchange, whether it be for work, shopping, production, or consumption, interacted with a central location, which traditionally was the market. In terms of residence, the model postulates that everyone would live in concentric rings around this point, the rings being differentiated according to the trade-off between what one might pay to travel to the center and the space that one was able to consume at different distances away from the center.[31] In this simple characterization, the cost of travel is all important in determining both how far one is able to live from the center and how big the city is likely to become. As we have emphasized several times, when the dominant form of transport was walking, the city could not get much bigger in area than 30 km^2, and this was the limit until the beginning of the industrial revolution. As soon as steam power developed to the point where railways were invented, the city began to grow outward, this growth being termed "suburbs" and the process "suburbanization." London, for example, grew outward very rapidly from the Napoleonic Wars, with its population increasing from about 1 million in 1800 to 2.6 million in 1850 then up to 6.5 million in 1900. In the administrative area we now call the Greater London Authority, it grew to eight million just before World War II, fell back a little in the late twentieth century, and is now back to its previous figure.[32] The density of population within its built-up area, in fact, has decreased since the beginning of the industrial revolution, but the predominant driver of its expansion has been falling travel cost. Railways came first, followed by the streetcar and the omnibus, then the underground railway—technologies that were all prior to the automobile, which only began to dominate suburban growth from the 1920s onward. To an extent, the automobile is still the dominant technology for moving people, although in the biggest cities, those greater than about four million, the main technology is now a mix of public transport, particularly roads, bus routes, and subways, which are only economical to build once the population of the city reaches about two to three million.

As we implied earlier, most people will be living in cities by the end of this century, but these will not all be big cities. There are many more small cities than large cities, and this will remain the case.[33] The idea that we will all live in megacities is fiction, despite the fact that such cities have become trendy, increasingly dominated by the young, while the suburbs and small towns tend to be associated with the graying population, at least in Western Europe and parts of North America. China is likely to follow within the next decade. The standard model of urban growth, however, which can be pictured as a simple diffusion of growth in concentric rings around the city center, as in von Thünen and Alonso's models, is increasingly one that pertains to the industrial past. As cities have grown since the middle of the last century, many have begun to fuse together, and the dominant pattern everywhere is of a polycentric physical form, with many cities merging together. According to the Nobel Prize winner Paul Krugman, the pattern now looks more like a plum pudding than a regular landscape of city centers hierarchically ordered and evenly spaced in such a way that they do not exist within each other's shadows.[34] By the late twentieth century, this fusion had begun to dominate many city systems in the West such as the northeastern seaboard in the US from Washington, DC, to Boston in the US, and from London to Manchester and Leeds in the UK, and from Amsterdam to Brussels in the low countries. Really big urban agglomerations, which are almost a merger of polycentric city systems themselves, are now fast developing, such as the Greater Bay Area from Hongkong-Shenzhen-Donguan-Guangzhou-Zuhai-Macao and the Shanghai-Hangzhou-Suzhou-Nanjing urban region in China.[35] The standard model is disappearing, for city systems are no longer simple in their structure, largely due to the fact that older cities built around older industries and slower transport technologies are being regenerated into polycentric forms where the old fuses and mixes with the new.

This fusion to create polycentric cities is, to an extent, independent of changes in the speed of transport technology. As cities have grown, even with the same dominant transport technologies, the space required for this growth has led to the fusion of older cities, particularly those that grew rapidly during the first and second industrial revolutions. Even cities with the most recent transport technologies begin to fuse if growth is large enough and they are close enough together, and this has little to do with our being able to travel further. It is purely a product of natural population growth or migration or both in volume, which, at fixed densities, implies the consumption of more and more land. In contexts where there is zero population growth, the way in which cities might fuse is only through the reallocation of the population to ever more distant suburbs. This can only happen if the limits on transport technology and the time spent traveling change. Thus, suburbs are the first direct consequence of the death of distance.

The city of the mid-to-late twentieth century, however, was still a physically connected form, despite continually increasing trends to economic globalization that generated many linkages to economic development outside the city, the region, and the nation. But with the growth of services and massive deindustrialization in Europe and North America from the 1960s onward, the bigger cities acquired more and more global functions. Into this nexus came information technology, and in the last quarter century since the invention of the World Wide Web and the global diffusion of the internet, cities are increasingly part of a global network—a skin of telecommunications where cost of access is no longer a major function of distance or even connectivity. In many environments, telecommunications costs have sunk to near zero, and as we noted above, pricing is no longer determined by physical considerations. When Cairncross developed her thesis twenty years ago, she speculated that "equality of access will be one of the great prizes of the death of distance."[36] The disconnect between physical locations and the extent to which locations are related by distance between them is now writ large. The old mantra of "form follows function" is fast becoming irrelevant.[37] By the end of this century, the disconnect will be complete, and insofar as the standard model will prevail, it will simply be the physical residue of a bygone age. Cities may well look as similar as they do today, but if they do, this will be because we continue to build them in the way they were because of a nostalgia for the past. As we have said before, when we look under the hood, the post-industrial city is already dramatically different from the industrial city.

RADICAL RESTRUCTURING OF THE TRADITIONAL CITY

As telecommunications costs fall and become less significant and as digital information becomes ever more important in the production and consumption of any manufacture and service, many if not most of the activities that take place in cities are losing their comparative locational advantages. The fact that in the past we have engaged in production and consumption in distinct spatial locations and at particular times is weakening in that a good deal of organization is now being achieved digitally, where distance and, in fact, time are now much less constrained than hitherto. The cement that ties us to workplaces, residences, and shopping centers and to all the other places that we usually have to visit to partake in those activities such as health, education, and entertainment is slowly dissolving, and this has clearly accelerated due to the COVID-19 pandemic. We do not have a clear view of the implications of all these changes, for the degree to which they will disrupt contemporary practices is largely unknown. In a complex world, where everything and everyone relates to everything and everyone else, the way existing and new network layers interact with one another is ever more difficult to figure out. Thus, it might appear that in the future city, nothing is fixed any longer in the way that it was in the historic past, even in the cities of the first and second industrial revolutions.

We need to make a distinction between products whose demand can be mediated by digital technology, that is, ordered online but shipped using conventional transport technologies, and those that are also distributed online. In fact, we have three categories: those that are ordered online, those that are distributed online, and those that are both. The best example of both are online booksellers such as Amazon, whose sales are always ordered online but whose books are either shipped using conventional transport distributors or simply sent online as e-books. Sometimes companies not only take orders online but also let the customer configure the product online with the final goods if they are material, being shipped in the conventional way. In the case of cars, the customer often goes to the dealership in the usual way to take delivery but searches and orders online. In terms of distribution, in fact, there are many options, ranging from home delivery to pickup points at convenient locations. The internet has enabled all of this, with new services emerging all the time that have spawned different types of delivery as add-ons to traditional distribution channels.[38]

This focus on online sales is generating a radical restructuring of the high street and the shopping mall, and many types of outlet are being massively disrupted by these developments.[39] This is particularly the case for bookshops, and to an increasing extent, this is true for anything that changes the travel patterns of the consumer from those where the consumer actually buys the good at the outlet to those that now operate home and workplace delivery. Supermarkets and various food outlets, for example, are increasingly being reorganized through home delivery. How these disruptions will play out is impossible to predict. There is a strict limit to how many products can be turned into digital bits and bytes, thus removing all traces of physical transport, although books are one, entertainment another. But many kinds of information such as financial services are subject to such change, and many have already gone down the digital route. Moreover, shifts in the delivery patterns due to online sales but where physical transport is still necessary are extremely difficult to predict. Whether these new patterns will diverge from those trends that continue to reinforce decentralized living or herald a return to inner-city living is an unresolved serious issue. We simply do not have enough information about existing patterns and the prospect of new ones to know what will happen. Nor do we know how our own behavioral responses will be conditioned in the online world. As these arguments have resonated throughout this text, the future is largely unpredictable.

Another feature of how cities are beginning to change involves non-fixed and flexible locations for anything other than sleeping. With new forms of transport, which we will discuss in the next section, particularly autonomous cars, then it may become easier to work within the car, and many already do so within different forms of public transport and private vehicles that individuals can hire. In fact, the internet makes possible a much more fluid configuration of economic land use, particularly office space, which can be reconfigured continually for different uses. Of course, the prospect of working from home has always been on the cards since the early industrial revolution when those wealthy enough had control over their own transportation and work time.[40] In fact, many futurists of the twentieth century, from E. M. Forster in his novelette *The Machine Stops* to Alvin Toffler in his book *Future Shock,* painted a picture of a highly decentralized world in which our ability to telecommute would become the dominant standard for the way we would live and work.[41] We would all live, they argued, in the electronic cottage, cities being as George Gilder once described them "the leftover baggage of the industrial age."[42]

Of course, none of this has come to pass yet, and there are many reasons why it is unlikely to happen, but most of this speculation took place prior to the invention of types of telecommunications with enough bandwidth to merge with the computer. Once this began to happen, the prospect for homeworking grew massively. Current estimates for North America suggest that in 2019, prior to the pandemic, some 70 percent of the US working population engaged in some form of homeworking once a week, with more than 40 percent actually spending more than half the week working from home.[43] These proportions are still growing, while the percentage of people working on the move—in coffee shops, cars, and so on—at 10 percent is also still rising. Again, the pandemic has intensified this picture in that new forms of homeworking can be economically more advantageous to managers and workers, and as yet, it is unclear what proportion of workers will continue to work from home in full or in part now that the pandemic has passed.

Although much homeworking depends very strongly on the use and processing of information, an increasing proportion of those working are now essentially part of the information industries. This cuts across many traditional categories of work, and although the manufacture of physical goods makes this kind of working impossible, traditional manufacturing is now a very small percentage of the workforce in the US and UK (about 8 percent in both cases). As we are now within the post-pandemic era (2023 onward), barring any dramatic reversals in our battle with the pandemic, we are beginning to live with COVID-19 while the mandate to work from home no longer holds. But as we will elaborate at the end of this book in our concluding chapter, our experience with these relatively unconventional forms of location and movement is in its early stages, and to an extent, we can already say that working from home and the internet technologies that are able to sustain it contain very strong forces that are already fusing work and home, threatening to change the form and function of the city in ways that were once unimaginable.

Other distinct digital functions that have locational consequences in cities involve health, education, and leisure. Health requires transport if anything other than online diagnosis is required, although the balance between physical interactions and digital consultation and dispensing is changing, with material transportation falling throughout the sector. Education and entertainment, on the other hand, are being dramatically changed by the new online world. The digital delivery of courses is well advanced, but there remains the need for some sort of face-to-face contact, simply for impressing the need for education in preadolescent schooling. The consequences for universities are different, and there is likely to be considerable disruption there due to online delivery of course materials. In terms of entertainment, there have already been dramatic changes in the delivery of music, which, like books, can be delivered as pure digital content. The impact of this on location patterns and transport is as complex and unpredictable as in other sectors, and this is made all the more difficult in that populations are increasing still, and individuals are becoming wealthier in most parts of the world, both of which lead to rising demand for all these kinds of activities. In the face of growth, changes in distribution have always been tricky to determine. To provide a sharper lens on these questions, we will return to them in later chapters, particularly in part IV of this book, where we will focus as much on organization and governance as on questions of physical transport and distance.

AUTOMATION, AUTONOMY, AND AI

It is fairly uncontroversial that the industrial revolution, certainly the first and the second, paved the way for extensive and continuing automation of human tasks that were traditionally the fruits of our manual labor. Automation involves technologies that define processes that reduce the degree and amount of manual labor. In this sense, they change our role in processes of production, and these began almost as soon as humankind began to populate the planet. The famous scene in the Arthur C. Clarke/Stanley Kubrick movie *2001: A Space Odyssey* of humans as apes discovering, for the first time, that bones could be used as tools and weapons represents as graphic a beginning as one might ever see.[44] This enhancement to the human condition, however, remained modest, despite successive inventions through pre- and classical historical eras, which have slowly enhanced our abilities. Only when steam power was discovered at the beginning of the industrial revolution did automation begin to take off. The notion of building machines that could control themselves emerged pretty quickly in the form of thermostats that could feedback information to correct routine processes without human intervention. From around 1750, wave upon wave of automation was piled on top of one another to create machines that enhanced our ability to lift and move objects of ever-increasing mass and size. Transport technologies up to contemporary developments in magnetic levitation, space rocketry, and such like have always represented the physical cutting edge of this automation.

To an extent, changes in transport technology reflect automation with respect to the ease with which we can travel, as reflected in distance and speed. Automated processes essentially reduce our own personal role in powering the processes in question and, in this way, augment our abilities to perform tasks that previously we were unable to enhance. Prior to the industrial revolution, we enhanced walking by coupling ourselves to other means of animal transport, in particular to the horse, and we developed wheeled vehicles of various kinds to enable us to travel at much greater speeds using such horsepower. But in general, although these modes of transport did lead to improvements in mobility, they were largely reserved for the small number of individuals who could afford them. As we have already noted, it was the development of the steam engine that led to the first step in the automation of transport, enabling us to travel at speed over much greater distances. Individuals are not removed from the process, however, for these methods of automation still rely on human operation, although in terms of those being transported, only a handful are required to service the movement of large numbers of people. In this sense, then, all physical transport depends on systems that couple humans to machines, although, as we will speculate, truly self-driving vehicles, which are currently being explored, might eliminate the human from any form of control in some of the almost completely autonomous transport currently being tested. Many think this unlikely, but we simply do not know. In fact, the prospects for autonomous vehicles look somewhat bleaker than a decade ago, but much depends on advances in AI, and we will pick up on these in our last chapter.

What the digital revolution has brought to these processes is our ability to articulate automation as being a function of information rather than of energy. In Nicholas Negroponte's phraseology, this marks the transition from atoms to bits.[45] Coupling the idea of algorithms to the search for pattern in data that is incessantly available, such as that which is streamed from a variety of sensors, provides environments in which physical objects can be controlled in countless ways due to the fact that, at any one time, the information that is captured and fed back to the behavior of the object enables the object to react instantly to the environment in which it finds itself. If the object can capture all the significant information necessary to control itself perfectly in the context for which it is designed, then we can approach complete automation quite independently of any human involvement. To an extent, complete automation will always be an idealization, for the data that is sensed must also be complete in some way. There should be no possibility of any unknown or extraneous events entering the environment, and the feedback must be well below the threshold for action. The list is long enough to convince that there will never be a perfectly knowable situation in other than the most routine elemental processes for complete automation to operate.[46] Even in that context, there will always be room for doubt.

The classic contemporary example of automation that we have already noted is the idea of autonomous cars—self-driving cars—which, if exploited on a large scale, would undoubtedly alter the pattern of traffic in cities. In doing so, this would probably alter the structure of land use and activities in cities themselves, where the key issue is exactly how autonomy in vehicles will change the pattern of travel. This is largely a second- perhaps third-order effect because the simple act of changing the driver of a vehicle from a human to effectively a robot in which the vehicle itself is the robot has no obvious impact on the distance traveled or the amount of time spent traveling. It is simply a complex substitution of human behavior for artificial behavior, where the algorithms mirror all the features for controlling a vehicle in its environment. But control does not reflect the behavioral motivations of the person traveling, where that person wishes to go, the stops needed to be made along the way, the attraction of the route, and so on. There is enough controversy anyway as to whether there will ever be an environment in which a car without a human driver can function autonomously.[47] This requires a situation in which there is no possibility for the environment to be disrupted by events that introduce data that the vehicle's algorithms cannot respond to—data that is beyond the functional experience of the machine itself.

Before we speculate on what autonomy in vehicles is likely to mean for the death of distance, we need to note that there are several levels of autonomy that have been defined, and each of these implies an increasing level of AI embodied with the vehicle itself.[48] Level 0 represents the status quo, where all vehicles are under the complete control of a human driver. These levels increase to level 5, where humans are no longer part of the equation and the vehicle is entirely autonomous. At this level, vehicles can function without transporting humans either as drivers or passengers, and this may be possible where the purpose is to move materials other than humans or animals. Freight traffic thus aspires to such autonomy. In terms of passenger cars, the six levels can be made distinct with respect to the control that the human exercises on driving the vehicle. Level 1 is referred to as "hands on," level 2 as "hands off," level 3 as "eyes off," and level 4 as "mind off," with level 5 being complete autonomy, with the passenger ignoring all aspects of the vehicle's behavior and environment as if the passenger were asleep. A good deal of the current hype focuses on the development of autonomous cars to level 4, but it is still unclear as to what this means. There are few who would hazard letting level 4 cars loose on the existing street system without any human presence, and in the last analysis, there is caution as to how such autonomy might be introduced into the simplest of all driving functions.

Once an autonomous vehicle is released into a wider environment, it is not clear how one can control a mixture of autonomous and non-autonomous vehicles, for such environments are never completely defined. There is always the potential for extraneous events—"black swans" in Taleb's terminology[49]— that cannot be anticipated, thus destroying the ability of any vehicle to remain in control of all its functions. In fact, this introduces the notion that vehicles might be connected in that they might be networked wirelessly with server stations scattered all over the terrain, synthesizing the massive volumes of data that pertain to all driving conditions across wide areas. Vehicles might be connected physically in platoons in certain simple situations where aggregate autonomy is maintained.[50] This introduces notions about wide area control that changes the picture of autonomy dramatically, but there is little sense that such control is even on the horizon. I might be forced to eat my words, but the idea that everything of significance can be captured in real time and processed so quickly that vehicles will always act intelligently with respect to their movement is far-fetched. In other systems, where humans are not part of the equation, such control is never complete, and these are much more routine than driving a car.

In fact, this discussion is focused here on how autonomy can alter the key questions of this chapter, which pertain to speed of travel and the amount of time spent. To speculate about this, we need to assume complete autonomy rather than anything less, which is plagued with ambiguity. Cars that are completely autonomous and which operate in environments with perfect intelligence will presumably allow passengers to partake in a very different travel experience than either a driver or a passenger has in a conventional vehicle. In this sense, the length of time spent in the car is of less relevance, and people will probably travel longer, for then the car can act as an office in ways that they cannot when driven conventionally. In principle, autonomous cars that are connected to operate without any possibility of accidents will allow people to travel further and longer, and in this sense, it could well be that this simply reinforces the trend to suburbanization. Autonomous cars might reinforce decentralization from the core city, for their cost is eventually likely to be lower than conventional vehicles, congestion will be nonexistent due to overall wide area control, while the notion of door-to-door transport enables all the traditional advantages of the car to be preserved. Autonomous cars are thus like individual seats on point-to-point public transport but with the advantages of door to door. With complete autonomy, then, the constraints on cost of travel and time spent would appear to relax dramatically, but as we have been at pains to emphasize, it is impossible to consider all these issues simply in terms of robot cars. Other behavioral issues pertaining to location, distance, time budgeted for travel, and other activities, as well as the density of occupation and the very modes by which people prefer to travel, will be as important or more so.

There are several issues we have yet to discuss in terms of the death of distance that involve the future of work, the way production processes might be automated internally, and the way future interactions using physical and electronic media and networks will play out. Many of these issues pertain to the IoT, which are central to the way computable devices and sensors are being embedded into the environment at every scale. In this sense, everywhere we look, computers and sensors are being embedded into buildings, transport infrastructures, natural environments involving ecology and biology—in fact, into every physical asset that we have in the artificial as well as natural worlds, including ourselves. Add to this, the behavioral implications of using networks built from computers and sensors to manage our cities across different scales, to engender greater levels of prosperity, to generate more sustainable environments, and to improve our work, then the picture of what the smart city actually looks like becomes highly complex. We can continue to peel back various layers, as we have been doing and will do so in future chapters as we begin to unravel the computable city. In the next chapter, we will examine what we can and what we cannot envisage with respect to the new technologies that now enable us to communicate with anyone in any place at any time.

Just as we began this chapter, with ideas about how space was being annihilated by time and how Karl Marx grasped this path-breaking idea as underpinning the transformations that drove his great theories of historical materialism, we must also give him the last word. From the notes he made in his first days in the British Museum Reading Room, which were only published in English in 1973 as the *Grundrisse*, he also sketched the most radical of ideas. He saw not only capitalism as simply the exploitation of labor for its surplus value but also the transformation of labor into machines that would force labor to take on a very different role. In short, he anticipated the destruction of the traditional role of labor by machine, by automation, which he did not consider wholly bad. In other words, what is currently happening with the world of work being radically changed by autonomous machines, by robots—indeed, by all forms of digital information technology—he more or less predicted as happening 150 years before it really became apparent. He said this many times, but in its clearest form as "the means of labor passes through different metamorphoses, whose culmination is the machine . . . set in motion by an automaton, a moving power that moves itself; this automaton consisting of numerous mechanical and intellectual organs, so that the workers themselves are cast merely as its conscious linkages."[51]

8 BUILDING CYBERSPACE

Cyberspace. A consensual hallucination experienced daily by billions of legitimate operators, in every nation, by children being taught mathematical concepts.
—William Gibson (1984)

THE VISIBLE AND THE INVISIBLE

In 1880, if you were to walk north up London's Tottenham Court Road, which runs from the end of Oxford Street to Euston Road, you would barely see any evidence of the mechanical and information technologies of that time. The underground railway was only just beginning to be built, although the first rail—the Metropolitan Line—passed close by along Euston Road.[1] The telephone, of course, had only just been invented,[2] but there was no evidence of its illustrious predecessor, the telegraph, in the form of the poles that dominated many cities and landscapes, particularly in the US. Toward Euston Road, which by then was dominated by the big railway stations linking London to the centers of manufacturing in the north, the environment would have had a distinctly rural feel. Had you been lucky (or not, as chance would have it), you might have seen Karl Marx and his German buddies on one of their famous pub crawls—there were eighteen pubs in all along the Tottenham Court Road—but this was in the twilight years of his life, and we do not know whether his penchant for beer had been finally quenched by then.[3] In fact, although the road was a relatively non-descript Central London street, theater land had begun to develop at the southern end of Oxford Street where the street rapidly merged into Soho, while information technology was much in presence at the northern end where William Cooke and Charles Wheatstone[4] had first demonstrated their telegraph system along the railway from Euston station to Camden Town in 1837.

If you had walked along this same road twenty-five years ago, you would not have seen much information technology either. The telephone wires would by then have been buried underground, and there would hardly have been any public-facing closed-circuit TV cameras (CCTV), for in 1995, the technology was only just beginning to be used for widespread surveillance. The internet had only just reached the point where it was about to be launched on an unsuspecting world by Tim Berners-Lee at CERN,[5] and the wizards in Urbana-Champaign, who invented the visual interface Mosaic and, for a brief moment, held the limelight with its conversion to the browser Netscape, had only just begun. Had you stood outside Karl Marx's last watering hole, now called "The Northumberland Arms," and walked east one block, you would have come to UCL, which was the home of the first internet connection outside the US built in 1973 (as we recounted in chapter 5),[6] and between the pub and the college, there were many buildings where academics and researchers were beginning to exploit the internet for multiple purposes. But to the casual walker, none of this would have been visible, and at the time, it was barely imaginable.

Unless, of course, you knew where to look. West of Tottenham Court Road, in the area they call "Fitzrovia," the General Post Office had expanded into a number of buildings in the 1960s. Near Fitzroy Square, they built the Post Office Tower, now the BT Tower, which was the highest building in Central London at the time.[7] This communications hub for satellite data capture was supported by a number of other GPO operations in the vicinity. If, in 1995, you had also walked south from the Post Office Tower, along Whitfield Street, parallel to Tottenham Court Road, you would have come across Cyberia, which advertised itself as an internet café[8]—a coffee bar where all its customers were sitting in front of PCs, surfing the internet, initially using nonvisual interfaces such as Gopher and Archie, soon to be replaced by visual browsers such as Mosaic, Netscape, and Internet Explorer. This was the online presence that was just about visible, but even this was not obvious.

In those days, however, even the term "online" was used in a rather specialist way, but the net had matured so quickly that by the mid-1990s, we had reached the point where those in the know could both advertise their own presence by constructing their own web pages and explore the small but rapidly growing treasure trove of web pages that were being created at dramatic speed. Cyberia badged itself as the first internet café. In fact, it was not, for there were examples in the late 1980s in places such as Seoul in South Korea and in the US where there was a blurred transition in internet access from private to public places. But Cyberia was sufficiently iconic and focal to provide real evidence that the internet was likely to become a dominant presence. It also quickly acquired superfast access because the internet service provider (ISP) Easynet, which also engaged in the venture, was colocated in the same building, and insofar as you could touch and feel the internet, this was it. Moreover, Cyberia began to attract a wide following and look-alikes were quickly being spawned in other cities.

Eva Pascoe, who set up the café with her partner who started Easynet, chose its location close to where she studied for her doctorate in human–computer interaction at nearby Birkbeck College. She had good links with University College too, where there was considerable expertise in networking and communications, and her specialism in visual interfaces meant that her vision for Cyberia was as much a social experiment as it was a place where you could get net access.[9] The design of the space was such that it gave a strong educational feel to the whole venture, and first and foremost, it was designed to appeal to students. Fitzrovia of course is adjacent to Bloomsbury, which is where there is a critical mass of academics and students, who Eva and her colleagues considered would be turned on by the existence of such a forum where they could hack and digitally converse to their hearts content.

In fact, it did not turn out that way, for her venue rapidly became a veritable honeypot for those who were intrigued by the notion that the internet was the most promising of all future media, as indeed it ultimately turned out to be. Next door, there were the Whitfield studios, where many of the Britain's popular musicians made their recordings, and because Cyberia also made great coffee, something of a rarity in London in those days, the artists began to popularize the café. A succession of stars—Kylie Minogue, Bono, the Motown groups, and so on—patronized Cyberia, learning quickly about the net and convincing people such as Mick Jagger and David Bowie to invest in the enterprise.[10] Internet cafés did not last that long, for by the end of the 1990s, the critical masses, which now embraced a growing public at large, were all sufficiently online that internet cafés became somewhat passé. But Easynet grew like crazy in the late 1990s, with the company and Cyberia taking over the whole of their building but then suddenly relocating to the city fringe. The company was instrumental in setting up broadband in the Hoxton area in the early 2000s, and their initiative came to dominate the entire area by the time Tech City was established in 2010. The growth of these start-ups can be traced to the early influence of Cyberia, but this is another story for another time.[11] The lesson here, of course, is that to reveal this kind of internet technology and explore its visibility in the city, one has to dig deep, almost literally, at the most local level.

One of our four themes at the core of this book is the question of how visible something as abstract as information technology actually is or can be. In chapter 2, we noted that more often than not, we cannot observe much digital technology in its physical network form because the networks that enable it to function are not observable. Such networks are now everywhere, often buried underground or in the ether, popping up in almost random fashion as masts of various kinds, data centers, containers for switching gear, and so on from which the multitude of networks that define our communications systems are almost impossible to unravel. Add to this the fact that usually we cannot directly observe what information is being passed along these networks, and even if the owners of these networks capture everything, which they often will admit to doing, no one has the ability to figure out how one network interacts with any other. The physical network and the flow of information through it can barely be reconciled, as one cannot ever map one to the other: the material channels that take the flow cannot be reconciled with the topology of that flow. Moreover, we cannot know what information originates from individual decisions and why we cannot get inside the decision-making processes that generate such information. Thus, if the flows of information and their physical channels cannot be linked in any fundamental way, then there is little hope for providing a satisfactory picture of how information technology underpins the city.

This might sound a little depressing, but we must make a start to get some idea of the way the city is becoming computable. All is not lost, however, because we cannot measure everything in and out of sight, for we do, at least, have a good idea of where and why networks are being put in place. A good place to start is around the time when the internet first became visible in any form—the time when Cyberia was set up, when the diffusion of the internet turned into a veritable explosion. Round the corner from the café on the corner of Tottenham Court Road and Torrington Place, in one of University College's buildings, a group of nerds were busy at work in the basement attempting the impossible: attempting to map the internet, which they defined as "cyberspace." This was the group that laid the foundations for Steve Coast's project on OpenStreetMap (OSM),[12] whose origins we will recount a little later, but Martin Dodge and his brother-in-arms, Rob Kitchin from Queens University Belfast, embarked on a program that built on the basic notion that such mapping would always be far removed from the actual geography of where the wires and the servers and the routers and even their clients were located.[13] The fact that so much was happening and that it was so remote from the general public provided a very strong rationale for such an exploration. Along with a few other efforts at such mapping, Dodge and Kitchin attempted to get to grips with the whole question of how cyberspace actually functioned and how visible it might be made.

MAPPING CYBERSPACE

We introduced William Gibson's definition of cyberspace as the introductory quotation in this chapter. He elaborated his description in his 1984 book *Neuromancer*, where he defined it as "a graphical representation of data abstracted from banks of every computer in the human system. Unthinkable complexity. Lines of light ranged in the nonspace of the mind, clusters and constellations of data. Like city lights, receding." Since then, he has admitted that "it seemed like an effective buzzword. It seemed evocative and essentially meaningless."[14] Yet, it stuck. Taken from Norbert Wiener's use of the word cybernetics in general systems theory,[15] which he defined back in the 1940s as "the art of steersmanship" that we noted in chapter 3, the term has gone from strength to strength, with the present focus on cyber-physical systems[16] representing a synthesis of the world of wires and the world of flows—where the physical and the topological merge.

Thus, cyberspace at the time was in fact the internet then defined as the network of networks. The purest form of cyberspace mapping was essentially based on maps of its topology with respect to how websites linked to other websites. If you open a page and find links to other pages, then their topological representation is what you are seeing. The fact that the page you are on links directly to another page does not mean that when you click on that page, the request to serve you that page goes down one wire directly to the physical facility where that page exists. It may go through many intermediate links, for the data gets broken up into packets that can diffuse all over the net before they reach the requested site, while the data that is sent back also goes in myriad ways.[17] Data may go halfway across the world to reach a site that is physically adjacent to the place where you are viewing the web, and therein lies one of the key problems of operating and designing such systems. Although the time taken to traverse the web is miniscule compared to anything we could do hitherto with respect to the transmission of information, this time is not zero, and therefore if you can link physically to two sites that are physically adjacent, the latency, or time taken for transmission, is at a minimum. The biggest problem, however, is how such links are constructed, for this requires information about the physical network and the topology of how physical sites relate to information flows, and we only ever have a very partial knowledge of this phenomena.

When Dodge and Kitchin began to map the internet, they built on the back of others engaged in similar pursuits. In particular, the commercial firm Telegeography,[18] which specialized in producing high-quality detailed maps of the net as well as handbooks of the key links between networks, owners, operators, and those who leased lines, provided an important source of data. Bill Cheswick, then at Bell Labs, began what he called the "Internet Mapping Project,"[19] and the group at the Cooperative Association for Internet Data Analysis (CAIDA) in the San Diego Supercomputer Center were also early into the game, producing astounding maps of the topology of network structure, reminiscent of fractal geometries that implied that the web grew from the bottom up, as it clearly did.[20] Most of these early maps, and indeed most of those that followed, are based on the notion that it is the topology not the geography that is really important: who links to who in contrast to where links to where. Our focus in this book so far has not been on one or the other, for we have argued that we need both topological as well as physical connectivities to get an understandable picture of every spatial scale from the global to the local.

There are, in fact, many different types of map that allude to cyberspace, but those that we consider here are those of cyberspace itself: as the physical connections of where the wires actually go as routes or where they surface as towers, switches, or data centers, or where the flows actually go or surface. The former are based on the physical geography of the space, whereas the second throw away the physical geography and just reflect the logical links or topology. Before the computer but once the industrial revolution had thrown up the telegraph and then the telephone, there were plenty of attempts to map the location of the cables that linked telegraph station to station. On the North American continent, these followed the rail lines, after which they began to appear on poles in denser clusters in cities, while the locations of these poles mirrored the pattern of settlement itself.[21]

Undersea cables were sparser and easier to chart, but here there was a degree of invisibility in that once the cables were sunk, there was always the possibility of them drifting, and were they to break, as they sometimes did, fixing them became a severe problem due to the fact that they were locationally not very well charted.[22] Once the telephone was invented and instantly became a more portable device—or rather a device of more personal dimension—telephone poles became standard, while the density of the phones themselves mirrored much more closely the density of population across the globe. Yet, in terms of the difficulties of invisibility that currently plague our geography of information technology, the telegraph and the telephone were well-known devices, whose location and networks could be physically mapped with ease. The fact that we still do not have good telephone maps is largely due to the fact that the locations of where these devices were used did not seem very consequential to the growth of cities. The telephone tended to lag rather than lead, although it remains somewhat surprising that there has never been a concerted effort to look at the geographical spread of the telephone to assess its impact on urban and economic development.[23]

Dodge and Kitchen in their book *Atlas of Cyberspace* plot three kinds of map.[24] The first is the traditional map of the physical geography of cities, regions, nations, and the entire globe, which presents what we know about physically embedded networks such as the telegraph, fiber optics, and such hardware, which essentially is where the wires go. This is what we can observe within limits, and sometimes we can augment this type of picture with flows of traffic of various sorts—data and information in the case of cyberspace. The second map is entirely topological, relational, the best and simplest being the original sketch of computers attached to the internet associated with the first message sent on the ARPANET by Larry Kleinrock in 1969 from UCLA. This famous diagram shows four mainframe computers at UCLA, UCSB, Utah, and the Stanford Research Institute with their four switches or routers linking UCLA, UCSB, and Stanford directly together in a triangular network with an additional link from Stanford to Utah.[25] This kind of map is pure cyberspace in that the computers and the switches might be anywhere: it is the links or network segments that matter. As we have noted, the tagline for SUN microcomputers in the 1990s said it all in the phrase "The network is the computer." The third map that Dodge and Kitchin present is a mixture of topological with physical. Back in the 1980s and 1990s, maps of the US were drawn that showed the actual locations of the computers at each node in the ARPANET, but these were topological links—direct links between them—which bore little relationship to the underlying physical terrain into which they were embedded.

The architecture of most communications systems, at one level, is quite simple. You have devices that are the center of the action such as computers in this case but, in the past, phones and telegraphs; these act as nodes or hubs in the network. You have network links, which have many channels for the processing of data between the nodes or hubs in question. But you also need routers to enable the data to be switched to the right hub, and these exist all over the network as intermediate nodes, which are staging points, so to speak. It is here that latency can be reduced between the biggest sites with the most traffic, but these systems largely have a bottom-up design, with network operators figuring out quite locally whether it is worth linking two or more hosts from each particular router. These elements can, of course, be mapped in geographical or in topological space. As well as the servers and computers that locate at various hubs, maps of more abstract conceptions such as domain names—uniform resource locators (URLs) or web addresses if you like—can be plotted, and at the city scale, these often correlate with the distribution of employment or population, depending on the types of networks being mapped. Dodge and Kitchin demonstrate just how important financial services are in their plots of these domain names in central cities such as London and San Francisco.

More abstract conceptions of cyberspace that do not have any physical significance per se can be developed. Before the dominant access point to the web became the browser, cyberspace was partitioned into various subdomains using software such as Gopher, FTP, Usenet, and a host of other tools for exploring the net.[26] Various infographics were developed to chart the way the city is wired, while the dynamics of the web—changes that are continually taking place in the size, scale, traffic, content, media, and such like, all associated with the internet—were being visualized in new and insightful ways. Of course, as we will explore in later chapters, the computable city also embraces the real-time city, and there are many different ways in which we can visualize real-time change. But there are also spaces within spaces in cyberspace itself. Gaming, for example, on networked systems was fashionable long before the internet exploded for public use. From early games called "MUDS"—MultiUser DungeonS (and dragons)—came internet gaming with massive visualizations that were clearly influenced in diverse ways by the fact that the gamers were playing their roles from different places in different cultures. Out of these came virtual worlds, which in themselves produce tangible geographical spaces but in VR.[27] Some of these virtual worlds took on a reality that was similar to our real world but sufficiently different to jolt one's perceptions, with the whole arena of VR and augmented reality defining its own corner of cyberspace. And last but not least, when we ask users to map their own views of cyberspace, we are in the realms of mental maps, sufficiently different from one another to reveal that there is not one cyberspace but many, just as there is not one city but many. The city, like cyberspace, has as many personas as there are people occupying that space.

THE PHYSICALITY OF THE INTERNET

At one level, the physical internet is very simple, as it just consists of hubs or nodes, which are the computers that process data, the switches that enable that data to be routed within the whole system, and then the links that constitute the conduits that take the copper wires and the fibers through which the data associated with computation at the hubs flows. We envisage this as an enormous network embedded in physical space, and while it is impossible to map in anything like its entirety—and probably never will be because its development is from the bottom up, and changes in its form and function are particularly rapid—we can get some sense of it from partial descriptions in very local situations where one does have enough information to figure out where its various pieces are located. In fact, world maps of its hubs mirror the distribution and density of urban population, revealing that North America and Europe have the densest pattern, while Asia-Pacific is fast catching up. John Matherly's map produced by his search engine Shodan,[28] which queries more than two billion devices that have internet addresses, is probably the most complete at the time of writing

(2023). Nevertheless, this is still only a sample, for the current estimate of connected devices is nearly twenty billion.

The architecture of the physical internet, however, is massive and variegated, rapidly changing as much because new data centers, switches, and computers as well as new networks are being developed all the time, while the functions of existing ones become obsolete or decline in importance. New hardware is also being developed continually, and when the pattern of ownership is forever changing too, our picture of even its physical infrastructure is far from clear. Try mapping the meaning of data flows, which are articulated at the level of users querying, and adding to information at the hubs as though one has a direct line of site from one's origin to that destination, and the pattern that emerges is chaotic. It is not that it is not understandable, for in principle, all this can be mapped, but in practice, the sheer volume of information that is needed to produce a coherent picture is simply beyond our collective means and, as we have implied throughout this chapter, will probably remain so. We do, however, need to grapple with this problem. Otherwise, we have no sense of what the growth of the internet might mean for cities in particular and for society at large: the implications are essentially indirect—based on second- and higher-order effects, we might say—that need to be pinned down both geographically and in terms of the social and economic functions that the internet enables.

It is worth noting that in computer and communications science, the idea of a network is conceived conceptually as a seven-layer model, which we noted in chapter 5. This need not concern us very much here because at the lowest physical level, the focus is about the ways in which the bits and bytes generated by computers are organized into packets that form the raw material that flows from one hub to another and are variously switched along the way. The basic layer, called the "physical layer," deals with the bitstream, and as one goes up the hierarchy, we hit the middle layer, often referred to as the "network layer," which deals with all the translations into the directions that route the data packets from origins to destinations. Higher layers pertain to the presentation and application of data that is moved by the network.[29] This model has been widely used to articulate the way computers can be linked across networks, but it is largely of conceptual value in thinking one's way through the myriad of issues involved in translating data into a form that can be transmitted over physical networks.

The most obvious evidence of the physical internet is the stuff that the network is made of—today, largely fiber-optic systems that are contained in the cables that link computer hubs together and can be seen as usually being buried under the ground. This is the material that is laid in large tubes that themselves rest in conduits, usually reserved for such cables underground, often occupying the same space as other forms of utility such as electricity wires, sometimes adjacent to sewerage systems but invariably, in the Digital Age, dominating a hidden network landscape. Once the fiber within the tube is revealed, it is usually marked with given colors that define its function and its ownership. This is best seen when roads are being dug up to install new cables or to fix existing ones, but the clearest sign of what lies underground is when the cables pop up invariably into the buildings that act as places where the hubs are located. This, in fact, means every place where there are computers. In a domestic context, once your internet connection leaves your computer as an Ethernet cable or as a Wi-Fi link, at some point close to the computer, it will encounter some form of router that sends your data onward—usually over a copper wire. Increasingly, however, these wires are being replaced by fiber for the last mile from the center where your message is being processed for onward transmission to its destination.

These locations also represent the largest pieces of the internet, although there are now many thousands of such centers, usually in anonymous buildings that contain massive amounts of equipment of various kinds, normally routers that switch data automatically. They also represent the points where network engineers can connect networks together to increase the speed of transmission or to reduce latency in the jargon of communications science. Just as there is a distinct hierarchy of physical network pieces—from no wires at all in the case of the most local wireless signals to copper wires through to fully-fledged fiber-optic cables, all increasing capacity as one rises up the network hierarchy—there is a hierarchy of hub locations too. These are from one's own computer or device, portable or fixed, up to ever bigger exchange centers where the routers are located, spinning off into centers that store data that is needed to keep the internet working and to house the information that we as users require and produce. It is convenient to think of these as being similar to the telephone exchanges of the past, which were full of operators switching calls but which are now almost entirely automated and much more anonymous than in the early days of telegraph and telephone technologies.

There is a distinct hierarchy of exchanges. Telegeography[30] has mapped what it considers the biggest—more than three hundred—that process the vast majority of data that is being shipped across the planet at any one time. The biggest exchanges sometimes define themselves using acronyms such as IX (Internet eXchange). The Frankfurt exchange (DCE-IX) is the biggest in terms of processing power, with its computation at some seven thousand gigabytes a second. This is followed by Amsterdam, various linked centers in Brazil, the Telehouse complex in London's Docklands, the complex west of Washington, DC, originally at Tyson's Corner and now in Ashburn, VA, and so on. Wikipedia gives the complete list of some five hundred IXs, but this must be taken with a pinch of salt because these are only the official ones, albeit in one sense the biggest and most prominent. There are many private locations where there is massive compute power that enables very specialist routing, and a good deal of the military traffic worldwide does not touch the mainstream internet. Andrew Blum's book *Tubes* provides a detailed travelogue of his search for the physical internet and captures much of the Ballardian landscape of these places.[31] They are a little like airport landscapes with the same sense of placelessness and the anomie that these locations portray.

The last stop in our description of the physical internet are the hubs that are often referred to as "data centers" that house the massive data systems and the data that the world's major computer companies collect, manipulate, and market. In this context, think GAFA—Google, Amazon, Facebook, and Apple—and the centers that they have spread all over the globe.[32] These are almost industrial in scale and have the aura of the past world of the military–industrial complex with the same, if not more, security and extensive surveillance. There is an even more elaborate hierarchy of these than there is of internet exchanges, largely because every computer user has his or her own data center, be it only something such as a flash drive or, as in the past, a floppy disk that contains material that is not stored on a computer itself. Data centers at different scales emerged, just as mainframes gave way to minis and then to servers and PCs, where PCs themselves became servers that were placed remotely from the client computer itself. In short, the client–server architecture hastened the development of remote computing, where the server held the data and the client computer was solely used to extract, compute, and display it. Put another way, the machine room became the server room with increasingly more such locations being added. In fact, currently, servers and centers are beginning to sprawl across the computing landscape and even in my own small research center—located as it is on London's Tottenham Court Road—our own servers are located in at least four places across the college, and as these are managed remotely by people other than ourselves, there could well be many more locations than this. There is a clear tendency to buy additional servers for each new project, with this phenomenon of "server sprawl" often becoming an issue.[33] The quest to limit such sprawl by the creation of new virtual machines on existing servers—by compacting servers if you like—is the computable or digital equivalent of reducing urban sprawl by moving toward a compact city-like form.

Internet exchanges and data centers do tend to be low rise and sprawling, often in the suburbs of key cities, but this may be changing as the growth of the net continues unabated. There are already proposals for creating data centers in skyscrapers, with one of sixty-five stories recently being proposed by two Italian architects. As the first skyscrapers where christened "cathedrals of commerce" in the early twentieth century, such proposals for high-rise data centers could be termed "cathedrals to data," but it remains to be seen if the landscape begins to change in this way.[34] Nevertheless, the geography of where data centers are best located depends on their size, which implies they need massive cooling facilities and therefore extensive and cheap power supplies. In the US, favored centers are in the Pacific Northwest in dry, mountainous climates with abundant power, places such as the Dalles,[35] Oregon, where Google built its first data center, and Prineville, Oregon, where Facebook has one its four centers. Google now has thirty centers worldwide, and these installations are becoming a distinctive mark on the landscape. Google's centers are increasing at a rate of about three a year. It is reported that data centers now consume more than 2 percent of the world's electricity, with a growth of some 12 percent a year. Questions of sustainability are becoming significant.

Once we really move offshore in computer terms, we think of the cloud, and this means that we have little idea of where data is stored and even where computation now takes place. The invisibility of the computer landscape is thus increasing ever more rapidly, as it means little to know where the servers and data centers are actually located as long as the speed of access and the bandwidth to store and retrieve data is fast enough. In fact, what "fast enough" means is still very pertinent to network engineers and to the profitability of many computer companies that can increase their returns quite dramatically if they are able simply to go into one internet exchange, install a new router, and then connect this to another router that serves the hub that they wish to link to. Clearly, this depends on the scale of the traffic, but it makes the point that geography is still critically important to the way the internet is configured. One of the great paradoxes of the computable city is that as things get more and more invisible, rootless, and apparently only accessible in network terms, the geography of the internet becomes ever more important. This suggests that we need a new perspective on what it means to be accessible in the Digital Age. Once again, the death of distance is not what it seems.

WIRED SYSTEMS FOR CAPTURING URBAN DATA

So far in this chapter, we have outlined what we know about generic information technologies for which we have no clear idea of what data is being captured, computed, and/or transmitted. This is entirely consistent with the computer as a universal machine, but in cities, long before digital devices were first used to collect data, systems were put in place to capture more specialist data. Transport is a classic case, as is weather data. In the case of transport, inductive loop counters, which involve placing such a loop across the lane of a road, enable vehicles to be counted due to differences in energy levels—inductance—before, during, and after a vehicle has been sensed. These, in fact, have been used since the 1960s. Even in the late 1930s, various types of counter formed by laying electrical lines across the highway were used to count traffic,[36] and more recently, many varieties of photosensitive devices have been utilized. This data provides good counts of flow, but as we will see in the next part of this book, these tend not to be as useful to transport engineers and planners as data obtained on movement from other sources such as household questionnaires and travel diaries. In fact, there are now a variety of much more indirect digital tools for obtaining good traffic data, but loop sensors are still the most popular, with about twelve thousand being currently used in the Greater London area, generating more than four million observations of traffic each day. Now, this amount of data could have been collected at any time in the last eighty years, but it would have been impossible to store and process without digital technology. This reinforces the key point that digital technologies in cities need to interact with more traditional technologies for the city to become truly computable.

Another fairly obvious technology that can be developed quickly and is now extensively arrayed in cities—indeed, in many kinds of spatial environment—involves data captured visually with cameras. CCTV was first developed during World War II by the Germans, who developed it to capture V2 rocket trajectories, but it only took off when ways of storing the data using video recorders were developed in the 1970s. Almost as soon as this was possible, it was employed to capture the incidence of crime in city centers in the US and UK, among other places.[37] By the mid-1990s, CCTV was becoming widespread for monitoring and surveillance of people and traffic in many private as well as public places. If you wander around the streets of London, the proliferation of cameras is remarkable. Estimates suggest that there are about four hundred thousand in the Greater London Authority area alone and something like two million in the UK compared to some 350 million worldwide. Of these in London, only 1,777 are the property of Transport for London, which maintains cameras for its traffic control activities, while the congestion charge area, which is about six square miles of the city center, is monitored by about a thousand cameras that are used for vehicle number-plate recognition for charging purposes. The toxic congestion zone (or the so-called ultralow emissions zone), which opened in 2021, is eighteen times as large, requiring an order of magnitude change in the number of cameras to police it.

Cameras and cell towers that capture mobile-phone signals provide a new street architecture for the smart city. Just as telephone lines went underground for the most part, particularly in city centers, cell towers began to appear to capture wireless devices. These are mainly attached to streetlights that are themselves becoming digitally controlled, but in places where such lighting is not required or too expensive as densities in cities fall, purpose-built towers have to be constructed. There are more than forty thousand towers of various sorts (including masts) in the UK, with about five hundred in the City of London (the financial quarter). It is estimated, however, that when 5G networks roll out, another half a million will be required, and this is certain to change the visual urban landscape of this technology substantially.[38] When we come to explore what we call the "twenty-four-hour city" or "high-frequency city" in the next chapter, we will speculate on how good the data is for capturing the location and volume of mobile-phone signals, for there is still some optimism that this might be a good way of obtaining better vehicular traffic data. Linked to other demographic data, this is a potentially rich source for inferring better data about socioeconomic structure and human behavior in the city.

There is much other specialist data being picked up by sensor technologies installed in the city. In later chapters, when we deal with data streaming, big data, and new public and private services generated and transmitted across the net together with a whole host of real-time information, we will tie many of these loose ends together, for a complete picture of the smart city cannot be produced until information flows are mapped onto physical networks. What we need to note, however, is the existence of specialist data centers introduced to control the functions of various activities from which urban data is being generated, particularly those concerned with traffic control, security, policing, weather, and the economy insofar as there remain centralized points of exchange. Traffic signals are more or less controlled automatically now, but there are still human operators in control centers watching for anomalous incidents and accidents that need manual override. Such centers use the data from various sensors such as cameras and loop counters to provide an intelligible view of the operation of the system, but this data also reveals anomalies that only human operators can fix by overriding the system. Financial management controls associated with Fintech and controls on the local weather have not yet reached such specialization, but increasingly, the specialist networks that are needed to control various functions in the city are generating data that provides an ever more complete picture of what is happening in the city. Dashboards and portals are being designed to provide these immediate pictures, and to conclude our foray into how information technology is being hardwired into the city, we will sketch the emergence of such portals and their use. These are barely hardwired but do represent the softer side of the smart city, essential to its planning and management, which we will pick up in the last part of this book.

Dashboards, as the name suggests, are windows on the functioning of any machine that originally enabled users to monitor their internal workings and thus their performance.[39] They continue to have this diagnostic role. Apart from the obvious role of the dashboard in a mobile vehicle that is central to its operation, as in a car, the wider role is within computing, which enables computer users to display and, if necessary, act on key performance indicators. At one level, traffic control centers consist of a series of dashboards, which essentially are the windows on this world, and enable operators to control traffic in real time if the dashboard throws up indicator values that are deemed problematic. Because so much data is now online and available in real time, that is, streamed and available at key points on the internet, dashboards are being constructed for entire cities that enable those who have an interest in the running of the city to monitor a wide variety of its functions. The first dashboards emerged from decision support systems in the 1970s, but by the beginning of the internet era in the early 1990s, these were being collapsed into web pages that contain links to data sources that could be displayed in meaningful form so that city planners, mayors, and various policy analysts could judge the performance of their city over time. To an extent, dashboards represent the public face of any city, and to date, they are more presentational than functional. Kitchin's Dublin City dashboard goes a little beyond this in that the data that is captured is converted into more analytic interpretations, with the data being displayed using the basic functionality of GIS as well as a variety of display media that seek to understand everything from house price change to traffic flow. We will return to dashboards in the next chapter when we reveal what we mean by the high-frequency city, but it is worth noting that dashboards are part and parcel of the specialist data and control centers that are fast developing in cities such as those used to monitor a whole range of emergencies involving hazards—from extreme weather events to issues of crime, terrorism, and security. The center built by IBM in Rio de Janeiro is always illustrated as the classic example of how information technologies are making the city smarter, but there are many other such developments that we will pick up in later chapters.

CONNECTED MOBILITIES

We could continue almost indefinitely digging into the infrastructure of the city in the quest to reveal how richly and chaotically connected everything is at the physical level. Ingrid Burrington's book *Networks of New York*, which she subtitles *An Illustrated Field Guide to Urban Internet Infrastructure*, is a great attempt at unraveling the physicality and the role of local geography.[40] But what has fast emerged in the last decade is the notion that much of the connectivity of the city is not passive or fixed but rather is mobile. Connections change continually as people, objects, and activities move. Most of this mobile activity has come from people and objects being able to communicate while on the move, where the default state, of course, is a fixed location. Systems that mix wireless and wired communications are now becoming widespread, particularly in transit where there are fixed sensors but where their clients and customers are mobile. In stark contrast and for a long time, retail activity in the widest sense of the word has been accessible from fixed points of sale, where customers use credit cards to make purchases of various kinds. To an extent, such activities predate the Internet Age in that until quite recently, the networks that enable such purchasing remained fixed and largely inaccessible to the user apart from simply registering themselves using some sort of ID card. Essentially these systems are noninteractive, whereas the new mobility based on handheld devices or on any computer that is not fixed enables, at least in principle, any user to modify their own behavior with respect to wired systems through the ability of their device to interact with the system. So far, there has not been much activity in this context, but increasingly with car-sharing systems such as Uber, there is the prospect of continual modification of a user's behavior with respect to technology that is accessed on the move.[41]

The biggest change in mobile connectivity, however, is still a relatively distant prospect but one that is changing rapidly. We talked about autonomous vehicles in the last chapter, and the notion of vehicles being connected while they are on the move is becoming a real possibility. Interactions with fixed locations where data about their movement is archived and aggregated, connections with other moving vehicles, again through more centralized servers, and last but not least, digital interactions with highways and buildings that they are passing by are all phenomena that are being actively investigated. Already, there are plenty of systems that enable vehicles to communicate with central servers, which dish out information about a vehicle's performance and condition and capture the same diagnostics as well as information about the condition of the vehicle's environment. This information is being used in diverse ways, but it can, of course, extend to all kinds of other information that reflects the mobile internet, not so dissimilar from currently the largest use of mobile communications from personal handheld devices that are used to access the internet in one form or another. When autonomous vehicles are introduced that internally capture data about their own performance and the environment they function within in its entirety, this information will be fed back to control the vehicles themselves. This may well be augmented by information archived in real time from other vehicles and from the intelligent highway itself. It remains to be seen whether the array of information technology within buildings themselves along the routes of any vehicles will be captured too and used to enable further, and probably even better, safer automation.[42]

The prospects for connecting up a myriad of networks, fixed and mobile, as well as active network users and passive but connected objects is developing quickly but also unpredictably. Yet, the biggest source of information technology that has its own mobility is actually that within buildings that we cannot see from public space. Much of our discussion of the invisibility of the internet is based on the assumption that we should be able to see it from public rather than private space. So, to redress the balance a little, we need to say something about what happens inside the nodes—inside the buildings where most of our computers are still anchored. The best way to do this, in fact, is to walk again up Tottenham Court Road, starting where we began in 1880 and where we retraced our tracks more than a hundred years later in 1995. If now, more than twenty-five years on after the internet really took off among the public at large, we begin our walk from Tottenham Court Road tube station, which is now a key node on the UK's fastest and newest tube line, we encounter a bewildering array of information technology. First, you would have to leave the tube by scanning your Oyster card, credit card, or iPhone using Apple Pay, which are all ways of purchasing travel on the tube and buses in London. Then, on the street itself, walking north, you would observe lots of CCTV. The cameras, most of them private but monitoring public places, are not easy to see, as you have to look upward to reveal them, and they are scattered in a relatively random fashion on lampposts, building eaves, above doorways, and so on. Walking up the street, you are likely to encounter police vehicles, most of which are equipped with automated number plate recognition (ANPR) used for security purposes, but also a way of checking if a vehicle has been stolen or taxed. In that part of town, you are within the congestion C-charge zone (and, of course, you are well within the bigger toxic T-charge area), and thus there are no cameras checking number plates until you get to the top of the street at Euston Road where there are two.[43]

At the start of the walk back in 1995, the street would have been full of electronics and computer shops, selling items that have traditionally been cheaper than elsewhere, but with the movement of sales to the internet, these shops have declined quite rapidly in the last ten years as internet purchases have become cheaper and often easier to execute. This is London's electronic quarter, which is like Akihabara in Tokyo or the Golden Shopping Centre in Sham Shui Po in Hong Kong, but change here is very fast due to the power of the net to drag sales away from these physical locations. Proceeding up the road and noting that many of the shops are of an ever more temporary nature—due, we think, mainly to the decline of the high street in the face of internet shopping—the cameras continue to proliferate. The pandemic has also played havoc with the whole picture of retail along streets such as this in large cities such as London. As we head toward Goodge Street and Warren Street tube stations, I am guessing that the number of computers inside the buildings is increasing as we encounter more and more academic institutes and media firms. The frequency of buses along this street, which are now segregated from private cars, is virtually one a minute, and again, like tube trains, 90 percent of payment uses the Oyster card system. We will note the data that these systems generate in real time in later chapters, but the massive connectivity occasioned by this use is only dwarfed by the fact that most of the people walking along the street are now looking at their mobile phones.[44]

The other feature of the street is the sensor technology and the network lines that lie buried under the road. In 2019, the road was dug up from top to bottom to widen the sidewalks and to convert the street to bus- and bike-only lanes. This digging revealed the multitude of pipes and conduits that lay under the ground, and the fact that none appeared to be damaged by the construction indicates just how robust this kind of technology is. Mobile-phone masts, however, are harder to spot for these relate to many rooftop aerials that are hidden from plain sight.[45] Telephones too of the fixed landline variety now merge with other cabling, and thus one has to infer a good deal of where this activity occurs. In comparison to a generation ago, everything has changed, but nothing has changed. The street feels the same as it always has, but the sheer scale of computation associated with the place is daunting, for it is impossible to make sense of it. I do not mean that it is unexplainable or even inexplicable, although at one level it is, but insofar as all this is changing our behavior in cities and the way we make locational decisions, the city in our contemporary age is very different in function, although perhaps not entirely in form from the past. Moreover, it behooves us to speculate on what this future might be like in the face of such change. This we will do in the following chapters.

9 HIGH- AND LOW-FREQUENCY CITIES

We are reaching the stage where the problems that we must solve are going to become insoluble without computers. I do not fear computers. I fear the lack of them.

—Isaac Asimov (1978)

SLOW CHANGE

Patrick Geddes, the father of British town planning, argued that "a city is more than a place in space, it is a drama in time."[1] Despite his quest to impress on us that cities evolve, most of our past thinking about cities has sought to cast them in a static world, frozen in time, pictured as being coherent and highly structured. But nothing could be further from the truth. Cities are continually in flux, often in crisis, and invariably in a state of extreme volatility. However, most historians of the city have sought to cast them in a timeless wonderland based on notions that cities change quite slowly from one equilibrium form to another and that they evolve in an orderly manner. Even our images of future cities never tell us how those cities might emerge from our current urban condition, with city plans being statements of an idealized future world that owes more to our aspirations than a reasoned analysis of how we might understand their form and attain a better future.[2]

The predominant image, then, is what we might call the "low-frequency city," evolving slowly over time from one generation to another, with all our effort at thinking about its form focused on a slow progression of changes that become easily absorbed into its fabric. This perspective still exists, despite the very rapid development, as we described in terms of its recent history of information and related technologies, that have been deeply disruptive to many forms of urban life since the industrial revolution began. In fact, the standard model that we presented in chapter 6 entirely reinforces this picture of a timeless world, and it is no surprise that urbanists from classical times onward have sought to describe the city as having a strongly monocentric spatial structure, with transportation being the glue that holds its various parts together. The nature of the glue changes, of course, as new technologies emerge as they have done over the last 250 years, but what we see remains largely the same.[3] The suburbs grow as transportation costs fall and new centers—edge cities—emerge, but there is still barely any sense of what will happen over the next hundred years as transportation and interactions become ever more invisible. Cities are now all about moving information in contrast to the past where they were all about moving people.

As we described in the last two chapters, as computer and communications have converged, we reached a point, once the Internet Age began in the early 1990s, where the city fabric began to be penetrated by all kinds of information technologies. Since their very origin and even before when many tasks were being automated using electromechanical devices such as the punched-card machine and the typewriter, our work environments were being populated by many such machines.[4] But it was not until the invention of the PC and the development of networks to enable us to access them remotely that there was any sense that computers would spread out into the city, into public and private spaces that were accessible to the population at large. In 1993, I remember seeing this world from the vantage point of my SUN workstation running the internet browser Mosaic, where we pulled up moving images in real time from webcams that had been installed to monitor traffic on the San Diego freeways. Then, there was just a glimmer that computers might be employed in rather unconventional ways to sense the wider environment.[5] The networked age, however, was, as many developments in computing hitherto, barely envisaged, and the notion that cities could become populated with millions, indeed billions, of computable and connectable devices was still a figment of our imaginations, even as recent as twenty years ago. In 1994, Bill Gates said, "I see little commercial potential for the internet for the next 10 years," following this up by admitting that if anyone has suggested to him then that internet addresses—URLs—would appear on the side of taxis by the year 2000, he would have dismissed them out of hand.[6]

The fact that computers as sensors and our ability to connect them has become a dominant technology in controlling routine functions in cities such as payment for transport, social interaction and messaging, control of domestic energy, and a host of related services from online shopping to banking is a phenomenon that has taken us by surprise. There is little in the short history of digital computing to prepare us for such a change in the way we think about cities. Our shorthand for this—the smart city—betrays more the sense of what we would like technology to be if we could embed it into the built environment to help us to act more intelligently than the actual reality of this happenstance. As we will see, the embedding of computers and communications into public spaces and many other urban environments that define the city may be anything but smart. A good deal of what is happening, albeit in the interests of making things more efficient, tends to increase inequalities and decrease inclusion, often lowering productivity, encroaching on moral and ethical issues, and introducing many more problems than the technology is able to solve.[7]

The introduction of computer technologies into cities in the effort to control its collective, often public, functions more intelligently has introduced time onto the urban agenda in a way that has rarely been the case in our past forays into understanding how cities work. The low-frequency city is, by and large, the way we looked at cities in the past, while the idea of studying how cities change on a routine day-to-day basis remained beyond our interests. This is not to say that there is not an elaborate means of managing and controlling the routine functions of the city—there usually is, as any focus on policing, local authority organization, daily delivery of services, and the functioning of energy and utility systems would identify. But there has been little sense in automating or coordinating such functions hitherto. If this is to be done, we need a concept of cities functioning routinely and over the shortest time intervals—not only day-to-day but also hour by hour, minute by minute, second by second. This is the high-frequency city, and to a large extent, the smart city is seen as coincident with these functions.[8] The smart city, however, is not just about high frequencies—elements in the city that change rapidly during the day—but also about the low-frequency city, and as we will show in the next chapter, much of our thinking about how we can use computers to simulate and predict the future, how we can plan for it as well, is based on computational technologies that are firmly part of the infrastructure of the smart city. However, at this stage, it is worthwhile being a little more specific about the high-frequency city, and to this end, we will begin to narrow down our conceptions of what the smart city is actually all about.

THE SMART CITY

The idea of the twenty-four-hour city has been discussed for fifty years or more, but it gained little traction, largely because data on the activities that dominate every hour and minute of the day were generally unavailable other than through casual observation. Historically, when people wrote about cities, much of what was portrayed were descriptions of city life, which, in fact, dealt with what happened over very short periods of time, but most of this was not abstracted.[9] It was hard to collect enough information at sufficient a frequency to begin to explain and pin down diurnal patterns. There are countless documentaries and novels about daily life in cities, and these provide a rich source of information, but this data cannot be abstracted to the point where it becomes general enough to theorize about how cities actually function. Abstraction, of course, is the first step in explaining how systems such as cities work, but making sense of high-frequency routine events and even one-off events has not been possible until comparatively recently. Neither had we the data in the past or the analytics to make any of this possible.

Hitherto, the low-frequency city has thus dominated our concerns. Most of our theories explaining how cities were structured in terms of their form and function were based on the standard model we introduced in chapter 6, which was informed by the gold standard of cross-sectional censuses, certainly in most industrialized counties, where they are carried out at ten-year intervals. Of course, much of this low-frequency data is now automated in that it is made available using digital technologies, which in turn are used to analyze its meaning. There is now a surfeit of data that is being automated in terms of its delivery such as house prices, income, unemployment, retail price data, and so on, but such data is limited in size, largely because it pertains to the population, and much is still collected through routine collection of a manual variety. In fact, the frequency of collection has been significantly increased, but it is nowhere near the new varieties of data that are being collected in real time using new sensors and networked technologies.[10]

What has changed this focus on the low-frequency city is the relatively sudden embedding of computable and connected devices into the built environment. As soon as computers reached the point where they became truly mobile, which was in the early 2000s, and as soon as they became the predominant way of accessing the network of networks—the internet— they formed a new layer or skin covering the city. This skin is now the dominant mode for generating data in real time that in principle provides not only a window on the twenty-four-hour city but also a means of inter- preting and planning for the city on relatively short timescales. As we have implied, time has come onto the agenda of city planning, while individuals and agencies who could now engage in some sort of city planning grew to become much wider than the traditional public agencies that had hith- erto represented the cutting edge of urban planning. Suddenly, cities were flooded with data from the sensors that came in the wake of ever-smaller connected computational devices. This provided an entirely new form of data about the city in terms of social relationships, organization at work, the detail of how one could move around the city, and the way one might access individual services through applications that one could download from the internet. It was almost as though we now had two forms of city: first, that based on the low-frequency, long-term perspective of how cit- ies evolved, and second, the high-frequency, short-term perspective, which dwelt on embedding computers into the fabric of the city to enable their control through massive streams of data generated in real time.[11] These produced pictures of how a city was operating largely in a routine sense, but also with respect to any events that might be picked up with sensors and streamed to devices that could aid in their analysis and interpretation.

To an extent, we could consider a good definition of the smart city to be one mirrored in the high-frequency city, with its focus on embedded com- putation, real-time streaming of data from its management, control, and patterns of social interaction. But this would exclude the low-frequency city, where there is as much intelligence from data that is not streamed in this way but is collected conventionally by manual interrogation of the population by the population. In fact, as more and more real-time stream- ing builds up data over ever longer periods of time, this real-time data becomes equivalent in its interpretation to low-frequency data in that it begins to supplement as well complement these traditional data streams.[12]

As we have implied throughout this book, a coherent and focused definition of the smart city is not possible, for many different audiences and commentators all have a different sense of what a smart city is. Here, we prefer to think of it as the operation of embedded computation in the built environment with the use of those same computers to make sense of this embedded computation using models of various sorts. In this sense, then, the same kinds of computation lie at the heart of both understanding and building the smart city. This does not mean that if we are dealing with cities where our focus is non-smart, we are not still in the domain of smart cities, for we clearly are. In this regard, then, the boundary between smart and non-smart is not the boundary between smart and stupid but rather a blurring of our thoughts and theories about cities more generally, where there is a fusion of computation and more traditional methods of data capture, explanation, and prediction.

It is, however, worth considering what others have said about the smart city, for there is by no means widespread agreement about its form. Definitions blur across a series of dimensions ranging across different kinds of embedded infrastructure, which from the Wikipedia definition includes "different types of electronic data collection sensors to supply information which is used to manage assets and resources efficiently." Other definitions focus on how computer technologies are being introduced into various sectors of the city, from "traffic and transportation systems, power plants, water supply networks, waste management, law enforcement, information systems, schools, libraries, hospitals, and other community services," again from Wikipedia.[13] Clearly, all these elements are being informed by new forms of computation, but most definitions of smart cities are simply lists of items such as this with no consensus as to informed and well-developed theories of what a smart city might actually be. In some senses, there is never likely to be a watertight definition, for, like cities in general, where there are as many definitions as there are people who live in them, so goes the smart city too.[14]

The smart city is, in fact, a term that goes back thirty years. It was first used by Gibson, Kozmetsky, and Smilor in their book *The Technopolis Phenomenon*, but the earliest term, which implied that computers[15]—in fact, electronic communications—were beginning to change the functions of the city, was "wired cities." This term was used by Dutton, Blumler, and Kraemer in their book of the same name,[16] which I am guessing came from James Martin's use of the term "wired society,"[17] which he used in the early 1970s. To an extent, the idea of the wired city some thirty years ago or more was predicated on the development of cable television. Although the first email had been sent on the ARPANET by Ray Tomlinson in 1971, by the mid-1980s, email was only just beginning to emerge in universities, which were connected to the ARPANET, and the idea of a wired city was much more resonant with those dealing with communications.[18] In fact, the early computer network, called "MINITEL,"[19] which was first rolled out in the late 1970s by the French Post Office was a videotext service delivering many different online products from news to online sales, and by the late 1980s, three million such terminals had been installed. Even fifteen years or so ago, the service was still fielding ten million calls a month before it was finally retired in 2012. Other services such as Prestel in the UK[20] and various related teletext services were widely developed by piggybacking on television, but it was not until the PC had matured to the point where the internet became widely available and Wi-Fi access was routine that the smart city actually arrived.

Other characterizations are significant. As computer graphics became dominant and as online maps became widespread, the idea of the virtual city, based on 3D visualizations for the most part, became another synonym for the smart city. The concept of the information city gained currency and this was associated with the notion of 'intelligent' and 'knowledge' cities where online access enabled citizens to achieve a level of intelligence in their decision-making behavior that hitherto had been absent—at least, that was the assumption of those pursuing the goal of making cities intelligent using new information technologies. "Cognitive cities," "creative cities," and "learning cities" are all terms that resonate with the notion of smartness and are variously used interchangeably with the term "smart." In Asia-Pacific, particularly South Korea, the term "ubiquitous city" or "U-city" has become fashionable. But at the end of the day, the generic term "digital city" is probably a better conception of the phenomena described in this book, for the term "smart" is, to an extent, a double-edged sword. It is used quite widely in American English, but in British English, the term is often used tongue in cheek to describe people who are too clever by half. In many senses, then, it is far too broad and ambiguous a term to describe the digital city, and for many, it has become shorthand for anything to do with the contemporary city.[21] Here we prefer to use a more abstract conception, reflecting the idea of the "computable city."

Our definition here is more focused, for we consider that the functioning of cities depends on their frequency of change. Cities function across a range of frequencies from low to high, with the high now being almost exclusively dependent on digital technologies. Thus, smart cities are high-frequency cities, but low-frequency cities can also be smart, for low merges into high when enough high-frequency signals have been accumulated to let us explore how cities change over the long term. In fact, it is best that we keep the term "smart city" in the background as convenient shorthand for the general tone of the phenomena that we are dealing with. In time, as this first wave of the public dissemination of new digital technologies passes, the term will probably become less controversial. In fact, as we implied in earlier chapters, once the wave passes and all cities take it for granted that these technologies are part and parcel of their fabric, the focus is likely to turn to other frontiers: to the embedding of technologies into inner space and outer space, into ourselves and into nature, perhaps, rather than into cities per se.

RHYTHMS OF THE CITY

Many commentators on what we do in cities propose that we think of the city as a beating heart. The biological analogy is not lost on us in that just as the human body wakes, eats, works, and sleeps, we can aggregate all this up to the level of the city using the same language to describe how it functions. In this conception, the city is like a living body whose heartbeat pumps the lifeblood through its networks—not only through the information infrastructure that we catalogued in the last chapter, but also through the physical infrastructure of transportation channels that move materials and people, as well as other sources of energy. These enable the city to function.[22] A heartbeat is the signal that blood is being pumped around the body, and a healthy heart works at something like seventy-five beats per minute, providing food in the form of blood that enables the body to maintain its functioning. You could almost say—to get some sense of scale—that the heart pumps at one beat per second.[23]

When we think of this analogy in terms of how a city functions, there are many different cycles, but the equivalent is usually taken as the flow of people moving from home to work, with the flow increasing dramatically as people wake and begin to move, decreasing during the middle of the day when people are at work, and increasing again in the evening as people return home. This, however, is only one kind of flow. There are a multitude of different flows at different frequencies, depending on where people wish to travel to and from and why they need to travel at all to pursue various activities. To compare the morning and evening peak hour travel with the body, the heart of the city pulses twice a day, once every eight hours, with the system closing down overnight. This compares to once a second for the body—a very different temporal cycle but sufficiently regular in the case of the city (and the body) to provide a basis for thinking about how all the other rhythms that define it are coordinated or at least related to one another. Of course, within us, there are circadian rhythms—the twenty-four-hour cycle that includes sleeping, diurnal rhythms that are synchronized with day and night, and longer rhythms that pertain to monthly cycles. As soon as we move beyond the diurnal cycles, then the human body is linked to its wider environment, and weekly, monthly, and annual cycles, which can be psychological as much as physiological, begin to intervene relating to the weather, the seasons, and other forms of geophysical phenomena. Add to this the cycles that we invent for ourselves that pertain to our own rituals as well as the elaboration of our work practices other than the time spent at work during the day, and we begin to see some of the complexity of the city in terms of these patterns.[24]

We can also classify these cycles in terms of the networks that make them possible, and to an extent, we have already done so at earlier points in this book. But we must also make a distinction between the cycles themselves and the way we can detect them. By and large, until we had electronic sensors of either the fixed or mobile variety, most cycles were hardly detected at all, and insofar as they were, the collection of data was usually based on manual means through human observation or, in the case of analog devices, by electromechanical rather than digital technologies, as we noted in the last chapter, where we catalogued things such as loop detectors for traffic, weather stations, and so on. The devices that have become widespread in detecting urban data that we can interpret as urban cycles are our own personal handheld devices: smartphones. These have become ubiquitous and are fast extending to devices such as watches and other smaller devices about the person that can be used to detect our movements and activities.[25] In fact, these devices are often used with the other forms of sensor: fixed sensors that can be activated by personal devices, for example enabling us to pay using apps from smartphones as we are now able to do in a wide variety of retail contexts. Fixed sensors that are permanently activated are the other devices that are being literally embedded into the built environment, and both these kinds of activator now provide us with real-timed streamed data that we can archive and even begin to interpret close to the real time in which the data is generated.

The key cycles that define the rhythms of the city relate to physical movement, in particular to the movement of people using light and heavy rail as well as public buses and other vehicular transport along fixed routes and natural tracks. In many cities, these are now automated for public transport, and this is almost complete in countries such as Britain where smart cards of various sorts as well as phones can be used to enable payment.[26] The data that is streamed from such sources is key to interpreting the performance of these systems. New forms of personalized and public transport such as systems that can call up transport from online requests such as Uber represent an elaboration of these systems in that mobile devices can be used to call transport that is already mobile so that very fast response times are enabled.[27] Transportation that needs to be planned, which is usually less frequent, such as that based on flying or long-distance trains, is now usually activated from websites. There has been a consequent decline in travel agencies that one used to visit to make such bookings, which, like so much retailing affected by the online world, is now largely a physical residue on the high street, where such shops have largely been turned into advertising forums or have simply disappeared.

Digital traffic itself such as email and web-page visits probably does define cycles, but these are much more complex to unravel. These pertain to other cycles such as the working day, and because these represent a mix of work and play as well as maintenance of systems that support both kinds of activities, it is not possible to catalogue what these cycles look like, for they depend on particular circumstances. Online retailing and banking of all kinds is probably clustered in the evening after work, but the picture is too complex to articulate, and there is not yet much data pertaining to these cycles. Social media data follows the same patterns, although we do have some sense that for short text-messaging services such as Twitter, the number of messages rises during the working day and reaches a peak in the early and mid-evening. The same appears to be the case for posting all kinds of data such as photographs on Instagram, Flickr, Facebook, and such media sites.

Physical movements that we introduced at the beginning of this section, such as the journey to work, extend to all other kinds of journeys such as health and educational facilities that have some parallels with the morning and evening traffic peaks but also take place through the working day. We noted online banking, but the general financial markets have for many years represented the most visible of cycles. Long-term cycles reflecting various components of the business cycle have been clearly graphed for many years, but shorter-term cycles based on second by second, minute by minute, and longer segments of the twenty-four-hour day represent key indicators that can be linked to regular and irregular events—to political and other crises for example, as well as to the regular reporting of the value of capital assets.[28] The movement of capital is harder to gauge, and the notion of an online economy, where price movements in diverse sectors are central to the way we exchange material and information goods, is only just beginning. In the near future and certainly over the next twenty-five years, we are likely to see quite dramatic examples of how financial cycles can be captured in this way, amid predictions and speculations as to how these will play out in determining and predicting answers to some very big questions about economic growth, patterns of investment, and the state of the local economy.

TRAFFIC AT THE HEART OF THE CITY

The diurnal pattern of traffic flow, be it by car, bus, or rail, manifests very stable regularities dominated by a slightly shorter morning peak and a longer evening peak in traffic flow when flows are aggregated to whole towns or cities. The morning peak is consistent with the dominant type of travel as being the journey to work and school, which increases more intensively than the return from work and school to home in the evening. which is a little more drawn out. People appear to engage in other activities related to their workplaces such as shopping or entertainment in the evening after work, and in some large cities such as London, there is a smaller peak in travel in the late evening before the city closes down for the night.[29] These patterns pertain to all identifiable modes for five days a week, while at the weekends, the flow increases slowly each day, peaking around lunchtime on a Saturday and a little later on a Sunday, when people are active in shopping, entertainment, and sports. This, of course, is the standard cycle, where we have to add seasonal events such as public holidays, which are more like weekend days. Summer holidays, usually in the month of August in Western European cities, and the opening and closing of schools tend to add to the richness of these cycles, notwithstanding that such patterns are quite easily explainable.

These are aggregate patterns across the whole city system. We can disaggregate them to the finest spatial scale where we observe such flows at particular locations on roads or routes or at fixed points such as stations and stops. As we disaggregate, we begin to see that the flow graphs become much more complex, losing their homogeneity. For example, using real-time streamed data from all those who tap in and tap out using radio-frequency identification (RFID) cards (or from apps on smartphones), we can observe the profiles of travel flows at different points on the system. Many if not most of the world's subways have now been automated for this kind of access, such as the London Underground (tube), which uses a payment system (smart card) known as the Oyster card. These profiles can differ quite markedly from the aggregate profiles, where all individual movements are aggregated, and thus averaged, displaying a degree of heterogeneity that does not mirror the overall aggregate picture but reflects the structure of activities visited over the diurnal cycle associated with different urban locations.[30] For example, only by looking at the data at a very fine spatial scale can one identify one-off events—which may follow a cycle other than the diurnal but are quite characteristic of regular events involving sports, theater visits, conferences, and so on. Moreover, once one focuses on locations other than workplaces and residences, different activities and land use generate different kinds of rhythms, where they all combine across all modes of transport to present a rich and detailed picture of flow, which is often hard if not impossible to unravel. Once we reach this level of disaggregation, the homogeneity and regularity of trip making collapses. It is still understandable, but within such patterns, there are all kinds of one-off events that are not predictable according to any cycle but are affected by breakdowns of the system, accidents, extreme weather events, and such like. These are all rolled into aggregate patterns that often damp out such events, leaving us with the aggregates that are the only things we can observe.

In the smart city, information technologies that combine mobile and fixed sensors and devices that work in concert are beginning to generate really big data that is essentially the data exhaust that comes from introducing such control systems, usually for payment, into various transport systems.[31] This is increasingly the case with public transportation, which is much easier to manage than individual and largely uncoordinated movements in cars and other personal vehicles. In time, most cars will probably be equipped with GPS receivers that sense their location, and when such systems become widespread, it will be possible, at least in principle, to obtain a complete picture of traffic flow at the finest levels of location and over the shortest time intervals. But until then, our pictures will remain partial. Even when transport systems become more fully automated, there will still be significant problems of location, where people walking, for example, resist their movements being tracked and where bikes and related alternative modes of movement might never be automated. And even if they are automated, there may be strict limits on what data is collectible, and there is likely to be the facility for the users to switch off such tracking. Questions of privacy and confidentiality reign supreme in such discussions.

The journey to work is, to an extent, still the baseline rhythm that dictates the beating of the city's heart, but there are many other patterns that pertain to how work activities, such as education, health, retailing, and so on, play themselves out, and these are often mixed with work-related activities. These cycles invariably do not coincide with the aggregate work journey profiles, for their starting times are more varied than work journeys, starting later and often ending earlier but also peaking at different points in the working day. In bigger and more complex urban environments, what we observe in terms of aggregate traffic flows are complicated mixtures of different cycles. For example, although we have a distinct morning and evening peak, the overall traffic volumes do not drop to very low levels between the peaks—that is, when the peaks are over—but are bolstered by other trip patterns reflecting education or visits to health-care facilities, as well as the ubiquitous activity of shopping for different kinds of goods and, of course, entertainment, which usually takes place from midday into the afternoon and evening.

Other transport modes, such as the recent proliferation of public and private bike systems, are automated from online payments using various sensor technologies. These enable any potential user to pay for their use over different periods of time, and such systems have become quite widespread in the biggest cities. In the center of London, for example, the public bike system, where these bikes are only available at fixed docking stations, reveals a pattern of travel that follows the rhythm of the working day, although the locational patterning is different from the journey to work using other forms of transport.[32] In the morning peak, the usual flow in big cities is from the suburbs to the center, whereas bike systems are much more local, in the case of London flowing from the edge of the center (the CBD) where people arrive on heavy rail systems to the real core. The flow is thus from edge of center to the core in the morning peak and from the core to the edge in the evening—simple enough to explain, but only understandable in hindsight. When it comes to walking, although 40 percent of people who move in Central London at any one time are walking, we do not have much idea of the actual rhythm, for this kind of data is not routinely collected, and it is early days yet with respect to piecing together such movements from mobile-phone data.[33] Alternative forms of transport such as Uber seem to follow the normal peaks, but surge pricing limits these somewhat, and there is much more volume at weekends than weekdays. It appears that Uber (in London) is more for entertainment and ad hoc journeys than for journeys to work. Essentially, we still do not know enough about these rhythms to make sense of them. To get a better but different kind of understanding, we need to turn to other data such as those that we can extract from mobile-phone calls in order to obtain more information about the frequencies of movement and travel in the city.

DYNAMICS ON THE MOVE

Even fifteen years ago, nobody imagined that most of the sensors that we might use to detect data streamed in real time would be personal devices—mobile phones that could capture many kinds of data associated with a variety of networks and related sensors. But now, the number of smartphones worldwide (at the time of writing in 2023) is a good bit more than half the world's population, some 6.5 billion, while as many again are other mobile devices. Smartphones in particular are universal devices, mimicking many functions such as telephone, camera, music players, satnavs, as well as the PC and laptop. In terms of usage, in the US, adults spend on average about twelve hours a day involving themselves at work and play with media of all forms, of which only half is digital, the rest being conventional TV, film, and video. Currently, usage of mobile-phone technologies in non-voice mode is about three hours a day, and this is rising by about 3 percent a year.[34] By the middle of this century, the dominant mode of interaction will be from personal digital devices, which will continue to expand to embrace many new functions that traditionally have been operated manually. Interaction with all forms of travel, for example, is likely to be fully integrated with smartphones.

Smartphones already have applications that enable users to pay directly for travel on public and private transit systems, and the raw data involved with any of these transactions is captured by the phone with respect to time, location, duration of call, type of application, and so on. Whether this data is accessible to the user or owner of the phone is problematic because the manufacturers of such devices, as well as the internet service providers (ISPs) enabling the service, often restrict this data, as it is valuable for marketing purposes. When the iPhone was first released, users were able to access some of this and retrieve their locational movements, but Apple soon embargoed this in a software fix.[35] Currently, it is necessary to strike a deal with the relevant ISP, and frequently such data is only available at substantial cost to those wishing to make use of it. In fact, almost 90 percent of all applications on Android phones generated data that found its way back to Google, and it is a safe bet that most of our movements are now tracked for those who carry a smartphone, notwithstanding the ethical and legal issues involved in such monitoring.

In terms of accessing data that is streamed in real time using smartphones, we need to note just how such phones work with respect to raw telephone and message data. Mobile phones act like two-way radio receivers and are very different from landlines that route messages down copper wires. The signals are routed to radio masts that receive and transmit with a message being passed to the nearest mast, which is located in a cell or zone over which the signal is captured. The data is then directed from the cell tower's base station to another tower that is closest to the phone being called, and that station essentially passes the call locally to that phone. When calls cannot be transmitted due to lack of signal strength or any other disruption involving capacity, they are passed to another cell—handed over—and in this way, the call is completed, albeit in a slower time. There are various protocols required so that calls can be completed in a timely manner, especially when calls are transferred between different types of network and, of course, to other systems such as landlines.

The data that is captured by the base station and, of course, archived as call data records (CDR) is fairly routine. Users are not known from the signal and are therefore assigned a random ID, while location, time, and duration of the call are captured, as well as other technical information pertaining to the routing. It is comparatively easy to work out phone usage by cell and sometimes by actual location, and there are a number of attempts to see how close mobile-phone call volumes are to other locational data such as the distribution of resident and working populations. But there are problems in doing any of this pertaining to data generated by mobile and smartphones. These relate to inferring substantive meaning to the origin and destination locations of such calls. Usually, the meaning of such location data has to be matched with the time of day and the duration of the call, with the assumption that if a user is located for any length of time during the working day making phone calls from any one place, this is their place of work (or less likely their residence). If the location remains fixed and is activated over time, it is assumed that the correlation with their place of work increases. The same sort of inference is made with respect to place of residence, with nighttime hours being used to figure out if the place is a residence or a workplace.[36] This kind of analysis is pretty crude and, so far, is indicative only of volumes of activity at locations. It may well be a long haul, as in all social media data, to reach the point where such data can be used to infer movements and locations in contrast to more straightforward but ever more expensive collection of data using traditional means (by asking people or getting them to record personal responses). In fact, we may never be able to generate stable and reliable observations from such data.

The jewel in the crown for this kind of analysis is to associate mobile-phone calls with physical transportation, that is, to define the origin and destination of such calls in relation to the origins and destinations of movements made on physical transportation systems. As with call volumes, such analyses are associated directly with the journey to work. If people spend time at work and then at home, based on the frequency and duration of calls, one can then infer that they make a trip between home and work or vice versa if the data can be matched to the temporal cycle. However, there are major problems in developing flow patterns as a proxy for real trips from CDR archives. In essence, the assumption that mobile-phone calls are a proxy for physical travel is highly contestable. It may be that, in time, we will discover that other applications on smartphones such as Tweets are a better proxy.

What makes the analytics so problematic is the fact that the diurnal flow of phone calls is not that close to the rhythms of materials and people traffic flows. Phone-call usage rises more slowly in the morning peak than actual traffic flow, and although it follows the same peak-hour profiles, it is different enough to suggest that although mobile-phone calls are correlated to a degree with travel, then the calls do not actually mirror the physical act of traveling. Of the many case studies where such data have been compiled,[37] there is quite a lot of variation in daily call patterns using mobile phones, and all one can really say is that there is sufficient variability to cast doubt on the validity of using such data as a proxy for trips, especially as our knowledge of physical trips anyway is often confounded by our not knowing the actual purpose of the trip. All one can say is that interpreting such data is a veritable minefield, once again showing how difficult it is to extract meaningful patterns from big data sets such as these.

In fact, in the extreme, when one drives, in most places, it is illegal to hold a device that one speaks or types into, and it is clear that most people who have such a device on their person do not yet have the technology for hands-free conversation, although this may change, of course, as digital car technology becomes ubiquitous. Last but not least, the actual physical locations—the origins and the destinations of mobile-phone calls—are often only available by cell tower. The number of cells is always a magnitude smaller than the actual number of locations we need, while the number of calls is a magnitude above the number of individual locations. The spatial resolution is usually quite low, with a cell radius of about 2 km on average in suburban areas, but it can be much finer at the level of 400 m in denser urban areas. In Paris, for example, with three thousand cell towers, there are about three million mobile-phone calls per day, which is an average of about a thousand calls per cell per day.[38] In fact, the predominant usage of mobile phones is no longer for phone calls.[39] Calls came eleventh in a survey of phone users, which gave the following ranked order in terms of frequency of use: text, 88 percent; email, 70 percent; Facebook, 62 percent; camera, 61 percent; news, 58 percent; online shopping, 56 percent; weather,

54 percent; WhatsApp, 51 percent; banking, 45 percent; and watching videos, 42 percent. For our next set of rhythms, we turn to social media, for as this survey shows, a strong and increasing usage of smartphones focuses on such patterns of contact.

SOCIAL MEDIA, CONTACTS, AND NETWORKS

As we outlined in chapter 5, as soon as we were able to access the internet with truly portable devices—meaning smartphones, lightweight laptops, and tablets—our ability to link with others entered a new phase. The 1990s saw the emergence of the passive web—Web 1.0—in which most activity was based on users searching for and viewing information, but by the early 2000s, this had evolved into much more active usage—Web 2.0—which essentially is defined as web usage involving user-generated content. Users are able to interact with web pages and, through them, with other users through a wide variety of media.[40] These involve straightforward query-response procedures, all the way through to augmented realities, virtual worlds, and elaborate graphic forms of human–computer interaction.

Essentially, social media originated from chat rooms, internet relay chat, bulletin boards, and simple message passing, often on the same machine or certainly the same network. This eventually reached the mainstream web, and various forms of messenger systems emerged, which when mixed with graphics, and the latest human–computer interfaces generated a variety of systems for making contact across the net with others who would declare themselves related to one another in digital terms. Social networks became the basis for these developments in media, which we quickly ported to smartphones, which became popular about fifteen years ago. The first systems were text based, with Twitter being released in 2006 to enable users to send short text messages originally no longer than 140 characters in length. In the case of Twitter, the length of text was doubled in 2017, and various video and graphics outputs can now be added and streamed with respect to the sending of any tweet. Followers of the user in question receiving a tweet can retweet it if they wish, passing other public and private messages pertaining to its content to their own followers and to anyone else who is interested in the media, and also relate it to other messages and users on the platform. The number of active Twitter users now stands at about 220 million worldwide, although the number has peaked in the last five years, and to an extent, it has become more specialized, with many users being interested and influential in entertainment, politics, sport, and related pursuits. Twitter is quite widely used by politicians and media stars to popularize their policies or their fame.

Every second, on average, more than six thousand tweets are generated, with this coming to some 360,000 tweets per minute, just over half a billion per day, some 190 billion per year.[41] Tweets during the working day, unlike mobile-phone calls, do not coincide as clearly with physical peak-hour traffic, and there is not much correlation with the journey to work. Tweets peak in the very early morning but then fall during the morning peak—which is largely the journey to work—and begin to rise toward midday into the afternoon, peaking again in the early to mid-evening. There are, however, distinctly different patterns in the largest cities that have much more pronounced social media locations in the form of restaurants and entertainment hot spots as well as much larger venues for sport and theater.[42] These attract people with strong active interests in these events over which they tweet more avidly than the general population.

The temporal rhythm of tweeting, however, is complex, for it relates very strongly to different communities and interest groups in cities, and only in the largest cities does the profile mirror other rhythms such as the journey to work. In these cities, however, when people wake up, before they set off to work, tweeting is stronger than in the actual physical traffic peak. In big cities, the number and density of tweets is closely correlated with the density of population, but this is also where distinct entertainment hubs, which are hot spots for tweeting, emerge. These are closely linked to stations and other places where people gather to wait for transport to other key tourist attractions and to places where there are major sports events. As with mobile-phone calls, extracting the actual links from the implicit social network of those who tweet to each other is highly problematic. Frequently, in fact, for about 95 percent of all users, the GPS is not activated when people tweet, and thus there are no unique locational identifiers in the corpus of most tweets.[43] Even though the total number of tweets is enormous, if only 5 percent are geocoded, the links to other users are problematic. It depends on guessing the location of those other users, on mining their tweets for hints of location, and on extracting geography from the process of retweeting messages, all of which are indirect and particularly suspect ways of figuring out the geography of tweets.

Nevertheless, at a macro level, at the level of whole cities, countries, and continents, the patterns of the social network based on the links implicit in tweets produce impressive maps of global connectivity.[44] There are many other similar sites—for example the Chinese site Weibo—from which similar patterns can be extracted, and the analysis of textual content for geography and other patterns, although in its infancy, is rapidly increasing. All of this, of course, is a kind of guesswork. To generate data about the social network, this is almost like taking two steps back to go one step forward in that the data we want can be produced through direct observation and questionnaire, despite some practical difficulties. But in the case of using social media to do this, we have to guess the data before we can even see if the data is meaningful for the quest in hand. In many senses, this might be worse than having no data at all, and it puts the notion of data from real-time streaming in a particular perspective with respect to how far we can use big data in learning about the smart city.

We can say a lot more about the kinds of rhythms that define the high-frequency city, but many of these face the same problems we have identified with extracting geographies, flows, and related patterns from mobile-phone calls and short text messaging. Many large cities are now covered with a skin of sensors that monitor pollution. For London, this data is streamed in real time and made available publicly as open data relevant to exploring ways in which pollution might be managed and reduced. Much of this data is being used to make predictions about future air quality and is essential for the diagnostic monitoring of safety levels. To an extent, the profiles of this data follow the overall traffic flow through the day, largely because many of the pollutants monitored relate to pollutants generated from road traffic. There are a variety of weather-related features that can be monitored in this way too, and microclimatic factors follow similar profiles. However, as one gets finer and finer measurements in terms of spatial scale, the data becomes more and more heterogeneous and, to an extent, volatile. Extreme events also characterize these series, implying that we need ever better measurements to gain a much more complete understanding of the implications of disruptions to these dynamics.

THE REAL-TIME ONLINE ECONOMY

Just as mobile phones can be used to activate countless physical systems such as paying for transit and accessing many social systems such as Facebook, they can be used, along with credit cards and many other varieties of smart devices, to initiate online shopping associated with many different sectors. Although the initial systems were analog, based on a mixture of manual processing and electromechanical sensors, the notion of paying in this way is little different from our current online systems other than the physical act of presenting the card to the seller. We have already noted that the online economy with respect to retailing is making a big impact on the physical location of shops and the distribution of goods, but there are also significant economies of scale from such automation involving the actual processes all the way from the manufacturer to the buyer. In fact, smart cards are being extended all the time to embrace other services, but there is the prospect that all this will collapse eventually, perhaps rather suddenly, into standard payment systems as part of smartphone technology. New patterns of distribution within the city linking buyer to seller and manufacturer are emerging, this being one of the few examples where it is obvious that traditional patterns and flows of good are being altered by the imposition of information technologies that make these changed behaviors possible.[45]

Other forms of payments and savings are part of the whole domain of the online economy, specifically, in this context, banking, where, like many bookshops, furniture outlets, and department stores, institutions dealing with capital flows, pricing of physical assets, loans, debts, interest payments, as well as tax and subsidies are moving to the web, with their physical presence in the city under continual scrutiny. The actual assets of the city— its land and buildings as well as their contents—is being affected by these changes in terms of their value, but their valuations and the location of their work are now largely virtual in terms of the networks used to price and change their ownership. As yet, we do not have a clear view of how all these changes are playing out, and due to the invisibility of the networks that are being created to project them into the virtual world, it is quite likely that we will continue to struggle in producing an integrated understanding of their impact on the future city. There is little doubt that the problem of their invisibility is likely to get worse rather than better as we begin to attempt to provide an integrated understanding of how these various networks are interlinked and synchronized and how they underpin the resilience or otherwise of the future city.

With the advent of the interactive web, there are a variety of processes that do not occur at anything like the speed of the highest frequency events in the city but are captured more slowly but nevertheless at much faster speeds than anything we have measured hitherto. For example, many traditional systems, such as the decennial population census, lists of house prices, land values, and even indexes of economic performance, have been updated monthly or even yearly in the past, and increasingly, online reporting and archiving is leading to rapid increases in the speed at which current data becomes available. House prices and land values are updated continually and can change daily, but as we disaggregate their scale to finer spatial units, then their temporal frequency declines. In a large world city such as London, house prices and land values, for example, can be seen to vary daily, but when one examines finer and finer spatial units, the number of changes inevitably decreases, as the number of such movements in prices depends on densities and on the number of buildings. The same is true of all objects in the city that have their scale determined by numbers of people, physical activities, and such like, which are in the millions at most, rather than real-time streamed data, which typically is studied in terms of hundreds of millions or billions of data points. Again, we are only at the beginning of thinking about the city in these terms, and here the interface between the real-time city and the online economy becomes critical.

A lot of this data could become open in the next decade but subject to tight confidentiality and privacy limits, while substantial amounts of data about the city are being created using individual populations as their own sensors. In this way, even equipped with mobile phones, citizens can become sensors, and a good deal of our measurement of the physical system is being aided by this kind of crowdsourcing.[46] OpenStreetMap is a magnificent example of how the wisdom of the crowd has in a very focused way led to a remarkable open-source free map of the world at the finest of spatial scales. This map is being tested and refined all the time as people whose knowledge of any local area are used to add and check detail that could not have been added or tested in any other way, certainly not by surveyors working in a traditional manner. Nevertheless, both ways are critical to constructing the best map. To an extent, all these measurements we have introduced in this chapter depend on contemporary information technologies that capture, create, store, and present such data about the high-frequency city. But we are still a long way from having any fundamental sense of how cities function in the short term and maintain their functional resilience. Our last example gives real force to this speculation.

THE ARRAY OF THINGS

Charlie Catlett set up the Urban Center for Computation and Data at Argonne National Laboratory, which is part of the University of Chicago.[47] The laboratory originated in the place where the world's first nuclear reactor was built. This was Enrico Fermi's Chicago Pile-1, which was established as part of the Manhattan project during World War II. But from the start, the laboratory's mission was much broader than high-energy physics per se, taking on projects of national significance that related to the broader implications of transport, environment, energy, and a related series of social goals, which included national security. More than ten years ago, the idea of the smart city was gaining momentum in many places, with those involved in the digital revolution becoming intimately aware that computation was spreading out into the wider environment. The prospect of computers of many sizes and varieties being embedded into the fabric of cities threw up a series of challenges that major centers of computing such as that at Argonne could well relate to. In fact, in the computational group that Catlett belonged to, grid computing had been developed by Ian Foster in the vanguard of cloud computing, and this established the reality of computation spreading out everywhere, particularly into the city in its many public places.[48]

This chapter has been largely devoted to incremental developments in high-frequency sensing without any notion that there is a grand plan to embed any integrated form of these technologies into the city. This is despite there being a variety of one-off initiatives to establish smart cities in the manner of new towns, particularly in countries that wish to showcase their economic prowess such as the United Arab Emirates and South Korea, where the new towns of Masdar and Songdu represent such iconic statements. Chicago is almost at the opposite end of the spectrum, where Charlie Catlett and his group have built a small but beautifully crafted network of sensors in the city in relatively slow, incremental fashion that is tailored to the particular conditions of the city and the many different projects that might avail themselves of such sensing technologies. These range from the usual kinds of pollution monitoring to pedestrian movement, the provision of information services on the street, water and sewerage provision, and a host of related infrastructures that need to be maintained, replaced, and enhanced. In fact, the group refer to the system[49] as the "Array of Things"—a fitness tracker for the city, almost a city Fitbit!

Many people now ask what the smartest city is, to which by now you must have realized that there is no answer. It is not that Masdar or Songdu are smart because of the way they are designed, top down. In fact, smart cities are built from the bottom up, and what is clever about Catlett's project is that his design for the Array of Things and its sensor network does not attempt to place sensors everywhere but rather to build an array of intelligent machines that are expensive and need to be placed in the locations where they generate the most useful data. So far, there are forty-two pilot locations that have been wired, and when the system is complete, it will have five hundred such sensor pods at roughly the same densities as cell towers associated with mobile phones but positioned in such a way that they reflect different intensities and densities of activity as well as areas associated with the most severe physical and social problems. Moreover, the system is designed so that the data that it generates will be in the public domain, accessible to anyone who wishes to use it and to design software that can add value to our urban experience, to city living. This will let anyone who wishes to build new analytics based on the raw data that the Array of Things is able to generate, making the city more sustainable, efficient, and hopefully more equitable.[50]

The nodes or pods will measure "temperature, barometric pressure, light, vibration, carbon monoxide, nitrogen dioxide, sulphur dioxide, ozone, ambient sound intensity, pedestrian and vehicle traffic, and surface temperature." The system will also be extended in time to "monitor other urban factors of interest such as flooding and standing water, precipitation, wind, and pollutants." These measurements can then be used to inform citizens and policy makers about the livability of the city with respect to pollution, congestion, flooding, heat islands and weather conditions, and, at the most local level, safe places for walking and sitting. The data is open and can be accessed in a portal, which is already operational in various formats that enable even the most nonexpert of users to extract it in a form that they can visualize and interpret. In fact, the portal is now already open, and data which is being generated routinely such as crime statistics and traffic volumes can be downloaded in a straightforward fashion. The system of sensors is also being designed so that they blend nicely into the environment with the overall aesthetic and functional design of the system under the advice of the Art Institute of Chicago. Unlike most technology in cities that is produced in an ad hoc rather muddled fashion with no sense of overall design, Charlie Catlett and his team insist that for smart cities technologies to be acceptable, their design across all dimensions must be world class.

The sensors themselves represent the latest in edge computing, designed specifically to enable an intelligent kind of sensing in their wider vicinities, particular to the project and not linked to any other sensor projects. The sensors are designed around security, privacy, extensibility, and survivability as key concepts in such a way that they generate a kind of sensing that is robust to many signal degradations. The system is called "Waggle," which reflects the way in which the sensor works like a bee that waggles. This is to ensure that bias in selecting mobile and fixed objects as well as atmospheres associated with pollutants and climates are measured correctly. Energy use has been a prime criterion in their design and deployment. Privacy and confidentiality are key to the data that is generated, and there are a series of organizational arrangements that make personal data anonymized and secure. With such a well-organized system, it is of little surprise that other cities wish to adopt the technology and partner with the laboratory and the city of Chicago. Projects are underway at Palo Alto, Seattle, Portland, Denver, Detroit, Chapel Hill, and Syracuse, while there have been more than a hundred expressions of interest from cities in the US and internationally.[51]

Part of Catlett's Array of Things is the notion that massive amounts of data that can be streamed and organized to enhance our understanding of the how the city functions can be tested in the wild, so to speak. It can then be produced as state of the art exemplars, which government and the private sector can then emulate in delivering it to those who can make best use of it. Catlett, in talking about the role of government and public agencies, argues that in cities, "most of that data goes unused because they don't have the [resources] to analyze it." The Array of Things, because it is coordinated, planned, but sensitive to the most local of issues and built from the bottom up, provides a way in which the public sector can begin to mobilize its vast information sources and resources to the benefit of its citizens.[52] The model that Catlett has pioneered seems to be a much more robust way forward in building the smart city than any kind of top-down plan making of the kind associated with the smart new towns such as Songdu and Masdar, but that is a different story and one that we will only take up in the last part of this book. Before we do this, however, we will explore what computation means to the contemporary city and begin to explain how computers can be used to simulate and predict what goes on in cities, notwithstanding that increasingly those very cities will be composed of the computers that are the subject of their simulation. We will thus return to that great universal—recursion: using computers to study computers.

1 The General Post Office and the Royal Mail in 1830.

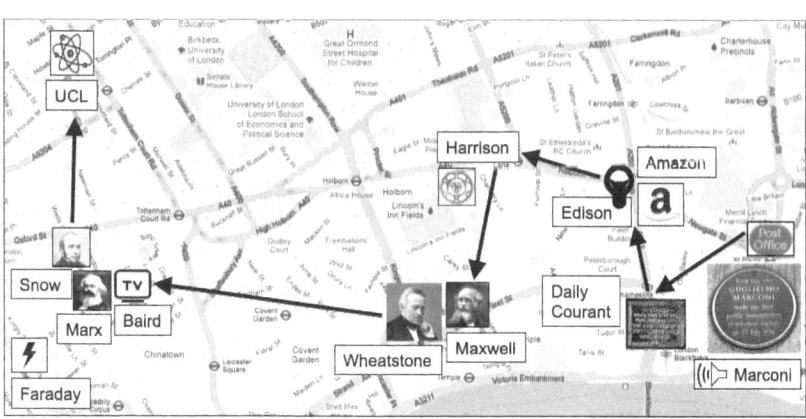

2 The information mile, walking west from the GPO at St. Paul's.

3 King's College London, where Maxwell first presented his wave equations.

4 Turing worked here next to the Manchester Baby in the early 1950s.

5 Colossus: One of the first digital computers at Bletchley Park in 1943–1944.

6 John von Neumann and the IAS machine at the Institute for Advanced Study, Princeton, 1951.

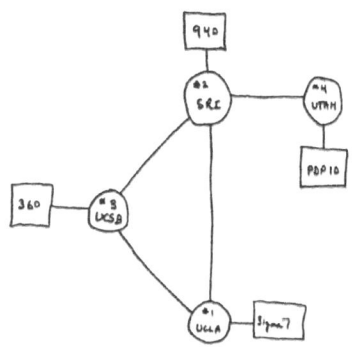

7 The first nodes in the ARPANET (internet)—UCSB, UCLA, Utah, and SRI—1969.

8 Bill Gates and Paul Allen at Lakeside School, Seattle, 1968.

9 Queen Elizabeth II sends her first email in 1976 from the Royal Signals Research Establishment in Malvern, UK.

10 Thomas Kurtz and John Kemeny with BASIC on PCs in 1984.

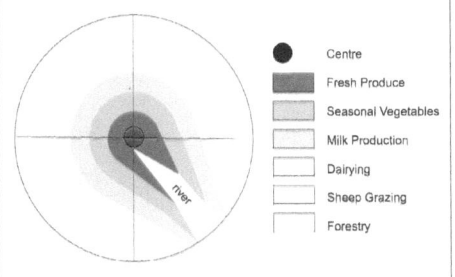

11 Johann Heinrich von Thünen and his model of concentric land-use rings.

12 Marx, Engels, and their families.

13 Jane Jacobs, vocal critic of architecture and planning, in NYC in 1961.

14 An interactive dashboard capturing traffic data.

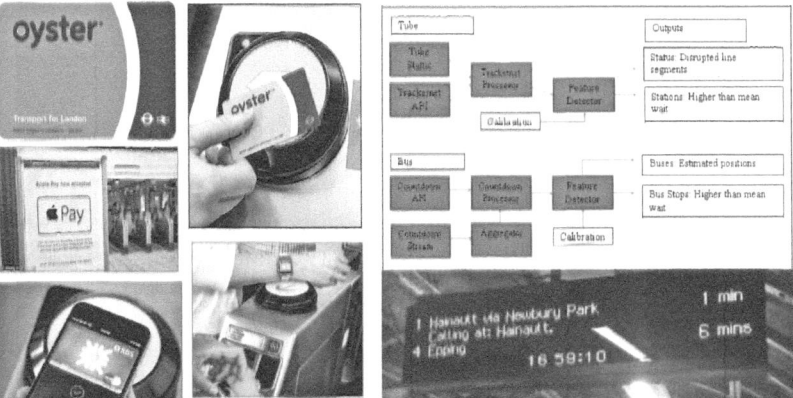

15 Capturing demand for travel linked to the supply of trains on the London subway.

16 Wassily Leontieff, inventor of input–output analysis.

17 Wilbur's simultaneous equation solver used by Leontieff for his input–output model in 1936.

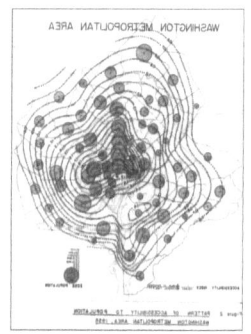

18 The CATS Cartographa-tron, 1959.

19 Hansen's accessibil-ity model, 1959.

20 Ivan Sutherland's Sketchpad, 1961.

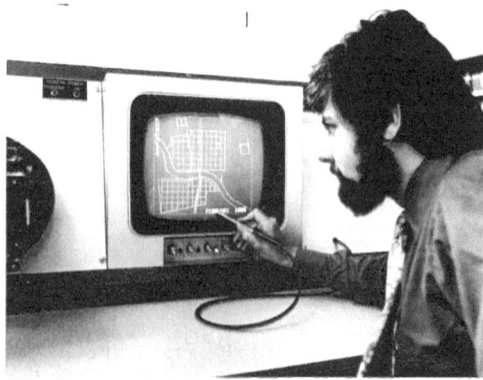

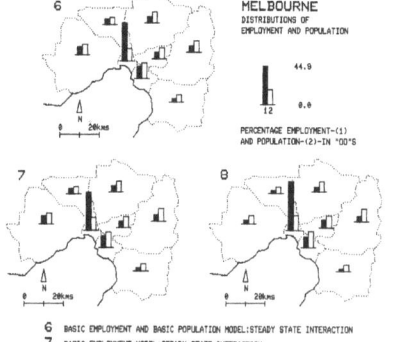

21 Kadanoff's interactive graphical simulations of urban development, 1970.

22 Early graphics screen dumps from a DEC VAX GIGI terminal, 1982.

23 Early line-printer graphics, 1960s.

24 Andy Warhol draws Debbie Harry, 1985.

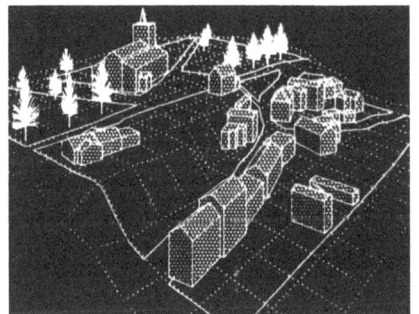

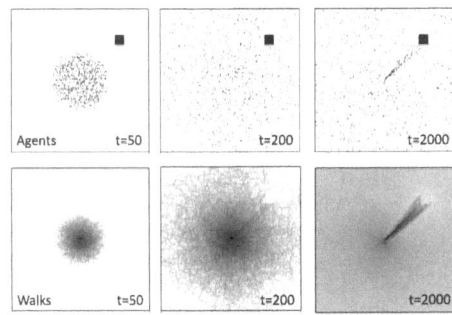

25 A simple 3D rendition of an English
village using AutoCAD, 1980.

26 Agent-based modeling of mobility.

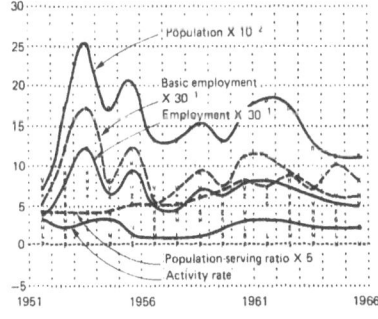

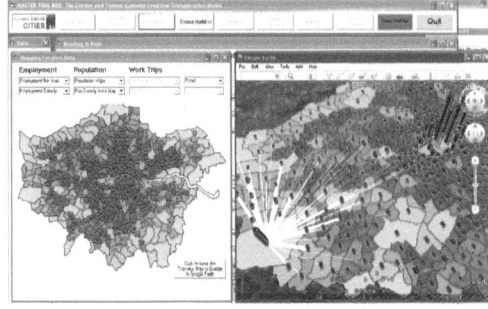

27 Urban dynamics with Forrester-like
graphics, 1970.

28 Desktop urban simulation of land use and transport integrated with Google Earth for 3D rendition
of map data.

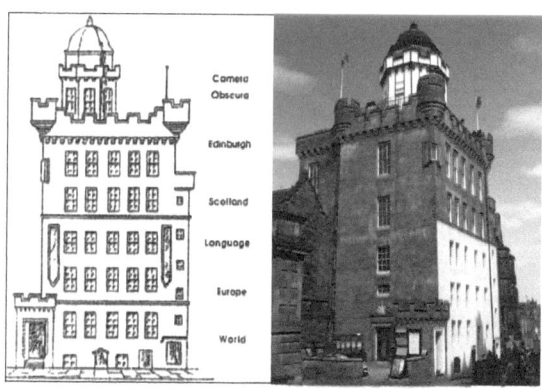

29 Patrick Geddes' Outlook Tower: The idea in 1910 and the reality in 2013.

30 John Paton-Watson and Patrick Abercrombie discuss the Plymouth Plan in 1943.

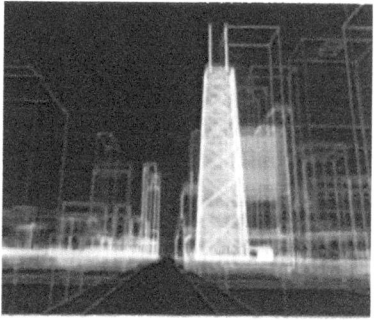

31 Wuhan Citizens Home: An entire building for participation using city models, 2017.

32 The first 3D digital city models by Skidmore, Owings & Merrill, 1984.

33 Embedding Virtual London into a virtual room where avatars act as participants in the design process.

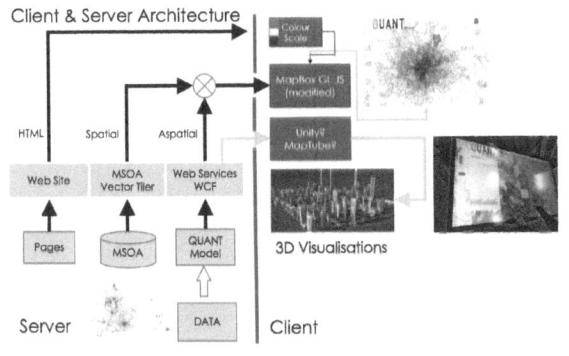

34 Models constructed from web-based services, big data, and fast computation.

35 Singapore: Building an intelligent island, 1988.

36 Geospatial server farm at Harwell, UK.

37 Cooling pipes inside Google's data center in Douglas County, Georgia.

38 Modeling riots and policing using the London touch table.

39 Augmenting models of physical plans with computable touch-table technologies.

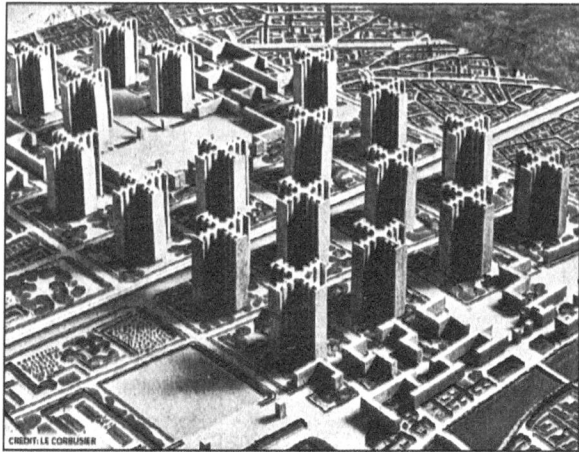

40 Ideal cities: Le Corbusier and his plan for Paris, 1925.

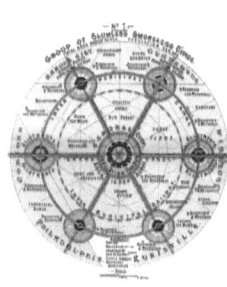
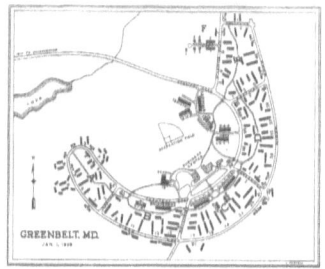

41 Ebenezer Howard's   42 Masdar Smart City UAE, 2011.   43 Greenbelt, Maryland, 1938.
garden city, 1901.

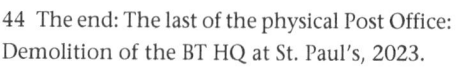

44 The end: The last of the physical Post Office: 45 The future: Social networking.
Demolition of the BT HQ at St. Paul's, 2023.

III MODELS AND COMPUTATION

Computers were first used to increase our understanding of the way cities worked by developing simulation models that could generate predictions to help us to explore how we could achieve a better quality of life in future cities. These models were based on traditional forms of scientific computation that dominated applications until they were dramatically enriched by computer graphics and GIS, which made such models much more accessible to both urban scientists and policy makers. The recent focus, however, has been on the high-frequency city and the embedding of sensors into the city, which is yielding big data, and this has changed the emphasis toward developing models that are much closer to the real-time functioning of the city. Into this nexus has come the development of new ways of interacting with cities using a variety of virtual and augmented realities that enable professionals to engage with their clients and publics in ways that are much closer to the real issues than anything we have had available hitherto. In part III of this book, we will explore these ideas, illustrating the arsenal of tools and range of data that new information technologies are bringing to our overall understanding and planning of the future city.

In chapter 10, we outline the history of the first simulation models built for land use and transportation that date back to the 1950s. The experience with these early models turned somewhat sour for many organizational and political reasons. It took the development of new ways of representing the city, particularly the development of GIS, to restore the focus, and we deal with these methods and tools in chapter 11. The kinds of models that pertain to the high-frequency city, which lie at the core of the smart city and the rise of big data, are presented in chapter 12, and we conclude with an outline of how new graphics technologies such as 3D and VR systems are being developed to improve urban design for the future city in chapter 13. Arguably, this focus on the high-frequency city is now being augmented once again from a fusion of different models and tools developed from the time simulation models were first introduced, and new developments in urban science that we have hinted at already are beginning to inform the practice of modeling. Much of this will prepare us for widespread computation across many temporal and spatial scales, which provide the emerging focus, increasingly referred to as the metaverse, that we discuss in part IV.

10 SIMULATION, MODELING, AND PREDICTION

All models are wrong, but some are useful.
—George Box (1979)

BEFORE COMPUTERS

Wassily Leonticf was born in 1906 in Munich to well-to-do Russian parents. Almost immediately, they moved back to their native St. Petersburg, where Wassily was educated until the Russian Revolution ended their middle-class lifestyle in 1918. It was clear that Wassily was a bright student with a keen eye for his father's profession of economics, which he studied at the University of St. Petersburg, soon to be renamed the University of Leningrad, as was the town itself in 1924. While he gained his master's degree in economics there, he was politically active against the Soviet regime, supporting the Mensheviks against the Bolsheviks, always sailing close to the wind, but avoiding the fate of many during those turbulent years.[1] However, in 1925, due to ill-health, he was lucky to be allowed to make the trip to Berlin to seek medical help, which, in fact, it turned out he did not need.[2] He stayed in Berlin and registered for a PhD degree, which he presented in 1929 with the title *The Economy as Circular Flow*, which, by then, he realized, would constitute his life's work.[3]

Leontief reasoned that economics was a systems science that required many sectors in the form of industries to be articulated in terms of the network of relationships that held them together, in balance, or, in the emergent semantics of the day, in equilibrium. As one sector could depend on any other, either directly or indirectly through an intermediate sector, if you were to predict the impact of an external change or shock to one sector, then you would need to take account of the ripple effects that would happen as the change in that sector made itself felt on all the other sectors to which it was linked. In this way, the ripples would work themselves out, the entire panoply of effects being referred to (albeit much later) as the "multiplier effect." Leontief observed that most economic analysis was not of this nature—it dealt with partial changes—and thus he argued that only by treating the entire economy as well as its imports and exports could a sense of how the economy worked be understood. In short, like Samuelson,[4] his contemporary, who we noted argued the same for the standard model in chapter 7, he believed the road to economic salvation lay in general equilibrium.

From 1927 to 1929, he nurtured these ideas at the University of Kiel, but he wanted more, much more, and in that year, conscious of the rising Nazi threat to people like himself, he emigrated to the US. He secured a job at the coveted National Bureau of Economic Research (NBER) in New York in 1931, but he only stayed long enough to impress his talents and ideas on the economists of the day. This led to an appointment as an instructor at Harvard in 1932, where he was able to convince the university authorities that to initiate his research program, he needed assistance as well as some rudimentary calculating machines. This, he argued, was crucial to begin his lifelong quest to model the general equilibrium structure of the US economy using what ultimately came to be called "input–output analysis." In essence, his work with various problems in demand and supply convinced him that one needed to move beyond the kind of partial analysis, which was the norm, to a fully-fledged general equilibrium theory of the economy that would enable a much better understanding of what was related to what. Although Leontief is credited with much of this analysis, there were historical precedents for such equilibrium thinking. In the 1880s, Leon Walras, for example, among other economists, produced a specification for how the various sectors of the economy might connected,[5] but it was the work of the French physiocrats in the mid-eighteenth century who provided the initial rationale. Theirs was a theory largely about an agricultural economy as might have been expected at the time, dominated by land but constructed so that the various inputs of land and labor were balanced in some way, which was again akin to economic equilibrium.

Leontief built on these contributions, but his focus was on the actual structure of a model economy—in his case, the US economy, which was largely industrial and service orientated by the 1930s. To construct his model, like Walras and others before, he formulated the relations between the various industrial and services sectors using simple linear equations, which were configured in such a way that a solution to the model was a solution to these equations. He chose two points in time, 1919 and 1929, for which he had reasonably good but highly aggregate data on interindustry linkages, which defined the data required for the system of equations. But the biggest headache that Leontief anticipated almost as soon as he formulated his model in the mid-1920s was how to actually solve the system. Solving forty-two simultaneous equations, which was the number of industrial sectors he defined, was an impossible task using manual methods, for it would take something in the order of thirty million calculations to do so. Something else was needed, and that something appeared from other sources at both Harvard and the MIT—in fact, from the remarkable developments that were taking place in the development of electromechanical calculators just prior to the development of the digital computers that ultimately emerged in our magic year of 1937.

Down the street at the MIT, John Wilbur, who had worked with Vannevar Bush on various calculators from the late 1920s, had devised a mechanical equation solver, which could handle nine simultaneous equations quite easily. Leontief got to know Wilbur and decided to trim his problem to the machine that Wilbur had devised.[6] Forty-two industrial sectors were aggregated to ten, and this meant that the nine-equation solver could handle the ten-equation problem immediately. Leontief could have never solved the problem without Wilbur's device, for to solve the reduced system of ten equations, something like half a million manual calculations were needed, and this would have taken some two years. Leontief described Wilbur's machine as being "made of tilting plates (representing the unknowns) and steel tapes (representing the equations to be solved)." Somewhat apocryphally, Leontief, ever the wit, recalled, "You could really change the coefficients slightly by simply sitting on the frames and if they did not give too much, this meant the solution was relatively stable"[7]—a classic demonstration of how a complex system can be simulated not by a digital computer (but this would come quickly on its heels) but rather by an analog machine—where physical components represented the abstract elements of the mathematical structure to be solved. Somewhat amazingly, it was almost as though those who anticipated the digital computer began by demonstrating simulation using analogs but in the firm believe that the digital would replace the physical very soon. Leontief worked in that ferment of ideas just before World War II, when all kinds of ideas about computation were in the air. In one of the most prescient insights into analogs and real systems, Leontief said, "The economic system of a country is like a big computing machine . . . and very often computers break down. Find out what's wrong with the computer with a good analysis."[8]

Leontief's work was so close to the computer that it might even be regarded as one of the first steps toward using computers to simulate social systems, for almost as soon as Wilbur's equation solver was used, Leontief began to think about bigger systems, and this inevitably meant digital computers. It took until 1944 for Harvard to acquire the Mark 1 machine built by IBM and designed by Howard Aitkin[9] as part of the cluster of developments at the end of the war in Harvard, Penn, the MIT, and Princeton. Leontief, whose project on simulating the US economy was growing ever bigger, immediately availed himself of this.[10] Yet, ideas about how to simulate cities and local economies go back to von Thünen, as we recounted in chapter 7, and throughout the nineteenth century, there were speculations that what society required were simulation models that could be used both to understand social problems better and to enable informed predictions. To an extent, it was the industrial revolution itself that forced this pace, as we have implied in earlier chapters. The steam engine and the development of the railways formed the first major impetus, and it was no surprise that attempts were made in practice to figure out how the role of distance, travel time, and travel cost were being changed by innovations in these kinds of transport. As we argued earlier in our remarks on distance, as soon as the first industrial revolution began, there was not so much a death of distance but a transformation, and as we have been implying all along, the future city will very much still be structured around distance and geometry, but it will be distance transformed and time compressed in ways similar to the manner in which the city has been transforming itself for the last two hundred years.

FLOWS AND GRAVITY

The first passenger railways built in 1825 between Stockton and Darlington and in 1829 between Liverpool and Manchester initiated a period of railway building of manic proportions.[11] It was not dissimilar in many ways to the race for developing new bandwidth that was part of the dot-com boom, which led to the collapse or bust in the early 2000s, with far too much bandwidth having been created. This was true of the railways in Britain at that time.[12] Favorable economic conditions and a growing middle class with savings to invest powered the boom in railway construction, which began in the early 1840s. There was an almost complete culture of laissez-faire during this period, with entrepreneurs simply having to get an Act of Parliament passed for permission to set up a railway company. At one point in 1846, there were 242 bills before Parliament, and the number of miles contracted grew dramatically, with more than 9,500 being proposed in that year alone. Most of these proposals never came to fruition, but the growth was rapid, from some 125 miles in 1832 to some 13,000 by 1870. In some senses, the 1840s was the decade when the physical backbone of industrial Britain was created. Although the network, which was more or less complete by the 1880s, represented the kind of diffusion of technology that all the industrial revolutions since the first have brought and continue to bring, much of the network was underutilized, as it took time for urban development and population growth to catch up. Nevertheless, this was the period when some farsighted thinkers began to ponder the nature of transportation and whether there were any regularities that could be exploited in the planning of such infrastructure.

In 1836, the UK secretary of state responsible for Ireland, Thomas Drummond, suggested that a commission be set up to explore the feasibility of building a railway in Ireland centered on Dublin. In their second report, published in 1837, a series of maps were included, which were produced by Henry Harness, a lieutenant in the Royal Corps of Engineers, whose expertise in surveying was used to take data about the movement of people and goods in Ireland and plot these to show the flows of activity from the most significant places to every other. In short, Harness produced the first flow map,[13] which is not so dissimilar from the many flow maps that are routinely produced by traffic engineers and cartographers of social media today and which have also found favor in showing the flow of information on many different networks—email traffic, Facebook links, web page logins, and so on, which we will describe in more detail in later chapters. Harness's maps showed exactly what one might expect: increasing flows of traffic as one got nearer the biggest places or, to put it another way, declining amounts of traffic as one moved further and further away from a given place to any other.

This observation that the amount of traffic declines consistently as one moves away from a place to more distant places is implicit in the standard model we introduced in chapter 6 based on von Thünen's idea that rents and yields associated with different agricultural products decline as one moves further way from the center of the market, while transport costs increase.[14] This, in fact, is Tobler's Law: "everything is related to everything else, but near things are more related than distant things."[15] It is a basic principle underlying flows in cities and regions, and it has even been incorporated into Leontief's model, where it is used to examine international trade at the country level. Essentially, it is an analog of Newton's second law of motion, which suggests that the force (or flow) between two objects (e.g., planets, cities, etc.) increases in proportion to the mass of the two objects and decreases in proportion to the distance (or cost) between them. This is the so-called gravity law we noted in our first chapter, and there is some speculation that only a matter of years after Newton introduced it in his *Principia* in 1687, it was used to describe social movements. However, once the construction of the railways began, there was a vibrant discussion about the configuration of the best network in terms of direct connections between places and the density of sets of branch lines, which was dominated not just by investment opportunities but also by questions of efficiency and time traveled.[16]

About the same time that Harness was preparing his maps of actual flows of people and goods in Ireland, John Herapath,[17] a mathematical physicist, was beginning to develop a strong interest in railways. In examining traffic statistics for the Grand Junction Railway in the 1830s, he figured out that if you looked at the actual flows of rail traffic from the data he collected, the flow of traffic did indeed fall off regularly with distance from the larger stations. In fact, his model was analogous to Newton's, except the decay of traffic followed a more intuitive law in which the falloff was proportional to the traffic at the point where it was observed. Herapath did not take his ideas any further, for he became interested in the question of direct and indirect flows and spent the rest of his life observing the way railways developed with respect to their timetables and such like organization.

However, in 1846, Henri-Guillaume Desart, an engineer, also examined railway traffic data in the form of passenger travel in his native Belgium. He came to the conclusion that the falloff in traffic on a much more comprehensive network followed a curve that was much more akin to Newton's second law, where the effect of distance was essentially a power law.[18] In fact, there is really very little difference between what Herapath was theorizing and what Desart was testing, but these early contributions demonstrated fairly unequivocally that there were important regularities to be observed in city and transport systems, the implication being that such regularities could become the basis for theoretical relationships used to construct a variety of simulation models that might be used to understand the system better and to make conditional predictions. This of course would be a hundred years in the future, but the fact that this kind of thinking began in the first industrial revolution represents a straw in the wind: ideas about how one can observe regularities in social phenomena would eventually become the basis on which cities and economies and societies would become computable in ways that would continually condition our behavior and inform our predictions. Indeed, the new urban science that is currently the fashion embraces this kind of social physics, generalizing this to city size and scale. What has been true since von Thünen is that this kind of science did not merely speculate on these types of relation, but rather it gathered data and, using the rudiments of scientific method, focused on testing hypotheses about the form of cities reflecting these ideas of gravitation and scaling. When the digital computer eventually did arrive, the scene was set for rapid empirical applications.

THE FIRST COMPUTER MODELS: CHICAGO AGAIN

There were several applications of gravitational models prior to 1937, but all of them were based on manual calculations. In 1858, Henry Carey had speculated in his three-volume work *Principles of Social Science*: "Gravitation is here, as everywhere else in the material world, in the direct ratio of mass, and in the inverse one of the distance. The greater the number collected in a given space, the greater is the attractive force there exerted, as is seen to have been the case with the great cities of the ancient world, Nineveh and Babylon, Athens and Rome, and is now seen in regard to Paris and London, Vienna and Naples, Philadelphia, New York, and Boston."[19] Thirty years later, Ravenstein, in his analysis of the patterns of migration between towns and counties from the 1881 British population census, implied that the flows he observed between any two places followed universal laws of gravitation, where the flow of population varied directly with the push-and-pull effects of the places in question and the distance between them.[20] Like Carey before, Ravenstein never actually stated the formulas directly, and it was left to his successors to make the formal connections. In fact, Ravenstein had been trained in Germany under the well-respected cartographer August Petermann, and like his mentor, he had many other interests, particularly in cartography, and much of his work on forecasting population as well as migration was dominated by visual representation.

It was not until the 1920s that the formula was directly stated, and it was a business economist, William Reilly, who first articulated the model for problems of determining retail activity at different shopping centers.[21] In 1929, he took Newton's equation and suggested that you could compute the point between any two centers—in his case, shopping centers—where the flow of traffic to either center was the same. This breakpoint could be used to compute the boundaries of the hinterlands between shopping centers, and in this way, one could sum up the activity associated with any shopping center within its hinterland. This was all done manually. It was assumed that the gravity model was applicable, but there was no attempt to ensure that it fitted the data. It was applied in its raw form as an inverse square law. Data on flows was absent, but Reilly built his model so that he could predict, albeit manually, the future patronage of shopping centers if a developer were to propose a new center or an addition to an existing center before it was built. Up until this point, Newton's model had not been applied to predict flows but simply to argue by analogy and by observation that flows appeared to follow his law. In fact, computer modeling of traffic and land use began without any formal theories being made explicit, and much of it was developed on the fly as the applications necessitated. Other developments propelled these applications as well, and to set these in context, we need to return to what was happening with respect to the structure of cities in terms of the ongoing industrial revolution.

Toward the end of the nineteenth century, another major transformation of distance began. Mass production emerged, and one of the most iconic manufactures was the automobile, which was invented simultaneously in several different places. Its clear utility over rail and bus transport quickly made a mark, and by the time of World War I, it was being widely produced for sale to the aspiring middle classes in North America. In particular, Henry Ford's factories producing the Model-T in Detroit led the way, and by the mid-1920s, about one in seven people in the US owned an automobile— about one in four households. The demand for road space grew dramatically, and cities began to be reconfigured so that work and shopping trips associated with individual drivers could be accommodated. At the same time, the city center began to lose its appeal as a location, and out-of-town factories and shopping malls began to appear. World War II lessened these forces a little, but particularly in North America, these trends continued to reinforce themselves to drive activities out of town. By the late 1950s, one in three owned a car, which meant that most households had access to individual transport. Now, about 90 percent of US adults own one or more cars, almost total saturation, but recently this has dropped back a little, and in many Western countries, a peak of ownership appears to have been reached.[22] More of this later, for it is in the vanguard of a sea change in the way we interact, communicate, and travel in the Digital Age.

In the 1920s, the development of the automobile industry was paralleled by new forms of roadway, particularly boulevards called "parkways," which were largely designed around car transport. The first expressways, which were grade separated, were built in the early war years in California, but it was not until the 1950s that serious highway building began to cater for the automobile. In fact, the US Interstate Highway System was not planned and put in place until 1956, partly by Senator Al Gore Sr. who cosponsored the Federal-Aid Highway Act of that year.[23] Nearly forty years later, his son, the former vice president, a strong advocate for the information superhighway, which brought together the ARPANET, NSFnet, and the World Wide Web, as we outlined in a previous chapter, was popularly accredited with the establishment of the internet. Transportation studies largely based on finding out where people traveled to (called destinations) and from (called origins) were begun in the 1940s using questionnaires from sample surveys of households. Rudimentary transport planning based on making trend projections of future travel were developed from these data, supported by various Federal agencies such as the Bureau of Public Roads. The first transportation study, the Detroit Metropolitan Area Traffic Study, began in 1953, putting in place the basic structure that came to be widely adopted by most subsequent studies.[24] Douglas Carroll, who had already developed ideas about how trips varied with distance and travel time,[25] was appointed director, and when the study ended in 1955, he was appointed to the much bigger effort, the Chicago Area Transportation Study (CATS). It was here, over a seven-year period, that transportation modeling and forecasting cut its teeth, leading to a flurry of efforts that continued throughout the 1960s.

CATS was significant for several reasons.[26] First, it pioneered a massive household survey that required state of the art computing to store and analyze the data. Second, it introduced a staged sequence of moving from data to analysis to forecasting and then to the evaluation of alternative plans for the city, all tied together within the land-use transportation planning process. Third, it introduced new forms of traffic distribution involving shortest routes computation and assignment as well as various forms of distance decay model, which Carroll pioneered both himself and with his colleague, Morton Schneider. And fourth, it succeeded in establishing the planning process as part of the organizational structure of the city of Chicago and its surrounding counties. In fact, at its height, more than 350 people were employed at CATS, and it used several different types of computer and computation. Irving Hoch developed a fifty-sector input–output table to support the land-use activity forecasting, while Schneider ran his traffic flow and assignment model on an IBM 704, which he had to gain access to in Cincinnati where he journeyed for many a weekend to avail himself of powerful enough computing. In fact, for many of these developments, Carroll had to improvise to get the most appropriate facilities so that his team could access the Datatron computer used for large-scale analysis of the survey data. He had "to get the Policy Committee to resolve the problem of obtaining the upgraded electrical service for the computer, which was initially prohibited by the state because the building (in which CATS was located) was leased. This Datatron electronic digital computer would eventually occupy a bank vault in the basement of the building. It was installed only after dismantling the front stairway and would have its own and the only air-conditioning in the building."[27] This was not so unusual a circumstance in the early development of computer models. Fifteen years later, in developing similar models in the UK, my colleagues working at Bedfordshire County Council could only gain access to the computer—an IBM 7000 series machine—at the end of each working day, once those involved in using it for payroll and other administrative functions left the office. Just as Leontief observed when he solved his basic ten-sector input–output model on John Wilbur's simultaneous equation solver, you could figure out what stage the model had reached by watching the way the lighting as well as input and output operations on the computer would oscillate through the course of computation,[28] matching, of course, the sequence of operations specified in the program.

CATS was quickly followed by transportation studies for Philadelphia (Penn Jersey), Pittsburgh, Seattle, Upstate New York, the Tri-state, and several others, all of which pioneered variants on the standard transportation model. By 1960, this had evolved into four stages in which trips were first generated, distributed to a network using gravity models, broken down by transport mode, and then finally assigned to the network or networks so that capacity and congestion could be determined and resolved. The earliest statement of the process was presented by Alan Voorhees[29] in 1955 in his paper "A General Theory of Traffic Movement," but almost immediately, the notion that land use and transport could not be forecast separately from one another emerged. This came to be enshrined in a new logic, with the first land-use transportation models such as Walter Hansen's accessibility model for Washington, DC,[30] constructed as early as 1958. This established the momentum for a new wave of land-use transport models that had begun by the mid-1960s, stitching these predictions more firmly to the land-use planning process, redevelopment in cities, and policies for dealing with urban renewal and urban sprawl.[31]

Many features that we now take for granted in building computer models of cities and related social systems were developed and adopted during these years. If you work with cities and if you are at the forefront of developing plans for future cities, the spatial context is axiomatic. Where land use, transport routes, and economic and demographic activities are located, how much activities in different parts of the city cost with respect to land, housing, and commercial use, how social inequalities manifest themselves differentially across the city—all these are elements that need to be seen in context; they need to be visualized, and the basic medium, as we have already implied many times in this book, is the map. Visualization is thus central and essential to understanding, simulating, and predicting spatial outcomes. As we recounted in chapters 3 and 4, computer graphics, in terms of 2D charts and pictures and rudimentary "maps," first appeared along with the birth of computers themselves. We noted Jay Forrester's demonstration of a bouncing ball on an oscilloscope on Ed Morrow's show

See It Now in 1951, and this heralded a long line of such demos that eventually became serious scientific extensions to many kinds of computation.[32]

CATS was no exception, and there was a strong momentum to visualize the kind of complexity that was contained in the detailed household surveys that generated data about trip making. The most significant features of the city that CATS could demonstrate was the flow of trips from suburbs to central city, and to this end, they developed what was called the "Cartographatron" at the Illinois Institute for Technology.[33] This was a cathode ray tube that could display trips between every origin and destination but over real time because the computers driving the visualization needed to process the trip data sequentially. It could display about three thousand trips per minute, covering the whole region's ten million daily trips in about four hours. We will return to these kinds of visualization later when we come to chart the emergence of GIS, which provided and continues to generate a strong momentum for computation in the city, but it is worth noting that the idea of plotting maps using display tubes became a key goal of visualizing urban data inspired by CATS. Ten years later, some 150 miles south of Chicago at Urbana in the University of Illinois, Leo Kadanoff and his colleagues developed similar technology to display a very fine-scale database of urban land parcels and streets for the town of Kankakee located on the edge of Chicago.[34] Kadanoff was a well-known statistical physicist, whose interests in graphics and cities led him to explore Jay Forrester's systems dynamics urban model,[35] which we will introduce below when we examine some of the detail of the first great wave of land-use transportation models that began in the 1960s. But for now, we need to note that computers for storing and manipulating urban data as well as enabling simple models of trips and economic activities to be simulated came onto the agenda in the 1950s, but there was a real push to visualize the results of numerical computation that finally came to fruition almost as soon as the PC was developed.

THE OPTIMISTIC ERA: LAND-USE TRANSPORTATION MODELING

The start of the 1960s saw the momentum that had been established in the development of transportation modeling ratchet up another notch as it embraced more comprehensive concerns involving land-use planning. The provision of highways was one matter, but rapid decentralization from the core of US cities, the growth of out-of-town retail centers, and the early establishment of large service centers and factories on the edge of the city, all served by freeways that enabled the population to travel more or less anywhere by car, became the dominant urban policy response. It was widely regarded that this kind of complexity required powerful computers to generate predictions for both the short- and long-term future, and out of the early transportation studies came a flurry of proposals for different types of land-use transport model. Nobody suggested that computers could not or even should not be used for such purposes or that their predictions might be flawed. After all, this was the decade when the US first landed a man on the Moon, and as President Kennedy mandated in 1962, "We choose to go to the Moon in this decade and do the other things, not because they are easy, but because they are hard."[36]

The model that took pride of place and became the dominant approach through the 1960s and into the 1970s was built by Ira Lowry for Pittsburgh.[37] His model was extremely successful because it merged two important traditions. First, it was based on assuming that cities and urban growth evolved from basic industries that drove prosperity, establishing an exporting capability that determined the ultimate wealth of the city and the locations and amounts of other services that were dependent on basic activities. Second, it enabled simulations of employment and population that were tied together by this economic base mechanism to locate these activities in small zones of the system using gravitational analogs that lay at the core of transportation models. Thus, the structure of the model was based on an input–output (economic base) function within which gravitational functions that enabled movement and flow patterns relating employment to population and vice versa were tightly stitched. Lowry's model set the pace, pulling together the long standing notion that gravitation was one of the key forces that determined which activities were located where. In fact, a somewhat simpler form of gravitation in which flows were implicit was quite widely adopted based on population potential, akin to Walter Hansen's accessibility model for residential development in Washington, DC, and this represented the basis for spatial allocation in Lowry's initial model. George Lathrop with John Hamburg, who had worked on CATS, developed variants of the gravity model for western New York State in Buffalo and the Niagara region, while "TR" Lakshmanan ("Laksh") with Walter Hansen, working for the consulting group that Alan Voorhees had set up, developed what became the most widely used gravitational model for retail planning, originally for Baltimore, building to an extent on Reilly's breakpoint analysis developed in the 1920s.[38]

In parallel, different styles of model other than the dominant transportation focus were pioneered. Econometric types of model based on systems of linear equations provided a popular approach, as developed in the EMPIRIC models, which again focused on allocating different types of population, employment, and housing to small zones of the city. John Kain, who would go on to develop a series of housing market models with a strong focus on market clearing based on explicitly microeconomic ideas in utility theory, began with similar econometric models. He swiftly moved, however, to rule-based economic optimization as embodied in his NBER model, initially built with data for Pittsburgh. Other housing market models such as that developed by Arthur D. Little for San Francisco provided further support for the microeconomic approach, and by the end of the decade, it was clear that both macro or aggregate and micro or disaggregate approaches defined the range of model types.

Two other types of simulation model emerged during these years. By then, it was clear that most attempts to simulate cities during these early years articulated the system to be modeled as being in some kind of equilibrium. As we have implied in previous chapters, most of our thinking about cities prior to the middle of the last century tended to see the city as a well-defined rather static affair with a strong polar structure, which has been reflected in our descriptions of the cities from the classical era. By the time economists such as von Thünen began to explain this structure, it was widely regarded that cities grew slowly in their overall structure and could be considered, for all intents and purposes, to be in equilibrium. The notion that urban change was chaotic, always out of equilibrium, or far from equilibrium, as the complexity theorists of the late twentieth century would say, was way beyond the horizon that dominated our thinking about cities then. Yet, despite this notion, it was widely agreed that building models that could enable the dynamics of urban change to be simulated was a worthy quest. After all, planning was about prediction, and the future and its mandate was to produce a better future. Thus, temporal change was regarded as important, despite the inability of the field at this point to do much about it with respect to the computer simulation models that were fast emerging. Data and theory were key limitations, despite the widespread sentiment that dynamics was important.

In fact, Lowry himself noted in somewhat detached fashion that his model created an instant metropolis, implying that what was being created was a somewhat idealistic version of the real thing—a city that could adjust immediately to changes in the forces that determined its structure. It was suggested that his model could be extended to simulate the changes in the form and function of the city, simulating the increments and decrements that gave rise to a changing urban structure over short periods of time such as five or at most ten years. The CONSAD Research Corporation picked up Lowry's model, with John Crecine extending it in this way. It led to other models being developed in the same spirit, such as Stephen Putnam's model for the much wider region of the Northeast Corridor from Washington to Boston. These quasi-dynamic models represented a quick fix to the problem of urban dynamics, but elsewhere, more realistic approaches were being proposed.[39] In the University of North Carolina at Chapel Hill, Stuart Chapin and his colleagues proposed that the actual land development process could be simulated. In a rather far-seeing project, his team built a model of the process of how land was developed for residential use in typical small-town America—the town of Greensboro, NC—where the growth of car ownership and the rising prosperity of the middle classes was characteristic of the urban sprawl that was driving development everywhere. Chapin's model was essentially one that explained urban development in terms of typical indicators such as accessibility, land availability, terrain, utilities provision, and so on. Tacked on to the end of these processes of how development took place, they hooked up a probabilistic model using Monte Carlo simulation, which enabled them to predict different scenarios for future long-term urban growth.[40]

The other distinction in these early models was, in fact, more fundamental. Some argued that the true role of such computer models was not to predict the future but rather to invent it—to design it. In fact, this idea is gaining ground once again in present times. The idea that optimization based on selecting the best location of land use and transportation routes should be the focus rather than their prediction (usually on the basis of past trends) defined this distinction. Optimization theory in the form of linear and nonlinear programming had developed very rapidly since the end of World War II in the emergent field of operations research, and there were thus ready-made sets of computational algorithms that could be adapted to enable land-use location to meet some predefined objectives such as those based on siting activities in places with the greatest accessibility, the lowest cost for terrain development, and so on. In fact, although various optimization models were proposed, such as Kenneth Schlager's land-use plan design model for southeast Wisconsin,[41] and despite the obvious excitement for the idea that one could define models that could generate plans, even in those early years, there was the feeling that the most effective planning needed to be somewhat less formalized. The prevailing view was that the process of making plans should be based on using predictive models for second guessing and exploring the future from past trends but set in the context of the overall policy process being the optimizer, rather than the model being used to generate the plan. In short, the policy process was dominated by planners and engineers, and thus optimization was their prerogative, not that of the models. The other way, based entirely on formal optimization, even then was too automated for it to be easily adapted to city planning, which was and will always remain a manifestly political process. Only now has this perspective been resurrected in our newfound concern for weak AI, although it is clear that some of the original objections to optimization in planning remain.

There is, in fact, a further blurring of these viewpoints that complicates the use of models in any human context, and this is reflected in the extent to which we can consider real-world human problem solving and human behavior to contain elements of optimization. Economic theory still persists in developing an ideal type of rational behavior in which individuals optimize some criteria subject to known limits, often portrayed as optimizing their utility, which embodies their preferences subject to constraints such as the income they have. The more ambitious of these models during this era sought to take an explicitly economic viewpoint based on the location of land use, which optimized various utility functions on the premise that this was the way, in fact, that locational decisions were actually made. This rational decision model implied processes of optimization that were embodied in John Kain's NBER model[42] that we noted above, but they emerged in the late 1950s in the Penn Jersey transportation study, whose research director, Britton Harris, developed a variety of models based on linear programming that embodied many of the elements of residential choice consistent with the microeconomics of the land market.[43] Thus, by the end of the 1960s, there were predictive models that mirrored how the city had actually developed—models that enabled planners to determine how the city could best be developed using optimization—and a third set that used optimization as the basis for how actual human decision making took place in reality. Needless to say, these models were not necessarily inconsistent with one another but simply embodied different perspectives on how the form and functioning of the city could be articulated.[44]

ENTER JAY FORRESTER: DYNAMICS, VISUALIZATION, AND DATA

We have already come across Jay Forrester as one of the key developers of the Whirlwind computer system, which began toward the end of World War II at the MIT's Servomechanism Laboratory using funding from the US Navy for tracking missiles. The project was based originally on analog machines, but once the ENIAC came onstream at Penn, Forrester's team began the construction of a digital computer with random access storage, along the way inventing the magnetic core memory for which he held one of the original patents. Forrester's project had fizzled out by the mid-1950s, but not before it morphed into the biggest missile defense project of the early Cold War: the SAGE system, which was still in operation in the late 1960s. SAGE and Whirlwind before it were the first projects at the MIT's Lincoln Laboratory set up in 1951, which spawned a variety of technologies, among which was the PDP series of computers that eventually became the basis of the Digital Equipment Corporation. In terms of early digital computation, the environment in which Forrester worked was one of the most influential ever, with a variety of famous names emanating from the laboratory, associated with everything from radar to game theory.[45]

Forrester himself returned or rather moved on to the study of servomechanisms in the wider field of systems dynamics that he pioneered. In 1956, as Whirlwind was finally wound up, he moved to the Sloan School of Management (at the MIT), where his work on supply chains in industrial applications led quite quickly to a formulation of dynamic systems— he actually called his approach "systems dynamics"—that were applicable to any system, where feedback, both positive and negative, was instrumental in governing trajectories of change.[46] His focus was mainly on business applications, but he developed a particular language called "DYNAMO" that enabled applications to be built almost but not quite interactively and whose trajectories were displayed graphically using line-printer technology. In some respects, his graphs paralleled the maps called "SYMAP" developed using line printers. These primarily came out of Harvard, and we will elaborate on these in a little more detail in the next chapter once we broach questions of how we represent the city digitally using geographic information.[47] We should not doubt, however, the innovations of that time, for Forrester is credited too with one of the first computer graphics—the demonstration of a bouncing ball on an oscilloscope that he produced for CBS TV in 1951, which we noted earlier at the end of chapter 3.

Although this is perhaps only incidental to the focus of this book, industrial society from the early nineteenth century onward came to consider cities as being undesirable places—polluted, poor-quality environments, cramped and manifesting all the sins of those contemporary times. The predominant response from the late nineteenth century for almost the next one hundred years was to steer development away from the central core cities to the suburbs and back to the countryside. There was a brief interlude in this portrayal in the 1950s and 1960s, when suburban growth, which was built on the back of increasing living standards, seemed to stamp cities in a more favorable light. At least, our interest in cities as engines of growth and vibrant places took a turn for the better during these years. Since then, the old idea based on the city being an evil sort of place reasserted itself until the late twentieth century when things began to change. Since the millennium, pollution is being reduced quickly, and cities are becoming more livable, with their quality of life improving quite rapidly. Today, cities are very much back in fashion in terms of the vibrant environments they can offer, creative growth, innovation, and places that act as crucibles for upward mobility of all kinds.

Cities are thus once again regarded as the motor of growth and of increasing prosperity, but although the interlude in the 1960s raised the profile of urban problems and suggested that solutions could be found using new technologies, there was still a dark side that formed an ever-present undercurrent. It was into this context that Jay Forrester proposed a model of urban dynamics that focused not on the relatively prosperous decentralization of activity to the suburbs and the consequent new transportation systems that were being built but rather on the problems of the inner city. His urban dynamics model was built on the back of his ideas about systems dynamics,[48] particularly the notion that urban change could be simulated as a series of feedback loops between key variables such as employment and housing, which could lead to both positive and negative effects on income and employment. The context that Forrester used for his model was a stylized version of inner-city Boston in the late 1960s, where problems of race, poverty, and segregation were as significant as those of sprawl and transportation within the city's wider region. In fact, he flew in the face of establishment thinking about cities in that his avowed knowledge of the inner city encoded in his model was culled from conversations with the mayor of Boston rather than any painstaking analysis of Boston's urban problems. This was hardly regarded as consistent with the state of the art and was certainly somewhat tangential to the problems of modeling complex systems.

As if this were not enough, four other things made Forrester's model significantly different from the wave of simulations that had emerged during this decade. First, his model did not have any spatial context: it was focused on a typical inner city with no transportation or movement to the suburbs and, in this sense, ignored the considerable body of urban theory that had been assembled so far. Second, the model, like most of his systems dynamics models, generated exponential growth, which ultimately was capacity constrained—in this context, population growth was limited by the land available in the inner city—and he so tuned his model that population would overshoot this limit, oscillating around a steady state but then suffering a catastrophic collapse as certain variables pertaining to employment began to decline. Third, his model was visually very evocative. His line-printer graphics were extremely convincing, and the fact that he could generate them interactively as the model ran was especially enticing to decision makers, who wanted to explore the future of their city in ways that could be communicated easily to their peers and publics. Last but not least, the model was not based on a proven database from any particular inner city. In short, the data was made up, plausible to a degree with some stylized facts but very different from the other simulation models that were being built at the time, where there was a strong emphasis on calibrating and running these models with real data. In this sense, the model stretched the conventional scientific traditions of testing any theory or its model against data, and thus it could not be evaluated prior to its use in prediction.

In fact, the model received quite wide acclaim, despite its rather suspect fundamentals. It was more proof of concept, and it led directly to this style of modeling being used to examine world dynamics[49] as embodied in the Club of Rome's "Limits to growth" report. This report gained widespread attention pertaining to the future of the planet, where explosive population growth was forecast to dominate.[50] The model also led to a number of attempts at making it spatial by adding zones such as suburbs, but it proved difficult to extend in this way and, in the end, largely remained a demonstration of how one might approach dynamics. In fact, earlier in this chapter, we noted the physicist Leo Kandanoff's work in urban modeling at much the same time, and he, among a number of other commentators, made extensions to the model, helping it to become part of a wider tradition.[51] One other feature of Forrester's work must not go unmentioned. In chapter 4, in our history of how computing became personal, we described how Forrester had installed a terminal in his home, which was online to his computer at the MIT. The fact that he was a computer engineer, had worked on servomechanisms, had invented his own language (DYNAMO), and was a widely respected and influential scientist, one of the elite of the MIT, made all this possible. It is worth quoting from the preface to his book *Urban Dynamics*, where he says. "Without the immediate accessibility provided by a personal computer terminal in my study at home, with day, evening, and weekend service, this exploration of urban dynamics would probably not have been undertaken." He wrote this in November 1968, and I wonder if this is one of the first references to the term "personal computer," albeit that he was talking about a terminal rather than a stand-alone machine.[52] In fact, John Mauchly of ENIAC fame used the term some six years earlier in a conference address, where he said, "There is no reason to suppose the average boy or girl cannot be master of a personal computer."[53] Of course, Vannevar Bush of the MIT also implied such a possibility in his 1945 *Atlantic* paper "As We May Think."

To an extent, Forrester being at the forefront of computing was almost bound to be influenced by computer graphics, and his focus on business in the Sloan School reinforced the notion that his simulations should be visually and immediately accessible for businesspeople. It is therefore somewhat ironic, given his focus on interactive computing and his installation of a computer terminal online to the MIT in his home, that a colleague and other member of the Lincoln Laboratory, Ken Olsen, the founder of DEC, who we quoted in chapter 4, went on record in 1977 saying, "There is no reason for any individual to have a computer in his home," but like many such speculations, it was probably taken out of context. However, Olsen was almost made to eat his words when his own company, soon after he expressed this sentiment, began to develop personal computers.[54] By the end of the

1960s, immediate accessibility and interactivity with respect to computation was very much in the air. My first foray into this world was on a visit to
the Systems Development Corporation, a spin-off from the RAND Corporation (Research And National Development), in Santa Monica in May 1970,
where a team was developing an interactive model for school districting.
Using various optimization models operated by a user whose learning
experience of the nature of where best to place schools in their catchment
could be improved through trial and error, the algorithm made suggestions
as to the best locations. This captured all the notions about how human and
machine might interact.[55] These interactions were not represented on visual
display units, for this was too soon for such technologies to have been developed, but users would scan a list of data, be primed to reflect on the problem,
and then decide where various locations could best be identified from their
intuition about the problem. The machine would then whir away and eventually spit out the best solution, with the user invited to modify this solution
once again in iterative fashion until some convergence took place. I seem
to remember that the solutions were positioned on the line-printer page
according to their locations, as if the page were a map, the widely used con-
vention for plotting such maps at that time, as we noted earlier in chapter 4.

THE REQUIEM

The long boom that began when World War II ended was drawing to a
close in the 1960s. The central city was increasingly regarded as an undesirable place in which to live and do business, and all the action appeared
to be in the suburbs. Problems of race and poverty plagued the inner cities,
and as the 1960s wore on, these became critical. It was hard to add such
issues and features to these models, with many beginning to consider such
models as being peripheral to the main concerns of urban planning. The
love affair with computer models and our belief in predicting a more desirable urban future using systems and computers thus began to turn sour.
In fact, the US experience was rapidly diffused to the rest of the Western
world—improvements to these simulation models, in particular in Britain,
Australia, and, to an extent, Latin America, appeared, but there too the
mood turned ugly. Many policy makers and urbanists argued that the systems approach in general and computer models in particular were part of
the problem rather than the solution. This view is perhaps a little strong,

but there is no doubt that in building the first wave of computer models, the model builders and those who were tasked to use them in public policy found it extremely difficult to grasp all the richness of the city that was a clear influence on location and, of course, on plans for the future city.

There were various voices that sought to critique the movement, but none were more vocal than that of Douglas B. Lee, who sought to dissect the emergent field and the short history in terms of its relevance and the quality of the modeling experience to date. His paper, entitled "Requiem for Large-Scale Models," published in 1973, argued that the experience to date had been horrific.[56] The model builders invariably failed to appreciate the size of the quest they had taken on, and what is more, they did not really learn from their experiences in that they genuinely believed that bigger was better rather than less is more. If some of the early problems of data could be solved—that is, acquiring the relevant data—and if the speed of machine processing improved, then the promise of these models might be recognized. Lee argued otherwise. He believed that although much had been learnt about computer models during the period with respect to their construction and their use and abuse in policy making, the experience suggested that the models that had been built were largely inappropriate.

Lee's argument was not that mathematics, computation, and science were irrelevant to understanding and thinking about future cities, but rather that the way the movement had emerged and the highly applied policy context were such that almost any computer model would fail at this particular point in time. Although he also implied that such modeling was always likely to be flawed, he did at least leave the door open to future work, and although his statement that these models "begun in the early 1960s and largely abandoned by the end of the 1960s" was not, in fact, correct, for such models continued to be built, his analysis at the time was at the low point when the community itself was fast becoming disillusioned with systems analysis in public policy.[57] This was even more significant in applied areas involving the military, corporate government, and management science. The peculiar nature of urban problems and our inability to make inroads into their solution is best summed up by Horst Rittel's articulation of them as "wicked." Instead of yielding to a solution, wicked problems actually become worse as one begins to address them, with the most obvious solutions becoming part of the problem.[58] This was captured in Melvin Webber's immortal phrase "Why is that we can get to the moon when we can't get to the airport!" This was a mere ten years after President Kennedy spoke his immortal words.

Specifically, Lee argued that the urban computer models of this era manifested seven deadly sins. First, such models exhibited "hypercomprehensiveness" in that they tended to try to explain too much of the urban system, embracing issues and problems well beyond what our theory and intuition was capable of handling, despite the fact that what was included appeared necessary as well as sufficient. Second, he accused such models of "grossness." In fact, although comprehensive, they were often too coarse to pick up the detail that was needed spatially, topically, and temporally. Third, such models manifested "hungriness," for data primarily, when the data required was frequently not available or only available in a form that did not fit the requirements of the model. His fourth point involved "wrongheadedness," by which he meant that many of the components of such models involved relationships that were hard to identify and ended up generating repercussions in model outcomes that were full of unintended consequences. Fifth, a related feature was their "complicatedness," which meant that as the models tended to embrace more and more components, their potential interactions—and our inability to explain them—veritably exploded in size. Sixth, Lee described them as showing "mechanicalness," meaning that there was an obsessive focus on detailed massaging of the computer inputs and model relationships so that outcomes could be generated at any cost. Last but not least, Lee accused them of "expensiveness," and there is little doubt that their cost was unanticipated and often grew exponentially as the model builders and policy analysts embarked on their construction for which there seemed no turning back. To an extent, reading Lee's polemic fifty years on, there appeared to be no saving graces, with the entire experience being damned.

In fact, Lee was also wrong when he said, "the models, like dinosaurs, collapsed rather than evolved." In 1972, when he first wrote about these experiences, these models were being picked up in other countries, which had somewhat different cultures and goals for cities and urban planning. Certainly, their construction came to an end in the US, but more considered and less ambitious attempts were being made to develop the better features of such models elsewhere. Models were somewhat smaller in the UK for example, while their structures were being refined theoretically. The wide range of modeling types that characterized the US experience was not repeated elsewhere, and the movement did, in fact, consolidate itself. Even in the US, efforts such as those based on Forrester's model, and the development of variants of the Lowry model such as Putman's DRAM-EMPAL, kept the field alive. There was little doubt during the 1970s, however, that the experiences with the first models from the 1960s could not be continued, and it took nearly two decades until the start of the 1990s for the field to begin to recover. In fact, in 1994, Lee's paper was revisited in the *Journal of the American Planning Association*,[59] twenty-one years after it was published, and various contributions therein revealed that the field had, in fact, remained alive. It was beginning to pick up a little, the movement becoming more significant, as data, computing, and the idea of the smart city were all evolving from that date, which marks one of our milestones in computing in general: the invention of the World Wide Web. Lee, however, was unrepentant, saying, "That LSUMs (Large Scale Urban Models) are alive and well may be fine for the modelers, but is it of consequence to anyone else?" A qualified answer must be "yes," but it is complicated, and we will only discover this in the chapters that follow, where we will take our story further. There is much more we need to say about simulation, modeling, and prediction in the 1970s and beyond, but we will weave these developments into other perspectives in the chapters that follow.

So far, our focus on these early models has been on systems that treat the entire city in terms of its spatial extent, and larger cities at that. It might seem that in the age of the mainframe computer, where big strategic problems about transportation in cities dominated, the idea of modeling how the city evolved in real time was simply a nonstarter. This was not, in fact, the case. During World War II, there was a strong emphasis on real-time simulation, beginning with radar, gradually accruing an arsenal of mathematical tools dealing with trajectories and their optimization that found their way into the emergent and vibrant field of operations research. The RAND Corporation, as well as a variety of spin-off companies from the defense industries, began to evolve these ideas, and by the 1960s, there were various tools that could be employed to figure out better ways of scheduling routinised resources for emergency services such as police and fire in the bigger cities. In the late 1960s, the RAND Corporation was tasked by New York City to deploy some of these tools to help the city manage its routine, in particular emergency services, and a significant amount of resource was put into developing software for this kind of urban organization.

Yet, this experience too was marred by problems of implementation.[60] The models themselves were experimental to a degree that made their outputs remote from the detail of how such services were actually managed in the field. In short, although these routine tasks as to how to define the safety and health-care needs of the population were relatively well defined, the complexity of how to actually deliver them was highly opaque. The bureaucracy of delivery and the social conventions and practices dominating such activities obfuscated the processes that needed to be simulated in all their detail to enable good solutions to be generated in real time. The kind of problems the RAND Corporation in New York grappled with were clearly "wicked," in Rittel's terminology, in that very often the simulations actually made the proposed outcomes worse than those that they were designed to improve.

These and similar experiences in the wider corporate world and in military applications have taken longer to be seen in perspective, and thus our collective experience of using this kind of science in public policy (and we believe in many industrial and commercial applications too) is quite clearly as flawed as that involving urban models. We will not detail here these kinds of problem that clearly now characterize the smart-cities movement, but there is a feeling that the same sort of experience is being repeated. Overselling models that are short of the mark in terms of their scientific credibility, and models that blow up in the face of decision makers that embark on them in good faith often find that the problems are much more complex that anyone ever anticipated. The smart-cities movement is, of course, much wider than these experiences of half a century ago, but they are based on our collective inability to organize the deployment of information technologies at scale. Society is now replete with such technologies, but many do not operate very well, and the notion that such automation is actually decreasing rather than increasing our productivity is a collective view that is growing in significance. But before we explore these ideas, we need to take our story about urban computing somewhat further and diverge a little to questions of how the city can be represented in digital terms. These ideas grew dramatically after the first wave of computer models, particularly as computers scaled down to the personal level and as computer graphics really took off. These we will describe in the next three chapters.

11 DRAWING, MAPPING, AND PAINTING THE CITY

Computer science is no more about computers than astronomy is about telescopes.
—Edsger Dijkstra (1999)

ART AND THE COMPUTER

From the very beginning, artists have always been in love with computers. David Hockney, one of Britain's greatest living painters, adopted the iPad almost as soon as it appeared more than a decade ago, having experimented in the early 1980s with paint packages.[1] Andy Warhol used an Amiga 1000, one of the most graphical of the early personal computers from Commodore, to sketch Debbie Harry (Blondie) in his New York studio[2] in 1985. A year before that, Steve Jobs released the most graphic of all machines at the time, perhaps of all time, the Macintosh, displaying all its artistic features, from handwriting to the design of fonts, to impress on all of us that computers went well beyond simple number crunching. His 1984 Orwellian video[3] portrayed the soulless nature of the rest of the PCs as Big Brother, which intentionally bore some resemblance to the company IBM and their competing product, the IBM PC.

We have already noted the earliest computer graphics in the form of Maurice Wilkes's animation of a highland dance routine in the University of Cambridge's Computer Laboratory[4] in 1950 and Jay Forrester's bouncing ball, which he displayed on an oscilloscope at the MIT's Whirlwind laboratory in 1951. But computer art, indeed graphics of any kind, took a long time coming, as its production was a painful process for many years. The problem was not that scientists could not write computer programs to generate artistic designs, but rather that the medium of display and, behind that, of storage and retrieval was simply not well adapted to the way computers were designed. Their architecture was based on tiny memories that could not easily be expanded to embrace the requisite capacity to store a picture and to manipulate it on the fly. Artists always need to experiment with their drawing and painting, and the early computers were simply not accessible enough to enable this kind of interactive iteration between human and machine to take place.

Yet, from the very beginning, art and the computer beckoned. Echoing Edsger Dijkstra's quotation that introduces this chapter,[5] we can say that computer art is no more about programming a machine to draw than it is about representing pictures in a digital medium. As soon as the computer was invented, we began to speculate that it could be used to produce new inventions, of which art would always be the greatest. As in all our technologies, computers open up new vistas and enable us to push the limits of our human capabilities forward. Although the early idea of drawing pictures on an oscilloscope demonstrated that computers could extend our numerical processing into the visual realm, this was hardly art. But right from the beginning, there were those who really did extend their own artistic capabilities in this way. Ben Laposky, who originally was a sign painter from Cherokee, Iowa, is accredited with the first images that represented swirling patterns reflecting, to a very large extent, the underlying dynamic motions on an oscilloscope used to create the swathes.[6] These kinds of image were by far the easiest to produce, and he named them "oscillons," also referring to them as "electronic abstractions." During the 1950s, computer art slowly moved from these primitive displays to line plotters, which were able to generate a much more constructivist-like geometry, where the embryonic movement came to be dominated by the production of much more regular shapes. These early manifestations were often labeled as "graphics." They were almost automated industrial designs, but the fact that the movement produced a new generation of digital artists, very different in their style and contributions from more mainstream non-digital artists, led many to suggest that these developments were not art at all.[7] If, however, you believe, as Marshall McLuhan and Andy Warhol both did, that "art is anything you can get away with," then this really did start a revolution that is only now

coming of age.[8] The revolution, in fact, was also part of the first developments in computer graphics and came to embrace much wider concern for
the way we interact with computers that included sound and text as much
as numbers and pictures.

It is worth going back one step and considering exactly how computers
produce pictures because the history that we outlined in previous chapters
illustrates that the hardware of the computer, as well as the imagination of
those using it, are closely intertwined with the development of art, graphics, and the visual medium. The oscilloscope represented the most obvious
way of representing what an artist produces, which, at the end of the day, is
by the stroke of the pen or brush on paper or canvas. You can see this on an
oscilloscope, and you can also see this even more clearly on a line plotter,
and these media dominated the early years. What was less easy to present
were areas of paint, produced, of course, by a brush but in such a way and
at such a resolution on the canvas that a picture was often composed of
regular blocks of color. These were filled in, so to speak, at a scale that gave
the impression of solids. To get a computer to produce such realism in the
early days was almost impossible, although by following the time-honored
way of filling in the image with strokes of the brush, such imagery could
eventually be constructed.

The easiest way to produce blocks of color was by overprinting the image
using the computer's main output device, the line printer, which dominated computers until Xerox PARC invented the laser printer in the 1970s.
Line printers essentially printed rows of text and/or numbers, which were
the usual outputs from machines, whose central tasks in the early days
were various kinds of numerical and transactions processing. By controlling the print head, one could overprint the image by moving the line
printer back one space and using another font to increase the density of the
image, this character occupying the same space in question on the paper.
In this way, one could build up pictures of solid color, invariably black and
white, which, when overprinted, produced gray tones. Color, as far as we
are aware, seemed to be out of the question using line-printer technologies,
although there were various one-off devices constructed more in the realm
of photography and Xerox copying during the 1950s and 1960s, implying that color was not far away. The key to overprinting was based on the
notion that the picture could be represented on the computer, and this
meant that the computer had to hold the picture as 2D text (usually part

of the program), simply using the line printer to produce the text in literal fashion. Printing images with more than two dimensions on the flat plane was more or less impossible unless the image was sufficiently clear at line-printer resolution, which was rarely the case.

Big images could be produced by partitioning the pictures into segments, printing these separately, and then arranging the pieces to form a large paper display. The archetypal image produced in this way in many computer centers—hardly by artists but more by the nerds and the high priests who ran the mainframe machines—was of Marilyn Monroe, but images ranging from near-professional-based maps to cartoons such as Mickey Mouse were the norm.[9] From the early 1960s, in computer mapping, this technique was quite widely exploited to print images that sometimes were then colored in—augmented from the computer printout—and there were even attempts at making movies from these outputs, where each map associated with a particular sequence of temporal intervals represented a frame in the anima-tion. We referenced some of these in previous chapters, where we noted the mapping style called "SYMAP" (SYnagraphic MAPping).[10] We will say more about this later, but it is important to note that these kinds of mapping led directly to the development of GIS at the Laboratory for Computer Graph-ics and Spatial Analysis, which was established at the Harvard Graduate School of Design in the mid-1960s. It is sobering to think that even by this time, computers were penetrating areas that traditionally had nothing to do with electronic hardware per se but everything to do with how we might use computers to picture the world in which we live—further evidence, as if we needed it, of Turing and von Neumann's original insight of the digital computer being a universal machine.

The key issue in getting a computer to produce anything in the visual realm is to have some means of storing the image that is produced, ideally so that once the image is produced, a user can begin to manipulate it. The image might be constructed alongside its continued manipulation, and this implies that any user could continually work at improving and changing an image through its basic programming. The most basic way is to produce an image that can be in any spatial dimension but then to transform it into 2D form in an area of the computer's memory that is reserved for the image and which can be addressed by the user. The display screen is the most obvious area of the computer that might be associated with the pictures that can be stored, with the user interacting with the screen directly from the keyboard or with a light pen or mouse—indeed, with any device that can enable interaction. In the early days of computing, the memory was usually an add-on to the main machine, often called a "frame buffer," but as soon as the PC was invented, this memory became directly addressable to the display screen, largely, one suspects, because of the strong imperative in personal computing for developing games, which were highly visual as soon as this became possible.

The other advantage of the display screen being directly associated with memory is that the pixels (picture elements) that compose the screen provide a coarse resolution to color polygons that were constructed from pixels, which generate blocks of color. This is raster graphics in contrast to the more geometrically precise line plots based on vector graphics. In essence, both types of graphic were constructed from drawing lines, and it was only the media of display and storage that was different. As the computer revolution progressed, screens became ever finer in resolution, but memories expanded much faster, and really high-resolution images could thence be stored and manipulated. All the tools of graphics could then be deployed to draw and color images, to transform them into different shapes, to render them with shadows to inject a degree of realism, to enable hidden line elimination, and so on. Computer graphics provided not only a new and important use for computers but also the very means for interacting with the computer in that the interface itself became graphic.[11] GUIs are now the norm,[12] and once these developments were in train by the late 1970s, several different varieties of spatial information system were developed for the city. Maps were the first to become digital, closely followed by the development of GIS, which provided the basic medium for spatial analysis, the kinds of simulation we illustrated in the last chapter, and the analytics we will discuss in the next. Three-dimensional representation, VRs, and related visualizations became significant in the 1990s—ideas that we will catalogue and present below when we will deal with the virtual city. Now, however, to set the scene, we must illustrate how maps became computable through their origins in computer cartography.

MAPPING THE CITY: THE ORIGINS OF COMPUTER CARTOGRAPHY

Maps go back to prehistory. Cave paintings that date from the middle Paleolithic, some forty thousand years ago, suggested that astronomical observations were encoded in star maps by our ancestors. Over many millennia through the Stone Age, more local maps slowly emerged, the clearest being the cave paintings at Catalhoyuk in southern Anatolia (modern-day Turkey) dating from about 7000 BCE.[13] Maps on clay tablets were soon transferred to other media. Paper maps, however, did not really emerge until medieval times and were really only produced in accessible format once printing had been invented in the early modern era. In fact, by this time, considerable thought had been given to how maps might be projected, and once they could be printed in any quantity, the science of cartography began in earnest, spurred on as much by colonial conquest consistent with their military origins and the quest for territorial expansion, which dominated the history of the world from the Middle Ages onward.

As we have noted, the first computer maps were generated using line plotters, and during the mainframe era, there were many such maps produced, but most remained behind the closed doors of the military-industrial complex. Jay Forrester's bouncing ball on the oscilloscope in 1951 was part of the precursor to the SAGE missile system, whose operation and control was aided by many rudimentary computer visualizations in the form of missile trajectories at a global scale. These applications used vector geometry in contrast to the other extreme of raster mapping based on line printers that began in 1965 with Howard Fisher's development of SYMAP at Harvard.[14] It is worth noting that an even more rudimentary mapping predated SYMAP, and this was Edgar Horwood's use of the line printer simply to print numbers pertaining to the geography of the map, which was represented by an array of points of longitude and latitude, which were themselves scaled to the coordinates of the line printer. At much the same time in Chicago, Duane Marble was developing a variety of network maps that would dovetail in various ways with the ongoing effort in the CATS that we noted in the last chapter. His focus was very much on maps as networks rather than as geometric projections or on thematic mapping, which was the prerogative of systems such as SYMAP.[15] But although the excitement of attempting to represent the city in these different ways was evident, the field was still massively constrained by technological limits posed by the devices used to store, represent, display, and hence communicate the cartography.

New methods did come out of this mix of media, but it is important to stress that new developments for constructing, manipulating, and displaying maps came from those working very close to both the hardware and the software. In fact, it is not really possible to disentangle all these different strands, for the quest in general was to drive forward computer graphics. One of the milestone developments, as much to do with developing ideas about computer-aided design as it was to do with mapping, was pioneered by Ivan Sutherland, whose work at the MIT in the late 1950s led to the first interactive software for enabling nonexpert computer scientists to create designs for a variety of systems from buildings to aircraft.[16] Along the way, computer cartography was part of Sutherland's arsenal of applications, and once the Harvard Laboratory was up and running, some obvious synergy developed between the two groups. Sutherland's software, Sketchpad, was developed on one of the forerunners to the PDP series of minicomputers that originated from the Lincoln Laboratory at the MIT in the 1950s (spinning off into the Digital Equipment Company), also spawned by the link to Forrester's Whirlwind and the SAGE missile projects. In fact, Sutherland's supervisor was one of the great pioneers of information theory, Claude Shannon, and his second advisor was Marvin Minsky, who founded the AI Laboratory at the MIT.[17] The links across computing and networking, then as now, run very deep at Harvard and at the MIT, of course, and elsewhere, but it is still hard to figure out precisely how all these developments coalesced to generate the new world of GIS. In short, the mix of ideas, personalities, and machines was and continues to be too rich to disassemble.

Sutherland did not simply choose an experimental interactive computer and then proceed to write software that produced the solutions to CAD problems in visual and numerical form. This was simply not possible, for he had to modify his machine quite extensively before he was able to write software commensurate with the tasks he set himself. He had to sort out the display, the way data might be input, and the functions used to interact with any design, and then he had to transform all this into suitable software, most of which was written from scratch by himself. On the way, he developed various geometric tricks to implement 2D and 3D representations, as well as developing an early version of object-oriented programming to handle the geometrical configurations that he intended to manipulate. At the same time, he was involved in other kinds of computation, relating quite strongly to the emerging ARPANET projects, which were partly responsible for his funding. In fact, he was developing software during an era when most people in computing did not speak of software as something separate from the machine on which it ran. His influence from those days was huge. Many of the developments at Xerox PARC in the 1970s, at Silicon Graphics, at Sun Microsystems in the 1980s and 1990s, and more generally throughout computer graphics originated from the group he subsequently assembled at the University of Utah, which subsequently had an enormous influence on the use of computers in the movie industry.[18]

Back at Harvard, the focus on thematic mapping developed using the SYMAP system by Fisher was paralleled by William Warntz's interest in modeling map surfaces using gravitational ideas in which action at a distance was key.[19] In these years, experiments were made in developing contouring through interpolation and in representing subdivisions of the map using vector graphics. Ways of visualizing surfaces as triangular networks adjusted to the physical form of the map, while various functions involving the topology of the map in contrast to its topography were developed. There was also a slow push toward interactive mapping using storage tube technologies such as that used by Sutherland but made increasingly accessible in the form of Tektronix display devices that were becoming ever cheaper, as we noted in an earlier chapter. Seeing the map of course is all important in representing the spatial extent of any geographical object(s), particularly in cities, where we tend to think about them in predominantly visual terms as much as in any other way. The glory years of computer cartography at Harvard lasted about a decade through the 1970s, but during that time, the rudiments of not only computer mapping but also GIS were established. It took, however, another decade until the mid-1980s for the many individual programming functions that are now central to GIS to be established, made operational, and integrated into organized software. What characterized the Harvard experience, however, was that the software that emerged was largely individual computer modules and programs, sometimes coupled together but never really entire GIS systems, with perhaps the exception of the Odyssey software, which was the nearest thing in the US in the late 1970s to a fully-fledged GIS.[20]

So, it was at Harvard during these years that the rudiments of GIS were put together. Such systems are essentially representational ways of describing the geography of any set of spatial units at any scale at any instant of time or over many time intervals. Such systems contain a language for representation—in the case of GIS, generically defined in terms of points, lines, and polygons— and a set of attributes of these elements, which provide the data defining the system. Instruments for operating on this data, which has particular spatial characteristics making it distinct from other database technologies, provide the core functions of GIS. All of these elements were put in place at the Harvard Laboratory over a very short period, and these provided the momentum for the development of many different kinds of GIS that we will describe in the next section. However, even though the Odyssey system was probably the most complete and organized system by the late 1970s, it was still largely nonoperational for routine uses and users. It took another twenty years for routinized GIS to emerge, and it took many different competing efforts to reach the point where these systems could be taken for granted. Of course, without them, it would have been impossible to manipulate the smart city in the way we are now able, and in this sense, these developments were foundational.

Before we launch into a description of the development of these systems and how they moved from mainframe to the desktop and then the web, we must note that the community that developed these ideas was an extremely tight one. Many of the key actors went on to found major software houses, particularly Jack Dangermond, who leads the biggest company worldwide, ESRI. He was a student at the Harvard Laboratory in the early days, and many GIS developments at software houses such as Intergraph, AutoCAD, MapInfo, and so on had strong roots in the Harvard Laboratory.[21] Moreover, the quantitative geographers, who were instrumental in developing various elements of the simulation models that we described in the last chapter, were intimately involved in computer mapping and GIS. The school from the University of Washington, consisting of Garrison, Marble, Berry, Tobler, Morrill, Dacey, Getis and Nysteun, among others, dictated the development of spatial analysis over the last fifty years, and many of their contributions are now deeply ingrained in the software and the theory that we use routinely to think about smart cities.[22] This list of luminaires is too long to report, but like computer graphics itself, the development of the internet, the development of workstation technologies, and the software that emerged from Silicon Valley in the last twenty-five years, there are many links between the personalities involved, which explain the genesis and development of this field.

Our focus, of course, is rather local, confined to North America, and to redress the balance, we need to note that these ideas spread globally with other contributions supporting the effort. In the late 1950s, David Bickmore, who specialized in producing school atlases, set up the Experimental

Cartography Unit (ECU) in Oxford to produce automated maps. The unit moved to the Royal College of Art in 1967, and during the same years when the Harvard Laboratory was developing SYMAP, the ECU developed some very sophisticated vector mapping software, which was quite widely used, even by Britain's national mapping agency,[23] Ordnance Survey, for experimenting with automated map production. At the other end of the country in Edinburgh, the University set up a program library unit, which began to develop similar mapping technologies to those at Harvard. Tom Waugh, the key developer, had been a visiting graduate student at the Harvard Laboratory in 1969, and back in Edinburgh, he developed a state of the art vector mapping package called "GIMMS." This was probably one of the first commercial packages anywhere, for it was sold by the university somewhat earlier than the Odyssey system and many years in advance of ESRI's ArcInfo to which it had similarities. The London–Harvard–Edinburgh link was important to the development of GIS everywhere,[24] and we now need to take up this story as automated cartography really began to fly in many fields.

GEOGRAPHIC INFORMATION SYSTEMS

Today, we associate computer graphics, computer maps, GIS, VR systems, and so on with widely available user-friendly software that you can buy relatively cheaply online. A lot of the more basic systems such as QGIS are now in the public domain, freely downloadable, and some are now packaged with other software such as spreadsheets and presentational systems, but this was not always the case. In fact, when people first began to think about integrated systems to organize problems where mapping was central to the activity of problem solving and design in fields such as urban planning, landscape design, and resource management, software was barely identifiable as a product, for systems were strongly linked but often idiosyncratically to the problem in hand. The notion that these systems might be rolled out routinely might have been a glimmer in the eyes of their developers, but there was no suggestion that such systems had any real commercial value. In fact, even if this had been the case, the systems were simply too unwieldy and not portable enough to be widely applied, and any pricing of them for others to use was purely on the basis of cost recovery.

The first GIS was developed by an Englishman, Roger Tomlinson, who, as a graduate student, had journeyed to Canada, where he signed up for a

master's degree in geography at McGill University in Montreal.[25] He apparently was the first to use the acronym "GIS," although the quantitative geographers Michael Dacey and Duane Marble spoke of geographic information
systems in their writings about spatial analysis in 1965. In fact, in 1962,
Tomlinson created the first identifiable GIS built on the sorts of principles
that we indicated above: maps were defined from points, lines, and polygons but embodying, at the same time, raster graphics, which were subsequently vectorized and arranged in layers. Because the system was designed
to inform natural resources management across Canada (hence the term
"Canadian GIS"—CGIS), it had a strong physical emphasis on map layers
that were rooted to the terrain and its topography, all part of the Canada
Land Inventory, which provided the database for the system.[26]

Tomlinson's work provided more proof of concept than an operational
system for routine analysis, but it did point the way. In particular, it threw
onto the table the fact that there were already different types of base maps
that were relevant for any GIS.[27] Cadastral maps had tended to dominate
our thinking about maps prior to the Computer Age, as these were produced
for military purposes, although throughout the nineteenth century, various
thematic maps were proposed. Network maps, particularly flow maps, were
also suggested by cartographers such as Charles Joseph Minard, whose map
of the decreasing size of Napoleon's army on the 1812–1813 March on (and
retreat from) Moscow has become a classic.[28] All these ideas were incorporated into various GIS, which gave each its specialist character and directed
attention to different applications, and although Tomlinson's work marked
the beginnings of such computerization, the origins of thinking in this way
about cities and their data go back much further to times before the middle
of the last century.

Let us return briefly to London and Tottenham Court Road. Instead of
walking north on one of Karl Marx's infamous pub crawls from the tube
station at the end of Oxford Street, we will walk south to Soho, where Marx
himself lived when he first came to London. A little to the west, but only a
few hundred yards, in 1854, John Snow, a local medical practitioner, made
one of the seminal discoveries of the mid-nineteenth century.[29] By mapping the location of the public pumps in the streets surrounding Broad
Street (now Broadwick Street), he realized that the spread of cholera was
associated with these pumps, and from this, he drew the clear conclusion
that its spread must be waterborne—a simple but clear triumph not only of

observation and theory but also of spatial analysis. Snow is often featured as one of the originators of thinking about cities spatially and statistically in this way, and he was the first in a long line of thinkers to suggest how populations organized themselves in space. This led to parallel developments in thinking about the role of distance in geography—another key theme in the development of the kinds of models we presented in the last chapter and the ideas about urban analytics that we will introduce in the next. One last digression. Marx lived at 28 Dean Street from 1851 to 1856 and would have observed the cholera epidemic that Snow catalogued and mapped so assiduously. It is a sad fact that Dean Street was not good to Marx, whose children suffered immensely from the diseases of the Victorian city.[30] His young son died there in the same year as Snow revealed that cholera was waterborne, no more than two hundred yards away. Whether his child died of cholera is not known, for data about diagnoses and the cause of death were notoriously inaccurate in those days, but it is a sad and poignant story.

Back to GIS. One of its most important features is the notion that the world can be classified and partitioned into distinct subsystems, which in this case are represented as map layers. One can thus build up a picture of the complexity of a city system by thinking in terms of spatial layers, which provide a way of at least comparing if not integrating very different data layers together to provide a picture of how the city, or rather its geography, functions. This is the model that was adopted in physical planning many years ago, going back to the late nineteenth century at least, when planners and landscape architects first began to define the physical constraints as layers that determined where one might best locate different facilities. The idea was adopted de facto in professional practice and finally made its formal appearance in the work of people such as Christopher Alexander and Ian McHarg in the 1960s.[31] Of course, there was little mention of GIS per se in this history, but quite soon, the idea of automating the process of combining map layers became a strategy for taking the data from GIS and focusing it on physical problem solving and design.

It was in Harvard that the layer model was institutionalized and perfected, albeit in the division of landscape architecture not the Harvard Laboratory, although both were part of the Graduate School of Design. There, beginning in 1969, Carl Steinitz pioneered the development of automated overlay analysis using SYMAP in the process, working with a gifted cohort of students, among them Jack Dangermond, who, as we noted above, now leads the world's largest GIS software house, ESRI; David Sinton, who developed GIS at Intergraph; and Lawrie Jordan, who founded ERDAS, a remote-sensing company, before its acquisition by Leica Geosystems. We will say more about Steinitz a little later when we introduce Geodesign, but he pushed the notion that overlay analysis was key to the integration of map layers. The technique, however, was not peculiar to Harvard or to GIS. It was widely used in many practical contexts where there was no automation and where its users were unaware of GIS. Out of Harvard also came ideas about how one might formally integrate such layers.[32] The Map Analysis Package (MAP)—which developed techniques of map algebra, ways to combine points, lines, and polygons using logical operations, and thence weighting combinations—was developed by Dana Tomlin. These developments ported basic GIS functions down to the PC level in the 1980s, while the general layer model was widely exploited in terms of the data structures common to many emergent GIS systems such as ESRI's ArcInfo. Indeed, ESRI developed a modeling routine in the 2000s called "Model Builder," which articulated the process of building a formal model as a set of layers where it was a requirement that the data be represented in layers before such integration took place. This posed limits on what this kind of GIS could do, but in general, the movement accepted that GIS was just one software strand in the wider process of understanding, simulating, and designing future cities.

By the time desktop GIS had been introduced following the release of Windows 95, the entire field was dominated by expanding GIS software, with various plug-in applications ranging from adding basic spatial statistics, enabling rudimentary 3D maps and surfaces to be plotted, incorporating location-allocation models, analyzing shortest paths in networks, and so on. From these basic functions, which incorporate much of the state of the art in spatial statistics, entire applications to transport, water resources analysis, business marketing, and so on have been fashioned, while currently, most GIS systems enable users to customize the product widely to the point where applications no longer have the look or feel of the basic products from which they are built. In the millennium, once the Internet became widely available for interactive work, which meant Web 2.0, GIS moved online, with internet map servers becoming dominant. Open-source GIS software such as QGIS emerged,[33] and now it is possible to develop such systems in the public domain, where many of the functions can be custom built from versioning software available at places such as Github, the current repository for software spanning many fields. GIS is now more defined by its applications, for it has moved into the mainstream and increasingly is part of the background to any problem-solving task. The move from desktop to web where most generic software is heading is best seen in online mapping, map servers, and web maps, and it to these that we now turn.

SEARCHING THE CITY: WEB MAPS AND ONLINE MAPPING

Maps are crucial to our everyday understanding of the world we live in, although as they become increasingly abstracted from the real world of our immediate experiences, they become harder to read. Reportedly, twenty-five million Americans cannot locate their country on a map of the world, and seventy-five million cannot locate the Pacific Ocean. The figures are even more scary when one notes that about two thirds of all those between eighteen and twenty-five years of age cannot read a road map, many because they are now reliant on satnav devices. There are lots of stories—many apocryphal, no doubt—of people ending up in the strangest of places because of their overreliance on GPS and related technologies.[34] Scale is key to mapping, but we avoid using the term "smaller scale" or "larger scale," as this depends on where one is looking from. At finer and finer scales, we pick up more and more detail, while we lose detail as we move to coarser and coarser scales. Clearly, as we get finer and finer, the amount of data increases often exponentially, and when we move the other way, we have to get rid of such detail, often aggregating and simplifying the map, which in itself is quite an art. Map simplification, sometimes called "generalization," depends on the purpose for which the map is being constructed, and there is always the tension of simplifying too much or too little.[35]

Maps are models, simplifications of the real world, and as with models, they are only useful for very specific purposes. Maps, however, are close to the real thing in that the degree of abstraction is largely based on how one visualizes the 2D terrain, but as the Polish-American mathematician Alfred Korzybski wrote in 1933, "the map is not the territory."[36] There are many playful consequences of this phraseology, with Lewis Carroll of *Alice in Wonderland* fame telling the story in one of his last books of the German "Mein Herr" whose exploits in producing a map with as large as scale as possible, one to one, was rejected because "the farmers objected: they said it would cover the whole country, and shut out the sunlight! So, we now use the country itself, as its own map, and I assure you it does nearly as well!"[37] Jorge Luis Borges, in his 1946 essay on "Exactitude in Science," told the story of a map of an empire[38] that was drawn at the same scale as the country but which was deemed useless by those who followed in the footsteps of the cartographers who originally produced this. There are strong analogies here to the idea of digital twins, which has become popular in thinking about the digital city,[39] and we will pick this theme up again in later chapters.

Maps are the simplest and most obvious of models, widely used for many purposes from simple navigation to understanding the location and variation of physical phenomena such as the weather. Almost as soon as the internet became widespread with the development of web browsers, maps were among the first artifacts to go online. The Xerox PARC Map Viewer came on stream in 1993 and provided rudimentary proof of maps being delivered to many users across the net, with the user actually determining what map or portion of the map they wanted to see.[40] Map servers in the mid-1990s were largely demonstration projects to show how data associated with the map could be pulled up, and these were in the vanguard of location-based services, which, as an idea, became hugely popular in the millennium. Maps, of course, are essential to the identification of such location-based services, but right from the word go, the idea of the map being the referent for layers and layers of data and eventually for the users to manipulate this data to some purpose became significant. This, of course, is the essence of GIS, and as might have been expected, commercial online map servers delivering GIS functionality began to appear in the late 1990s such as the Arc Internet Map Server from ESRI.

However, the most obvious of web maps were for navigation—for figuring out where you were and how you might get there. In 1996, three systems that became hugely popular for navigation, routing, and location-based services were introduced: MapGuide, MapQuest, and Multimap. By 2003, the first moves were made into 3D mapping on the web with NASA's WorldWind software.[41] We will say more about the 3D revolution in modeling cities below, but in essence, the third dimension opened the door to much greater realism, although, in fact, these developments have remained much less popular for navigation and location than the simplest 2D maps. This is partly because the third dimension can be quite confusing in terms of its visualization, for we are much less used to navigating above ground level, where most of our experience lies. In 2005, Google Maps was released, reducing the patronage of other map servers quite substantially at first, although with the massive growth of interest in using the simplest maps for navigation, the earlier map servers have held their own in the market-place. More particularly, all these map servers began to introduce additional content, which is now encapsulated in many different layers, ranging from traffic to points of interest to weather predictions and so on.[42] In fact, it is rare to find online maps that do not have the ability to be customized in some way by the user, as well as containing quite detailed archives about what happens where, particularly with respect to their use by tourists and other casual users. More professional GIS-like functionality too has been decentralized to the informed user, but this still requires some specialist expertise, at least in interpretation.

With the advent of Web 2.0, which heralded the interactive web, where users could edit, add, and delete online content, the idea of customizing maps grew in importance. In particular, Web 2.0 introduced the notion of crowdsourcing—the idea that the crowd (we as individuals or consortia) could add content and actually create their own data. In GIS, this was referred to by Michael Goodchild as "volunteered geographic information."[43] The best example of this is in the development of OpenStreetMap, which enabled any user to locate themselves in space and, using GPS and smartphone technologies, to add material, edit the map, delete what they did not agree with, and, in this way, generate ever more accurate mapping. In fact, as I noted in chapter 8, OSM was set up by Steve Coast at UCL in 2004, who acted as an intern in my own group, dropping out of his physics degree, a peculiarly un-British style of behavior, but like many illustrious people in software from Jobs to Gates to Zuckerberg, this seems to be the hallmark of people who pursue great software and killer applications. OSM is a remarkable testament to the power of the web and our collective ingenuity to populate the world with details tagging it to a map.[44] If you log onto the site, then in the densest areas of highly developed world cities, the detail is remarkable, but it is still only a fraction of what there might be. If you log onto the place where we started in chapter 1, the complex of buildings around St. Paul's that constituted the GPO, individual buildings and points of interest are clearly tagged, but there is still no reference to the place where Marconi sent the first wireless signal in 1896 or where Edison launched his first power station. This is not a shortcoming but simply a demonstration that much more can be added and, in time, doubtless will be. This, of course, is the power of a web that is open to all.

In 2005, Google Earth was released, and the 3D world that this revealed illustrated just how far computing had come in the preceding twenty years since GUIs first appeared on PCs. In fact, in 2007, Google Maps introduced a rather different visualization based on street views, which were recorded directly by the company, thus giving a much more realistic but very different view of places than offered by standard manipulation of the environment of the map using 3D functions such as zoom and pan.[45] Many of these systems quite quickly introduced functions that enabled users to modify content, collect and download spatial data, and even import the software itself into other systems. In the last decade, developments in mapping have been somewhat less dramatic, for there is a limited number of additional functions that can be introduced, and the major changes have been to augment the software itself, in 3D and other virtual domains, which we will present in a later chapter. In terms of mapping, Mapbox was founded in 2010 and now produces many open-source online maps for a variety of other services. Many map sites continue to adapt their software for different varieties of device, and some have moved mapping into other VR domains, gaming being a favored focus. Microsoft, Apple, and other large software houses have developed their own mapping domains, and the field is now quite crowded. But a major focus is how maps are being introduced into the public domain as open data and how this open data is being used by a very wide array of users with very different goals. We will explore these ideas, which relate to open data, crowdsourcing, and the organization of related software, in the last part of this book, where we will sketch how the digital city is restructuring many work and professional practices, not least the way we might plan the city and address its future problems.

GEO-POSITIONING, REMOTELY SENSED DATA, AND MAP HACKS

When the number of maps and their users was small, getting data into them was accomplished either by direct measurement or, in the case of cadastral maps, at fine scales, through photogrammetric and related survey devices, all of which required semi-manual, not automated, ways of capturing geometric data. But as the digital world became established, automation in cartography progressed through electromechanical devices that sensed map geometries remotely, capturing and storing such data in the databases that came to be associated with GIS and geomatic engineering. In fact, the kind of progress that we have recounted so far in this chapter with respect to automated cartography and GIS would have been quite impossible without such automation, just as none of this would have been possible without the internet, modern software systems, computer graphics, and a host of new technologies that were coupled together as being part and parcel of the digital revolution.[46] This is another example of the extent to which the software that we now have is composed of elements that originally were invented for different purposes—elements that can be put together in ways that we could never then have anticipated.

A large part of contemporary mapping now consists of data that is captured remotely using various sensors, the first of which were those from satellites launched in the late 1950s. The Soviet Union's Sputnik 1 went into orbit in 1957, taking the world by surprise but also establishing the basic principle that by bouncing energy off an object, radar (radio detection and ranging) could be employed to figure out the position of the object and, in this way, build up pictures of any remote object from the radio source. This was established initially for military use in World War II, but when the satellite era began with Sputnik, the idea of collecting Earth observation data from remotely sensed images of the terrain really took off. The electromagnetic spectrum is composed of a number of wavelengths that are all associated with detecting different features of the Earth's surface, where signals that are reflected, emitted, or backscattered can be interpreted as proxies for different objects that compose the landscape.

There are a number of satellite systems that collect data about the Earth's surface, producing images at ever higher resolution. Landsat was the first system of images collected at a spatial resolution of about 60 m, with the satellite orbiting the earth once a day, and this was launched by USGS and NASA in 1972. Clearly, these levels of spatial and temporal resolution can be traded off against one another, while the resolution of the electromagnetic spectrum determines what the sensor can detect in terms of objects that exist on certain wavelengths. Landsat is now in its eighth generation of satellite, and a very wide range of products is available. The French system, which yields SPOT imagery, was also introduced in the 1970s, and this has been focused on much finer-scale sensing, with resolutions down to 1.5 m in some applications. There are several other systems yielding imagery with different properties, but in general, the resolution on all dimensions is increasing, and in cities, very high-resolution imagery is now available to the point where building forms and footprints with natural vegetation could be quite easily extracted and used to produce 3D renditions of the landscape itself.

The other key system that is central to all we have been saying in this chapter is the GPS system, used to determine actual locations that are not fixed in space but usually associated with mobile devices.[47] The basic GPS system was introduced in 1973 by the US Air Force and consists of a system of thirty-three satellites in all, where four satellites are required to fix position and time. All four must be visible from the object whose location in space and time is to be calculated, three satellites to determine x, y, and z locations, with the time of the sensing computed from the fourth satellite. The GPS system continues to be maintained and improved by the US government and is open in a somewhat constrained way to the world, while there are Russian, Chinese, Indian, and Japanese systems that have greater accuracy in parts. The European Galileo system will be almost fully operational at some point during 2023. One of the key issues, of course, with GPS is that the signals depend on a clear line of sight (several thousands of miles) to get a good signal, and as we all know from our mobile phones, this is by no means assured. One of the key challenges is to improve this accessibility, as the mobile internet is entirely dependent on it. If we move to autonomous vehicles, some sort of connectivity will be needed to enable such cars to operate safely, and the reliability of such GPS systems is thus critical.

Remotely sensed images and geo-positioning are central to most of the map systems that are now used routinely online such as Google Maps, Bing Maps, and so on.[48] Users can switch between layers based on the traditional cadastral type of map and satellite imagery, with other layers such as congestion on traffic networks, points of interest, and so on being basic to the information that users need to identify location or aid navigation. Functions that involve determining shortest routes displayed visually and with respect to times taken on different transport modes are now built into such systems, where the location of the user, the origin, and the destination or many destinations is intrinsic to such calculations. GPS is also a feature of many social media such as Twitter, Facebook, Foursquare, and related networking systems, although in all of these as well as mapping systems, there is always an opportunity to switch such positioning on or off. As we will see in a later chapter, the quest to turn 2D maps into 3D, best exemplified by Google Earth, relies on GPS with respect to both the data itself and the usage of the various 3D systems that are now available in the various map interfaces that let users navigate in this augmented media. And last but not least, most of our contemporary map systems allow the user to customize their own maps, certainly selecting data from the available archive but, in some instances, adding their own data in the manner of OpenStreetMap.

Once mapping systems moved online and Web 2.0 technologies were established, map servers became quite common. Various methods for creating maps from the various atlases and archives centered around quick hacks of the websites themselves, sometimes called "map mashups," which combine two or more pieces of complementary web-based content or functionality to create a new web application.[49] These kinds of application involve a variety of ad hoc procedures such as scraping map data from the web, building new maps from these, and implementing basic GIS functions such as overlay analysis. The site developed by Richard Milton at UCL, which he called "MapTube,"[50] enables maps to be created on the fly from basic functionality pertaining to the map data itself. Taking a vector map stored in a common format and first converting this into a raster file that has a common format so that other raster maps can be directly compared to this is the first stage, and once the map is created, it is organized in an archive that the user can navigate and pull up whatever map content is desired. This, of course, depends on the number of type of maps in the archive, in this case in MapTube itself, and the system comes into its own only when a critical mass of maps is reached. Simple GIS functionality enables maps to be cloned and transformed as well as new maps produced. In fact, these kinds of system tend to merge almost imperceptibly into mainstream GIS and represent one of the major developments of the information society, with software building on software. It is remarkable how many developments in mapping, as in many other—in fact, most other—areas of software, build on a basic layer, often the current layer, thus adding more and more layers of complexity to the applications. The great danger, of course, is that few understand the layers on which current applications are built, with this being a classic example of how the world is becoming ever more complex. Mapping systems represent as classic an exemplar of this consequence of building layer upon layer as any software you will find.

There are literally so many systems now available that they are impossible to classify. The tools for elaborating quite basic ideas of online mapping and GIS are so widespread and relatively easy to use that most informed users of digital media can create their own mashups. We will return to these in the last chapter of this part of the book, chapter 13, when we explore the whole idea of the digital city, the city as built in 3D software of which the 2D map base is essential. But to give an idea of what is now possible, there is a rather effective application that is building the city from the ground up, by taking data about building plots and their sizes and volumes and associating as much data as possible with these geometries. In a sense, this is using the digital city—London, in this case—in terms of its 2D map base, which links data across many dimensions in such a way that the media is user-friendly to a wider variety of stakeholders who can adapt it to a variety of purposes.

The idea is to construct a detailed archive of everything pertaining to location at the building plot level, which is the level at which we can best appreciate the form and function of the city. Using data that has already been captured by remote sensing, photogrammetry, standard surveying techniques, and local sensing, as well as light ranging and imaging (LIDAR), a detailed configuration of the urban landscape is produced. A variety of attributes are defined for the building plots, and then the system is opened to the wisdom of the crowd. Like OpenStreetMap, we, the users and the occupants of the city, know the city as well as anyone else, and the idea is to archive our collective wisdom in filling in the detail. The project is pitched at the level of coloring the buildings, Colouring London,[51] and this enables a wide variety of the population, from children to adults and from those with barely any computer experience or any experience of maps to those who are immersed in social media, to form the crowd. The data collected pertains to many map categories, and there is scope to increase these and change them. Currently, the focus is on crowdsourcing data and converting this into information that pertains to the following: location (where it is), use (how it is used today), type (how it was first used), age (how old it is), size and shape (what its form and scale is), construction (how it is built), team (who built it), sustainability (energy, repair, retrofit status), greenery (whether there is greenery around it), planning (whether there is evidence of controlling change), and demolition (what this history is). Such a user-friendly website is able to provide anyone with access. Like OpenStreetMap, one has to register, and it is early days yet, but the quest is to build data from the bottom up. It is designed so that potential users will be attracted to "color the city" in terms of these attributes and that this, like OSM, will become the way we collect such data and build information about the city in the Digital Age.

In these last two chapters, models and maps have been two of the main computational themes that dominate the smart city, but our focus so far has been on low-frequency events rather than the higher frequencies that pertain to everyday affairs in the twenty-four-hour city. The models we have introduced are orientated toward a comprehensive perspective on the functions that determine urban form involving the location of land uses and the transportation that binds all the social and economic activities constituting the contemporary city together. The maps we have introduced provide the digital representations that support our urban models that have continued to evolve from the earliest days of computing. In this chapter, our foray into maps has been to suggest that routine and everyday change can also be sensed from maps, with many of the online systems being updated frequently and thus being useful for both high- and low-frequency events in the city. OSM is an excellent case in point, for as with any system of crowdsourcing, it is being updated all the time.

In the next chapter, we will transfer our attention to models and mapping in the high-frequency domain, although our focus on big data and urban analytics still cuts across our temporal spectrum. When we descend to highly routinized activities, which occur second by second, minute by minute up to the daily, weekly, and perhaps even the annual cycle, the spatial scale at which we focus tends to become finer than in terms of low-frequency events. The map begins to approach the territory, while our models become closer to real things, acting in some measure as digital twins. In the next chapter, the map, the territory, and the digital twin that emerged from ever closer correspondence with reality will provide a frame for our thinking about new notions of computation that are more data driven. These throw up new kinds of analytics that are beginning to form a basis in geo-computation for managing the smart city in the short term and designing its future form over the long term.

12 BIG DATA AND
URBAN ANALYTICS

The world is one big data problem. There's a bit of arrogance in that, and a bit of truth as well.

—Andrew McAfee (2017)

BIG DATA IN 1955

When we described what we meant by the high-frequency city in chapter 9, we implied that as we embed remote sensors into the city and as we use our own personal sensors to detect objects ranging from people to the IoT in real time, we unwittingly generate big data. Data is big because the sensors work continuously, and only when we switch them off is the data that they have generated judged with respect to its size. For sensors that operate by detecting changes in objects, second by second, over twenty-four hours, say, a single sensor can detect some 86,400 pulses per day. If we multiply this by the number of people with smartphones, which in Britain is about forty million, in theory, this generates some trillion numbers or a terabyte of data. Simple scaling through time and over the number of sensors soon yields volumes of data in the petabyte range, and it is this that is usually assumed to be the kind of size that defines the science of big data.

Big data, however, is not what it seems. We have always had big data, for big data is relative to the technologies we have available to unlock it and to store, retrieve, and manipulate it. There is a wonderful story about big data being entirely dependent on our ability to process it, which in turn depends on the capacity of the computers that we have available. Back in 1953, Joe Lyons and Company operated a popular chain of some 250 tea shops (and many other outlets such as hotels) in British towns and cities, and as part of their operation, they were at the forefront of digital computing in business.[1] They teamed up in the late 1940s with Maurice Wilkes (whom we came across in an earlier chapter as one of the founders of digital computing in Britain) at Cambridge, who was working on one of the world's first digital computers, EDSAC. In return for funding some of Wilkes's research, the company began the production of their own computer closely fashioned after the Cambridge model. LEO1 (meaning "Lyons Electronic Office 1") was first constructed in 1951, and LEO2 a couple of years later. By then, the embryonic computer group at Lyons had begun to take in computing jobs themselves to pay for their idle cycles from organizations as diverse as the UK Meteorological Office and Ford Motor Company.

In 1955, they were approached by British Railways, which wanted to compute all the distances between some 5,500 stations in Great Britain so that they could price their freight (and passenger services) efficiently and consistently. Now, 5,500 stations implies that there are some $5,500 \times 5,499/2$ symmetric link distances (more than fifteen million in all) to be calculated—assuming no one-way lines. This is a tiny data problem today, but in 1953, it was enormous.[2] This was big data by any standards, and although the LEO machine could just about handle this matrix of distances, the process was painful. Remember that this was an era when computers were still largely based on thermionic valve technology, when everything had to be input to the machine through an intermediate media—through punched tape and then punched cards in this case. Tim Greening-Jackson, who has charted the history of this railways project, says of the machine, "It had 2K 35-bit words of memory, implemented using mercury delay lines. Input was from punched card, and output could be either to card or to a printer. Programs were hand assembled (i.e. there was no separate assembler program) and the (decimal) op-codes were written on coding sheets. These coding sheets were then keyed as Binary Coded Decimal (BCD) onto punched-tape, which was subsequently converted to pure binary on punched cards by LEO itself. Each punched card could hold 16 instructions."[3] It is amazing that anything worked at all in those days, but it did, and it provided the kind of discipline in problem solving that still drives us to search for faster, smaller, and ever more efficient solutions to computable problems.

What the team did was to break the problem down regionally into parts, solve the shortest routes problem for each geographical bit, and then stitch the system back together. For example, Scotland had only three rail lines connecting the rest of Britain, and it was easy to see that this kind of partition could minimize the calculations. It is a completely manageable problem today without any partition, and there are applications out there on your phone and GPS devices that use similar procedures to compute shortest routes and display them on online maps in microseconds. But there was a more fundamental problem. No one had ever solved the shortest-route problems before. It was not until 1959 that Edsger Dijkstra invented his famous algorithm,[4] but this was 1953. So, the Lyons team simply invented it on the job, so to speak. When this story was discussed with a key member of the team, Roger Coleman, in 2012, he admitted he had never heard of Edsger Dijkstra!

There is another punch line to this story. When the job was completed, the team delivered a stack of printout to British Railways, so they could use this as a lookup table to price their freight. And in what is a typically British way of saying that the problem had changed, they never heard from them again! I suspect it might have been bureaucracy rather than anything more sinister—so much investment in ideas but no follow-up whatsoever. It is the story of invention in Britain from the jet engine to the computer itself, but we digress, and the British inability to exploit the technology they invent is legendary—another story for another time.[5] Whether the results were ever used, we will never know, but we suspect not. Yet, this is still a very sobering demonstration of the power of big data. Big data is relative to how quickly and easily it can be processed. In this sense, big data always stretches the limits of our computing power. What is small data today is what was big data yesterday, and absolute definitions based on volume are thus somewhat limited. Currently, we are able to process data volumes that are measured in terabytes—a thousand gigabytes—which can now be stored on an external hard drive that costs about $50, and even a petabyte drive now costs less than $50,000, with the cost rapidly falling.

Of course, it is limits on random access memory that really determine what we can process, as the LEO team was well aware sixty years ago. Once we pass beyond a terabyte, we are up against quite hard limits for most routine computing, and thence we move into the very specialist territory of big data and its processing, where data volumes are at petabyte or even exabyte level. This requires specialist skills as well as new forms of analytical modeling as the data volumes become greater. Volumes are thus still important, as they imply qualitative change in how we go about working with such data. In fact, there are much fuller definitions of big data that incorporate several key issues identified in the big data community. Big data is defined as being not only huge in volume but also high in the speed at which it is created, diverse in variety, exhaustive in scope, fine-grained in resolution, relational in nature, and flexible in terms of its extensionality and scalability. The so-called five Vs provide a quick shorthand for this—volume, velocity, variety, veracity, and value—while this shorthand can be extended to many more Vs.[6]

DEFINING BIG DATA

The definition of what constitutes big data is thus relative to our abilities to process it, as our example clearly shows, and it is processing rather than storage that is key to its definition. A simple rule of thumb currently suggested as a casual definition of big data is any volume that will not fit onto an Excel spreadsheet. If you filled the currently available spreadsheet from Microsoft Office, which has 1,048,576 (2^{20}) rows and 16,384 (2^{14}) columns, it would probably freeze, and no processing could be done, illustrating that it is not just storage that is an issue but also processing relative to storage. In fact, my own predilection for a definition of big data pertains to data that is streamed in real time. There are many data sets that are not temporally streamed that seem big in the volume sense, but it is only data that is truly incomplete at any point in time that can be big data in the streaming sense because that data can continue to be collected until the systems generating it are switched off; at any instant, the potential size of what has been collected already cannot be measured because temporal data depends on the future. In this definition, big data has no potential bounds; it is data that once the sensor is turned on so that it can be streamed, its volume is bounded by the point in time at which the sensor is switched off. This limit is not usually known in most of the applications that we will have recourse to explore here.

We can classify data by size in terms of its volume, but there are other criteria for data that can be generated automatically from sensors as well as produced more traditionally by probing the system directly.[7] Much of this data is produced for cities by census enumerators asking populations to respond to what they, the enumerators, wish to measure or by populations themselves being alerted in various remote ways to respond. Increasingly, traditional censuses are being automated but with responses still being generated by the population itself. As such questionnaires are expensive and time-consuming, they are rarely carried out with high frequency, with most nations administering a complete census of their population every ten years. Few undertake censuses of the population where the population is mandated to update information on a continuing basis. Traditional data from questionnaires is usually highly structured in contrast to data from sensors, and this is a major distinction in that with the rise of real-time streamed data, many new tools are being invented to mine such data in the search for patterns. This is in contrast to traditional data, where such patterns are already implicit in the way such data is structured. This is not to say that real-time data lacks any structure or order or that traditionally structured data provides deep meaning before it is produced. There is considerable variation in different kinds of data with respect to structure, but usually one has to work hard to extract this structure from streamed data that is collected automatically in comparison to traditional questionnaire-based data.

Although data that is streamed in time is big if the temporal interval is fine enough and the length of capture long enough, what might pass as small data where individuals or objects are modest in total, for example as millions of items such as the data generated from questionnaires, can become big if it is cross-classified in different ways. For example, if we have a large population of a nation such as the UK, which has sixty million people, and if the attributes of the population are broken down into a modest number of categories, tens or hundreds, then we can easily generate enormous data volumes. With sixty million people, assume each has an age that we can divide into, say, hundred-year cohorts and an imputed income divided into ten categories, such as less than $5,000 a year, $5,000–$10,000, and so on, up to, say, more than $100,000. This gives a total of twenty groups, and if we then cross-classify these data, we can generate a total of 60 million $\times$ 100 $\times$ 20 = 120 billion categories. Of course, the categories are mutually exclusive, and there can only ever be sixty million numbers in total, but in principle we are dealing with data that is in the order of billions. It is easy to see how we might blow data up in this way, notwithstanding that most of the cells in such tables may be zero. The data is said to be sparse, but to understand it, we have to treat it as big data. Data is as much about future possibilities as it is about the existing reality. We cannot dismiss categories that are now empty of data, for in the future, they may be full.

If we return to the Lyons Electronic Office and the railways project, then it is quite clear that when we look at linkages or relations between people or objects or both, we generate flows that might be physical or electronic or logical in nature but which nevertheless blow the system up to potentially enormous quantities of data.[8] The number of links between stations in the railway network of Great Britain in 1953, $5,500^2$, is not a particularly large number now with respect to basic computation, but if we were to divide our city into ten thousand zones or point locations, say, then there are a hundred million potential interactions. This might be a very modest representation of a 100×100 grid square lattice, and it is easy to see that we may need a $1,000 \times 1,000$ set of grid squares to represent a big city. This would generate some trillion interactions, which throws what might be considered traditional data into the domain of big data. When we explored the first great wave of computer models used in city planning—the land-use transportation models that we catalogued in chapter 10—the major limit was spatial representation.[9] Data itself on interactions was also a lot less available than it is today, although some of it was created by sample household interview surveys. But the limit was on the number of zones, and the spatial scale at which interactions could be represented and modeled was highly limited. Many of the newer models that now form the core of urban analytics, which we allude to in this chapter, assume the city can be divided into very fine-scale grids or pixels, and in this sense, although the data is not necessarily generated in real time, it is big in volume.

The computable city is now dominated by different kinds of flows that all manifest different aspects of the way urban populations and activities and the objects we use to engage in urban functions within the city are related to one another. The distinction between physical and electronic flows marks the transition from the material to the digital city, and as we have implied, physical flows have dominated electromechanical technologies as well as traditional mobility in the industrial city. The post-industrial city that is now to the fore is composed of a great cornucopia of flows of information that are generated directly from our formal physical networks as well as electronically from the multiple devices that we now have to communicate with one another. In this sense, we can picture the computable city as composed of layers of physical and electronic flows, continually changing, being related in diverse ways, with complex coupling that serve to keep the city functioning for the most part.[10] One of the significant features of the idea of the city being built up in layers of flows is that these layers are not particularly stable; they are changing continuously as new technologies are elaborated and invented and as different networks merge and converge with one another as well as collapse and disappear as populations switch their allegiances and business practices to other domains.

In previous chapters, we have noted that many flow structures and the materials and information that they encapsulate are rapidly evolving, while at the same time they are often invisible to direct observation. Even if we were to observe such flows in their entirety, knowing how they relate to one another is a massive dilemma. In some sense, our probing of them implies a quantum effect in that when we do get to observe them directly, they are continually morphing and transforming into new flow patterns before our very eyes. In short, so many new flows have emerged from the development of sensing in the high-frequency city—flows taking place at a fine spatial and temporal scale but, over time, eventually turning into flows pertaining to the low-frequency city—that we have only a naive and incomplete picture of how these are changing the city as it continues to evolve. Much of this activity is invisible, as we have pointed out several times, while new kinds of electronic and physical flow are emerging. Constructing flow patterns based on networks from this data is also problematic, for physical and electronic flows are converging with one another in ways that make it hard to know the bounds of inquiry when it comes to defining any flow that seems physical or electronic.[11] Many new kinds of flow merge both physical and electronic communications, and this complicates the picture. It is, once again, a symbol of the fact that the computable city is a continually evolving mixture of different kinds of flows, of material and ethereal types, and of the real and the virtual—themes that we will turn to more and more as we fill in the detail of what the future city is all about.

THE EMERGENCE OF URBAN ANALYTICS

More than ten years ago, Chris Anderson, then the editor of *Wired Magazine*, wrote a highly provocative article entitled "The End of Theory: The Data Deluge Makes the Scientific Method Obsolete." In it, he argued that many of our existing theories, from the hard physical sciences to the softer behavioral and social sciences, were largely flawed in that although they were the best we had for consistency, they were invariably imperfect and usually fell far short of the mark.[12] He quoted the English statistician, George Box,[13] who, in 1979, famously said, "All models are wrong, but some are useful"—the quotation we used to introduce simulation in chapter 10. Anderson went on to say that given the dramatic increase in data from real-time sources, we could devise much better predictions by simply exploring big data and mining the patterns therein—in short, that we could do better than science itself and that this heralded the end of theory. He did not really give any examples, but his argument appeared persuasive. The predictive models of cities that represented the first wave of applying new information technologies to cities using computers, which we sketched out in chapter 10, were those that Anderson had in mind as highly problematic, for as we noted, their predictive abilities were and continue to be forever limited.

His argument, however, seems to dismiss science completely and to imply that there are great discoveries to be made by mining big data directly. But this belies another quite different view about big data, which says that if you have great swathes of unstructured data such as that which is produced from much real-time streaming, and if you do not have some idea of what is within—and that means some theory—you are not going to get much out of it. This contradicts completely Anderson's views, where he says, "We can stop looking for models. We can analyze the data without hypotheses about what it might show. We can throw the numbers into the biggest computing clusters the world has ever seen and let statistical algorithms find patterns where science cannot."[14] Controversial stuff indeed. In this context, then, analytics (or here, urban analytics) refers to the tools and techniques—the methods—for getting into large data sets and exploring the patterns that one might find there. Such tools and techniques are the essence of data mining through searching over and over again in various systematic ways for explanations. In fact, Anderson is blunter than this, for he implies that we do not need to look for causation in data: correlation is enough.

Analytics, of course, means analysis, but data analytics goes back many years, at least to John Tukey, the father of data analysis, who, nearly sixty years or so ago, beautifully articulated the logic of all this in a famous paper. Tukey[15] did not say we needed to get rid of science to do this—he would have been appalled at the suggestion—but he did argue vociferously that we need intelligent approaches to data. Urban analytics is urban analysis, but it is more than this in that the emergence of big data and the high-frequency city gives it an added edge over what urban scientists and policy makers have done with computer models over the last fifty years.[16] A more recent comment by Michael Goodchild further emphasizes the bias to data analysis, where he defines urban analytics as "new kind of urban research, one that exploits the vast new data resources that are becoming available from social media, crowdsourcing, and sensor networks."[17] Contained in Goodchild's quotation is the clear and unequivocal notion that urban analytics is about the new world of big data—a world that has certainly emerged in the last decade with the continued miniaturization of computers to the point where they can be embedded into the fabric of the multitude of sectors and systems that we deal with in urban research.

Yet, urban analytics is also about how we manage, control, and plan the city system, and in this sense, it is unavoidable that we need models and simulations to enable us to do this, either directly or indirectly, in the development of plans and policies. We will spend much of this chapter on the new generations of urban models that have followed on from the land-use transportation models that we described in chapter 10 and are now part of the emergent field of urban science, but before we do this, it is important we dig into the data analysis tools that enable us to search for patterns in data. These tools again date back many years to the foundations of statistics and what came to be called "multivariate analysis" in the postwar years. The idea of mining data is one where the dimensions—the key components of that data—are identified and used to relate and correlate the attributes that define the data set. The kinds of patterns that might be found are only interpretable by those who are close to the subject matter, but there is the hope in unstructured data that patterns might be discovered that are unusual and insightful and, in this sense, add to our ability to explain and predict. Neural networks, which involve defining layers of hidden components that are matched to patterns in the data iteratively until a good correspondence with the desired output is achieved, are examples.[18]

There is, however, a more pragmatic reasoning that dominates this kind of data mining, and this is the idea that we might find regularities that cannot be easily explained but which are robust enough to enable us to control and manage or even plan our systems of interest to achieve greater performance. For example, there are many tools that enable us to introduce additional attributes that do not necessarily have any obvious or immediate meaning in terms of the data in question but are sufficiently robust as explanatory variables to enable us to use them in identifying important patterns. A good deal of this kind of analysis essentially lets us develop methods for relating many different patterns that appear to define an individual's preferences for example, and these preference structures or profiles can then be used to make suggestions with respect to the individual in question. We are all used to these sorts of systems, such as those that identify the item you may be interested in from online retailing sites—books are an obvious example on sites such as Amazon.com—but these are not supremely clever in any way. They are simply based on correlating masses of data over very large populations that enable us to look at what relates to what. Before the web and before we had computers powerful enough to search millions and millions of individual preferences, this kind of activity was easy enough to understand, but there was no way any of us could do much about it.

These kinds of tools used to provide structure to such large data sets are those that are loosely called "machine learning," although the term "learning" really relates to the way such systems optimize performance by continued iteration. AI is also being used as a collective term for these methods, but this is weak AI in terms of the way we defined it in chapter 1. We need to be careful in making a distinction between systems that are truly intelligent and those that look intelligent simply because they are able to amass very large quantities of data and produce patterns within them that we have never been able to figure out manually. In fact, this is essentially weak AI, in contrast to the much deeper tradition of endowing machines with intelligence based on mimicking human reasoning.[19] As we noted in our first chapter, the field of AI progressed very slowly for many years when the focus was on trying to build computers that could reason in contrast to the enormous strides that have been made in simply amassing data to reveal patterns that imply a degree of intelligence rather than systems that actually attempt to think. Many of the methods that dominate the way autonomous cars and suchlike are beginning to work are based not on any form of reasoning but on pattern matching. Language translation systems are a case in point, where the translation is made possible by the machine searching through billions of like examples drawn from the corpus of text defining the language being translated with that for which the translation is required. Speed and storage, as implied through Moore's law, is absolutely essential to the development of such systems.[20]

We will look at various examples of using such tools to figure out patterns in data, but our real focus in urban analytics will be wider than the data-science definitions that we have introduced in this section. In terms of computation involving cities, the quest to simulate many systems that define the city has continued fairly constantly over the last half century. Various model structures have evolved to address many of the problems posed by lack of data, lack of speed, lack of storage, poor theory, and our difficulty in focusing computation on models of systems that directly address the policy questions that have dominated the changing agenda of urban governments and urban planning since computers were first invented. What has happened since Douglas Lee wrote his "Requiem for Large-Scale Models," which we described in chapter 10 and which essentially marked the end of the first stage of what came to be called "urban modeling,"[21] has been the development of simulation models that have addressed many of the key critiques: these involve the scale issue, more open and more comprehensive models, models that have been more strongly rooted in stakeholders needs for using their predictions in various policy making, and models that have begun to focus on the high-frequency city as well as the long-standing low-frequency perspective with respect to how cities have been and are still planned for a longer-term future.

There are a number of features that make the development of these simulation models important tools in the armory of urban analytics. The first is the notion that we should attempt to simulate cities at a fine spatial scale and also move to thinking of social and economic activities at this scale as being defined by individuals. This was mooted many years ago for transportation, where an explicit microeconomic theory of transport preferences was evolved using ideas from utility theory, and from then on, these types of models were extended to deal with travel activity allocation, travel budgets, and household decision making, culminating in so-called agent-based models (ABM) of transport that now represent the state of the art. Urban applications also began a little later, and for the last twenty years or more, models of city development using cellular automata dynamics have become popular.[22] Microsimulation techniques as well as agent-based modeling now characterize these kinds of simulation. There has also been a move to dynamics, especially as many behavioral issues in such models involve changes in decisions over time, and ABM in particular is well suited to these kinds of development.

Extending such models to deal with major policy issues and to embed them in practical policy making has been exceptionally difficult. The field, however, is opening up quite radically at present with respect to engaging many interests that are not particularly expert in modeling but expert in knowing about cities and urban policy. Crowdsourcing not only for data but also for ideas about simulation is beginning to define the field. In the rest of this chapter, we will engage with these ideas illustrating how the city is becoming ever more computable. But we will show that many different perspectives dominate the current state of the art in urban analytics, and this makes it highly plural, with many models of the same system but from different theoretical traditions being developed. This is entirely consistent with the notion that there are almost as many theories and models of cities as there are serious urban scholars and professionals looking at the city. This pluralism defines the fact that cities are becoming ever more complex as they become more computable. It is consistent with the notion that the city is a rich, diverse, and chaotic ecology of desires, motives, and contradictions, and it can never be anything different. All this is consistent, of course, with George Box's hallowed phrase "All models are wrong but some are useful."[23]

PHYSICAL SIMULATIONS OF CITY DEVELOPMENT

Imagine you know nothing about cities other than through your experiences of living in them, but you are well aware that your friends live all over the city at different distances from you. If someone were to ask you to locate them, you might attempt to draw a mental map of where you considered each lived. What you might do if you were asked to be as precise as possible is to key into an online map each of the addresses of those you know and capture the position in terms of the latitude and longitude of each person identified. If you had more than fifty friends, this would become quite a laborious job, and as the number increased, it would become a real chore, but you could probably do a good job from your casual knowledge if you persisted. Once you had the locations, you could measure the distances between where each person lived, and you could work out the center of gravity of their locations: in other words, where you might locate if you wanted to reach all of them by traveling the least distance in total. Now imagine that you were called on to do this in a very large city, where you needed to know the location of every member of the entire population. Take London, for example, which at the time of writing (2023) had some 3.6 million households. Obtaining a good picture of locations would be a real effort, although not impossible by any means using modern GIS software, and there is now data available from many sources that would provide these locations. But if you wanted to compute the extent to which one household interacted with any other, then there would be 3.6 million × 3.6 million = 12,960,000,000,000 ($3,600,000^2$) possible links between them, and this is a really big number—something in the order of 10^{13} or a hundred trillion. You can soon begin to see that this is big data by any standards and that this arises quite naturally simply by describing the individual locations of people in a large city and the potential interactions between them. This was the sort of job that the Lyons Tea Company had to tackle back in 1955, albeit on a dramatically smaller scale but in an environment very heavily constrained by its limited computational resources.

One of the consequences of such simple demonstrations is that it shows our inability to deal with this kind of detail is still constrained by the power of the computers that we have available. We are unable to process this kind of data very easily using our current information technologies, even with the cloud, parallel processing, supercomputing, accelerated processing units, and so on. In fact, even if we forgot the interactions between the households, just working with a city with 3.6 million locations is difficult, and special techniques and representations are needed to deal with such problems. This was the issue when the first urban models were designed and constructed back in the 1950s and 1960s, and we catalogued these difficulties in chapter 10 where it was clear that to build the models at all, one needed to aggregate. Rather than dealing with each individual household (for which data was in any case very limited in the past), aggregations of households into populations were made (which is akin to grouping adjacent locations together into zones). For example, the limits on the number of model zones developed some fifty years ago were in the order of tens or hundreds. There are still limits, and very few land-use transportation models with more than ten thousand zones have been constructed, although this is changing rapidly. But the level of detail that can be captured spatially is still many orders of magnitude less than one might wish for.

Therein lies the problem. How do we design computable models at the requisite level of detail? For someone who takes a straightforward view of the city such as an informed stakeholder but nonexpert in urban analytics, then it might seem a very reasonable request to want to model the city at a level of detail that coincides with the way most of us perceive it. The fact that we cannot is quite an indictment of how far we still have to go in this field. So, taking up our story where we left off in the 1970s after the full force of Lee's "Requiem for Large-Scale Models" had hit the professions that considered simulation the only way forward, something had to give. The choice was either more and more detail in the traditional fashion using more and more computer power, which was quite painful, or a different means of representation, a different means of articulating the urban system. Both these strategies and various combinations of them were tried, and starting in the late 1970s, new forms for representing and aggregating elements that composed the city were invoked.

The simplest representations came from the emerging field of GIS, where the favored strategy was to divide the city into regular cells forming a lattice (often a grid of square cells) that could be adjusted to the required dimensions, thus detecting the appropriate level of detail. This was the notion of representing the city on a neutral canvas, which in GIS, as we saw in the last chapter, was often referred to as a "raster" in contrast to a configuration of irregular polygons, which contained the detailed geometry of the city associated with the concept a "vector." Earlier land-use transport models went down the route of defining the city as an array of irregular polygons, largely because data on regular grids was generally unavailable, the need to define data according to administrative conventions was paramount, and the need to define networks between different locations was important. With the development of GIS, which was much more strongly physical and spatial in nature than the first urban models, the focus was first on layers of data that needed to be represented in some sort of commensurate way, and thus the raster system was quickly adopted. Much of this GIS data also came from the first remote-sensing satellites, and the efficiency of pixel (picture element) representation came to dominate the way this data was delivered in map form. By the late 1970s, the notion of actually working with raster data directly, by either combining it or using it to simulate processes whereby the cells could change their state, had become significant, and the first simulation models of urban growth built around these ideas emerged.

In fact, the first models were somewhat earlier. Allan Schmidt, who we came across in previous chapters, produced a SYMAP animation of cellular growth for East Lansing in Michigan[24] in 1965. Stuart Chapin, Shirley Weiss, and colleagues at the University of North Carolina initiated a grid-based simulation of the growth of Greensboro NC about the same time.[25] Waldo Tobler developed a cellular model of Detroit[26] in 1970. There was then a slow march toward more physical representations of the city through the next two decades as more and more physical data was accumulated and transformed into spatial rasters. Coinciding with these developments, computer graphics made enormous strides, with the whole field availing itself of the massive new memories in computers given over to pictures that came in the wake of the development of the PC, the minicomputer, and eventually the workstation. Throughout the 1980s, new ideas about how one might model natural objects using ideas from fractal geometry hastened the development of generative models, which had an obvious affinity to urban development processes.

This led to proposals that regular processes of transformation that could be mirrored on 2D graphical screens might hold the key to models of systems that might display key features of living systems such as reproduction, diffusion, and perpetuation. A decade earlier in 1970, the mathematician John Conway threw out his famous challenge for the parlor game he had invented in the common room of his Cambridge College, which he called the "Game of Life." This game was represented on a grid, where each cell could reproduce if three cells were alive around a cell in question which was not alive, or the cell would die if more than four cells were active around it (through overcrowding), or the cell would not be born if only one cell around it was alive due to isolation. A live cell would thus stay alive if it had two or three live cells around it. Any other configuration of cells would remain the same from one generation to the next. Conway's challenge to the wider scientific community was to prove that this Game of Life[27] was self-perpetuating, and if someone were to take on the challenge and show unequivocally that this was the case, then he would reward them with a $50 prize.

Martin Gardner[28] advertised the challenge in his column in *Scientific American*, and within a matter of days, a group at the MIT led by William

Gosper had triumphed. This gave an enormous boost to the idea of self-perpetuating systems. CA had been suggested as far back as 1944 in the Manhattan project by Stanislav Ulam and taken up by the enfant terrible of the mathematical world John von Neumann.[29] But with von Neumann's premature death, the field had remained dormant until it was given this boost by Conway. Moreover, CA models also embodied the idea of emergence. All kinds of fractal patterns could be generated as cellular systems based on the application of rules that led to self-similarity at different scales were applied. The link to graphics was obvious, and to an extent, along with ideas about chaos, CA became the darling of the complexity theorists, whose notion that complex systems could only ever evolve from the bottom up, not the top down, really took off in the 1980s. Thus, CA became the most elementary exemplar for systems that demonstrated the power of emergence often in terms of fractal self-similarity as they grew in size, with patterns repeating themselves across different spatial but also temporal scales where the spatial element was only implicit.[30] The fact that one could also program such processes of growth on ever more powerful and accessible computers, on minis, and then eventually on PCs added to the momentum. The obvious step that came from GIS and PCs with large graphics memories was the development of CA models of cities, in particular of urban growth, and these then began in earnest in the early 1990s.

CA models of cities are very different from the land-use transportation models that were the first computable simulations of cities that we presented in chapter 10. The focus in CA is on temporal and spatial change, with the obvious assumption that cities are never in equilibrium but always in transformation. The simplest models articulate the city in terms of its land uses and activities, which define the state of any cell at any one time. In short, the urban landscape is one of regular cells, whose states represent the land uses or activities that are directly associated with the cell in question. Each cell may have only one activity, which can change as the automata transforms itself through time, but this depends on the level of resolution adopted. Sometimes, a cell may have several land uses or activities, especially if the number of cells is limited due to constraints on data and computing. In essence, the way a CA works is to define sets of rules for cell transformation. In the case of a city, where the cells represent development, these rules could be based on the way developers and those who ultimately occupy that development determine what the cell is to be used for.

The simplest example might be that an empty cell is developed for housing or remains vacant. Thus, typical processes would involve assessing the demand for housing in any cell and its potential supply, and would thus take account of the various decisions that need to be made with respect to things such as land ownerships, physical capacity for development, distance from other activities, and so on.[31] The most rudimentary form of CA would simply score each cell of the urban landscape with some potential for development that might be computed from its simple accessibility and then begin a process of diffusion from some cell in question that would generate the pattern of development through time. This pattern would, of course, be influenced by the potential scores of each cell, which, in turn, would be determined by the entire configuration of development in the city so far.

In this manner, a variety of CA models were constructed for large cities and for semi-urban, even rural, areas in developed and developing countries contexts. CA models are an order of magnitude simpler than land-use transportation models, and as such, they can be developed using very standard software that is applicable to many generic applications, not just models of cities. This is quite different from the earlier generation of models, which even in their present form are highly tailored to individual cities and require very detailed customization for the most part. The software for CA models is much more like GIS in that it is more formulaic in its implementation, although specific software packages have been developed for different contexts where there are intrinsic differences in the simulations. For example, from the mid-1990s onward, two sets of applications have become significant.

First, the SLEUTH model was developed by Keith Clarke[32] at UC Santa Barbara, funded by the National Science Foundation and the US Geological Survey. This has had widespread applications in the US and elsewhere around the world. The focus of the model is contained in the title "SLEUTH," which unpacks as a model dealing with *s*lope, *l*and cover, *e*xternal uses, *u*rban growth, *t*ransportation, and *h*illshade, the meanings of which are fairly self-explanatory. The model essentially filters out cells that reflect these six features that are used to determine how the cell changes state in terms of land use, and these form the transition rules that are used to drive the model forward. Essentially, the model is a growth model in that decline and redevelopment are harder to contain within the framework, while the model does not treat the transportation system explicitly as is the case in standard land use transportation models. The decision rules lead to cells changing state through diffusion, spread, and growth by breeding, and these are subject to the presence of existing land use, transportation, and the slope of the land in question. The model was originally applied to urban growth in the San Francisco Bay Area, simulating development from 1900 to 1990 using data based on old maps, remotely sensed data, and more up-to-date land-use data from the 1970s onward. Many other variants of the model have been developed and applied to different cities, but they all originate from the application of CA models to diffusion processes such as the spread of wildfire and the use of remotely sensed data from which land-use patterns can be extracted at different points in time.

Another class of models has been developed by the Research Institute for Knowledge Systems (RIKS) in Holland, which initially developed a CA model that simulated the growth of Cincinnati, OH. The model, developed by Roger White and Guy Engelen, went through several versions with packaged software under the general title "Geonamica" of which the variant for urban CA modeling is called "Metronamica." The model, like most CA models, is quite straightforward. The idea that local rules influence urban development—most CA models adopt the basic idea that cells are only influenced by the states of cells in their immediate vicinity—are relaxed, and the interaction between cells is more akin to a crude form of spatial interaction model. In fact, because the RIKS models have been quite widely applied to various European cities, they have also been steered toward more integrated sets of models interfacing with transport, demographic, economic, and environmental simulations, thus providing a decision-support environment for urban and regional policy analysis and planning.[33] Many other variants of these CA models now exist, and because they are so straightforward to apply, they represent a first step in simulating cities but mainly in those places where urban growth is significant.[34] Developed cities are much less likely to be candidates for such simulation, and thus the dynamics in many cities in Western Europe, for example, remain beyond the purview of such CA models.

The major problem with these models is that they tend to be simpler than their predecessors and are thus not well tailored to policy making. The earliest generation of urban models was directly the product of policy imperatives, mainly the growth of US and Western cities due to the impact of the automobile, the need for new highways and the decentralization of populations to the suburbs in search of lower densities, lower congestion, and more space. CA models emerged from very different origins than land-use transportation models. They came from ideas about representing the city in layers and from the availability of new software in remote sensing as well as GIS. Their theoretical foundations lie in complexity theory[35] rather than in a concern for the urban and regional economy and social physics that dominated the models we introduced in chapter 10. To an extent, although these models are much easier to implement, they are much less policy focused, and to make them relevant to the wider context of urban analytics, they need to be coupled with other systems that extend the meaning of their predicted outcomes. It is clear enough that we need to think about urban sprawl in cities and contain it in various ways or let it rip, or some combination thereof. But this requires a much more detailed set of functions dealing with travel cost, rents, land capacities, and behavioral preferences as well as issues concerning alternative modes of travel and styles of development than are contained in these simple CA.

SIMULATING CITIES OF INDIVIDUALS

What CA models show us is that we are able to simulate cities composed of many aggregates of individuals arranged in a regular lattice of cells. This enables us to move well beyond the very large but simplified spatial aggregates that dominated the first urban models. The logic, of course, of moving to ever finer levels is remorseless, and in the quest to simulate cities with greater and greater realism, there is an inevitability that we should focus on simulating individual behaviors. Although these developments are also consistent with and have been spurred on by articulating cities as complex systems, a new form of social simulation has emerged—ABM—where agents can be any form of object that interacts with other objects, usually endowed with some purposive behavior. Such behavior may be automatic and routine, but this idea opens the door to any kind of behavioral simulation that incorporates ideas about emergence, surprise, novelty, and innovation—all key elements in these new forms of model.

ABM is even more generic than CA in that the system of agents can be at any spatial and/or temporal scale, where an agent can be defined as any distinct object or grouping of objects, separate from other like agents. Such a system must be composed of enough agents so that they can interact in a nontrivial way, and this implies that ABM is usually dynamic in the sense that interaction between agents takes place through time.[36] There are examples of ABM, however, that attempt to simulate how agents relate to one another at a cross-section in time, but many models in cities that are agent based deal with movement and, in that sense, how agents move spatially in time. In fact, there are no strict criteria for defining an agent except that an ABM is likely to contain more than one agent (an almost trivial consequence except that in microeconomics, the focus is often on how a typical agent behaves). Most applications of urban ABM, however, deal with movement in space as well as through time, and the focus is usually on quite fine spatial scales, largely because agent behavior in terms of movement is at the small scale and the problems that generate policy questions are usually pitched in terms of local trajectories of movement.[37] In this sense, then, many urban examples of ABM deal with how pedestrians and/or vehicles (cars, which might include trains and buses) move in space over minutes and hours rather than anything much longer. These models thus merge into another class of related models that are built around micro-simulation that we will note below.

Pedestrian movement models came onto the agenda in the mid-1990s, with several applications to problems involving crowding, emergencies, and hazards involving danger such as fire evacuation, crime, congestion, and a host of related issues pertaining to the way in which human populations initiate local movement. Many of these models, like CA, were built around ideas of emergence associated with complexity theory. In the case of simulating how people move in small spaces, the way they interact, copying each other's movement, reacting by diffusing locally in crowding situations, and flocking together in spaces where the crowd begins to sense direction, are all features that have come to define how agents relate to one another.[38] Extending agents to different classes—for example agents as people, as vehicles, as elements of the environment such as buildings and streets, and so on—all characterize the way such models might develop, thus impressing the inherent flexibility of the approach. In fact, many of these ABM examples require extensive data resources if they are to be calibrated to real situations, and often such data is not available. They exist primarily because it is now possible to build plausible models that may lack data but which can be easily computed and visualized so that scientists and users can explore how such behavior might be best simulated.

There are several developments that incorporate the idea of ABM without being centrally wedded to the ABM concept. The land-use transportation model known as "UrbanSim" is an extensive urban model with implicit transportation that can be developed at the level of individual households that optimize their locations with respect to their preferences.[39] In some respects, this is ABM, but it is structured more like a conventional aggregate land-use transportation model than an agent-based model. In this sense, it is a blend of individual simulations with social physics. TRANSIMS (Transportation Analysis Simulation System), a model first developed thirty years ago at the Los Alamos National Lab[40] as part of the peace dividend from the Cold War by those with experience in battlefield tactical warfare, was agent based. Every household in the city in question was simulated with respect to their decision making in terms of movement through the city for work, shopping, education, recreation, and a host of other urban functions for which trips are generated. The latest application, which morphed very quickly into a European version, MATSim (Multi-Agent Transport Simulation), was first developed in Zurich by Kai Nagel, who had worked at Los Alamos on TRANSIMS, and this has a similar structure, where households decide their daily schedules, generating trips through their active day.[41] These trips are loaded onto the network, and the simulation takes place by ensuring that a feasible equilibrium traffic pattern is produced. These models can take many hours to run, especially if there are many thousands of households or trip makers in the city in question and each has to make decisions first without interacting with other trip makers, hence generating patterns of traffic that have to be restored to equilibrium using rational changes to their schedules. Despite contemporary computing resources, such models are very largely still proofs of concept, as they can take days to compute and are difficult to use in policy situations where rapid visualizations of outcomes are required once scenarios are posed.

One of the most ambitious theoretical agent-based models was developed by Rob Axtell and Joshua Epstein in their book *Growing Artificial Societies*. They developed a generic spatial model (which could be likened to a series of growing cities in an artificial landscape) from which they were able to generate a series of commonly observed relationships pertaining to emergence, city size, the distribution of incomes, poverty, and inequality.[42] These relationships tend to follow power laws, where there are very few individuals with the rarest of attributes who acquire great wealth and very many individuals who end up with very little wealth. There are connections here to new ideas about cities that are reflected in the development of complexity theory and scaling, but for the moment, it is enough simply to note that new approaches to cities based on urban science—which we have not really dealt with so far—are being fashioned alongside the smart-cities movement, and we will hint at some of these in the last part of this book. The final dimension to these new varieties of model, where the focus is on individual behavior, are termed "micro-simulation models." These are somewhat different from mainstream ABM in that the agents represent individuals drawn from probability distributions that are determined from data with respect to the explanations and predictions required. In particular, demographic models, whose distributions pertain to where people live and what stage they are in their life cycle, are sampled and provide the predictions that relate to the categories over which the various probability distributions are fine-tuned. Many of these models simulate the probability of an individual falling into some category or categories that determine outcomes, which are then synthesized to generate an entire population. Various mixtures of these kinds of micro-simulation and conventional ABM continue to be developed, with the field being currently quite vibrant, especially in terms of simulating social structures.

FUTURE ANALYTICS: DEEP STRUCTURE, MACHINE LEARNING, WEAK AI

All the methods and models that we have introduced so far are essentially based on theories that originate as plausible hypotheses about how cities work. They come from ideas about how the local economy is structured, how populations arrange themselves in cities to maximize their accessibility to one another within the resources they are able to command, and ways in which they can realize economies of scale by competing for space and occupying it at densities that are feasible but not necessarily optimal. The problems that emerge from such notions involve spatial and social inequalities, congestion, the costs of sprawl, segregation, and market failures, particularly in housing, all reflecting issues that urban planning has attempted to address since it became institutionalized as a function of government in the late nineteenth century. Some of these ideas have emerged by reflecting on and observing the way land use and activity structures and their associated populations evolve in cities, while the models that attempt to explain them are being continually tested in the attempt to improve their fit in terms of what we observe.

By and large, the traditions that have been developed have thrown up theories and models that are largely deductive in scientific terms, but a profound shift is taking place in the kind of analytics that is emerging from the world of big data and the high-frequency city. In the great data deluge that is happening, it is of little surprise that the focus has turned sharply toward the analysis of data in the belief that real-time streamed data will reveal elements of human behavior, new and existing, that we have not been able to observe or explain hitherto. Whether this is the case—and there is considerable disagreement and skepticism about it—we are only at the beginning of a search, for the deep structures that might well underlie the social world only manifest when we mine such data.[43] The methods that are now coming to the fore seek associations or correlations in data that signify patterns that are far more complex than the simple relationships that most of our theories about how cities work currently embrace.

In short, as we accumulate more and more data particularly through real-time streaming, patterns that might appear suggestive in small tranches of data only become important and inevitable when massive amounts of data are considered. The number of combinations of data attributes that one has to search through is enormous, but current computer systems and algorithms are now powerful enough to handle these possibilities, and the focus is increasingly on finding and testing relationships in data that become stronger and stronger as the amount of data accumulates. For example, it is well known that causal effects in systems often only become apparent after very long time periods have expired, while in the transition to such outcomes, no effects are seen whatsoever. In these kinds of systems, where it might take a long time for relations to become apparent, real-time streamed data over very long time periods is the only way that one can confirm the existence of such effects. As yet, we are nowhere near this, and it remains an open question as to whether any such patterns that are discovered will remain resistant to our unpredictable future.

The focus on searching for patterns is little different from any kind of search such as that we are familiar with from search engines such as Google. These systems amass data all the time by searching the web itself, that is, by searching for new websites but also by uploading great tranches of data from the conventional written and printed media to even live performances and films—indeed, every media one might think of. Systematic classification is required, which is always under scrutiny as new material arises, while as search continues, structure is given to the database in such a way that searchers become better and better informed. It is in this sense that the system learns. In terms of behavioral patterns that pertain to how people function in cities, patterns in flow data sets such as mobile-phone calls and related interactions are being associated with transportation movements. There is also the prospect that social media data might also provide insights into people's mobility, although as we have noted earlier, the problems of making unambiguous associations are severe.

It could well be that nothing like enough data has been collected in cities so far and that only when one reaches massive volumes will we discover new patterns that afford us much better insights into how cities might be improved. This view point is the one that Chris Anderson, who we introduced earlier in this chapter, argued, but it is by no means clear that big data will yield the big ideas that are needed to understand our cities better, thus leading to better planning.[44] We may need to develop systems such as those similar to IBM's Watson—a computer system that is able to compete with the cleverest human contestants at games such as Jeopardy, where such machines soak up data to the point where they are able to respond with the most likely answers drawn from the data that the machine has been provided with.[45] In time, there will be machines that seek out their own data when placed in a wider ecology of information, and to some extent, current translation systems fall into this category. In the case of these and most systems that produce logical outcomes from the questions posed, the processes are not intelligent in the sense of human thought but simply have large enough data and fast enough processing to produce plausible and usually correct answers to what would appear to be very human-like questions. This is not AI, or if it is, it is a weak form of AI, where all that is happening is that a computer is producing an answer to a question from a repertoire of answers that have been predetermined by the machine in organizing its data and which are then generated using various programming instructions that filter out the correct outcome. In short, a system has this kind of intelligence when it can turn on a light when darkness descends due to its being programmed to activate the switch when the overall light in the system falls below a certain threshold. This is not strong intelligence. It is basic automation.

Most of the systems that form part of analytics are of this nature in that they do not aspire to simulate general intelligence or strong AI. This is reserved for systems that engage in reasoning,[46] and in this sense, it was Turing who posed the first test. He argued that if, in any conversation, a human considered that a conversation with another source hidden from the human in question was the product of human thought—that is, if the human could not distinguish the conversation from that generated by a nonhuman—then the source was intelligent. There are many qualifications to this argument, but the point is clear: most algorithms that search for pattern in data, achieving this iteratively with some purpose in mind, may thus be said to be learning; they are not intelligent but simply testing and comparing and identifying answers to tightly posed questions within the limits of their knowledge base, which is big data. Thus, this form of analytics depends intrinsically on what the data might ultimately yield, and the tools to do this are only part of the arsenal of analytics that we might employ to understand and plan the smart city. In future chapters, we will return to these questions about machine learning and AI, but we will leave this until the last chapters of the book when we turn to bigger questions about how this new digital world will be organized. To complete our survey in this part, we now need to show how cities have been represented in more literal but still digital terms as 3D geometries, augmented realities, and virtual environments.

13 DIGITAL CITIES AND VIRTUAL REALITIES

Virtual reality was once the dream of science fiction. But the internet was also once a dream, and so were computers and smartphones. The future is coming and we have a chance to build it together.

—Mark Zuckerberg (2014)

THE DIGITAL OUTLOOK TOWER

The history of contemporary town planning, which began as a reaction to the evils of the industrial revolution in the nineteenth century, is dominated by the towering figure of Patrick Geddes,[1] a Scottish biologist who studied at the feet of Thomas Huxley and who was much influenced by the prevailing ideas of evolution due to Charles Darwin and Alfred Wallace. Geddes was a maverick in many ways, arguing that we should see evolution as the ultimate driver of ways in which cities develop but always wrestling with the tension that town planning was intrinsically a top-down activity that imposed a kind of order on development that, left to its own devices, could and usually did become socially divisive, if not chaotic. Among his many visions for the city of the future, he relentlessly promoted the idea that all cities should have exhibitions that would introduce the idea of the good city to its citizenship in such a way that it would engender public participation, spreading the word that cities should be built from the bottom up.[2] One of his most significant achievements was the Outlook Tower, a permanent city exhibition located in the Old Town of Edinburgh, which

provided the spark for a series of international exhibitions that he pioneered as part of the world's fairs that dominated civic life in the late nine-
teenth/early twentieth centuries. The artifacts in his exhibitions ranged far
and wide across natural and social biology but were always based on maps
and plans and models of the physical kind that impressed ideas of civic
responsibility on the informed public that these exhibits reached out to.[3]

The idea of communicating city plans in such a tangible manner continues to this day but in a considerably lower-key way than Geddes intended—
except in China, that is. A generation or more ago, just as the new China
was emerging from its long experiment with communism, if you were to
visit any of its big cities to talk with those responsible for their design, you
would have been treated to an elaborate display of 2D plans and 3D models of buildings and districts, somewhat like those that were displayed in
Geddes' Outlook Tower. These were carefully and painstaking constructed,
mirroring the rather old-fashioned way in which we thought about cities
in the nineteenth and early twentieth centuries prior to the Digital Age. In
1987, I remember going to the Chinese Academy of Urban Planning and
Design, which was part of the Ministry of Construction in Beijing where we
demoed a digital map on a PC—in fact, many of the participants there had
never seen a PC before, nor a digital map. On the same trip, visiting Kun-
ming in the southwest, not so many miles from the border with North Vietnam, our hosts, the planners, explained the future form of their city using
traditional maps and models in an elaborate suite of rooms that recorded
the top-down way that they had thought about planning since the time of
Mao and before. To China, in those far-off days, the digital revolution was
nowhere in sight. In essence, city planning was still about how one laid out
future functions using off-the-shelf designs, and the time-honored way to
communicate these ideas was through the traditional architect's model.[4]
The physical planning of the city was still rooted in the notion that form
follows function based on the long, unquestioned supposition that if one
got the form right, all else would follow.

Fast-forward thirty years from 1987 to 2017, moving geographically
from the hot southwest to the cold and deindustrialized northeast, and
you will find many big cities in China that all have remarkably elaborate
exhibitions given over to their future planning as well as their history. The
traditional architectural models of building form have not gone away.
Indeed, they have never been more elaborate, but they are now widely

complemented by digital models of the 3D kind through which one can fly, displaying all kinds of flows such as physical transportation and electronic media based on archives of mobile-phone calls, as well as a massive number of ever higher intelligent skyscrapers that dominate current city development. In Shenyang, a growing city of some six million but with a declining industrial base, their city exhibition consists of three floors in one of their municipal buildings. There, the history of the city and its future is mapped out in graphic detail using GIS, CAD models through which one can navigate, sand tables, and touch screens and related devices, where one can literally mold and manipulate the media to show different kinds of futures. Shenyang is proud of its heritage, being the early capital of the Manchu-dominated pre-Qing dynasty that ruled China albeit only for twenty years. The city has its own scaled-down forbidden city—a miniature replica of the Beijing palace complex that was at the heart of the empire until it moved to Beijing in 1644 and established the Qing dynasty. It is a proud city with grandiose plans for the future. It is as good an example of a digital Outlook Tower as one might find. Geddes would have approved.

What is so remarkable about these city exhibitions, of which there are too many to count in modern day China, is the fact that they play such an important role not only in advertising the city but also in drawing in its citizens, who seem to take great delight in picturing their cities in traditional ways but using the latest digital technologies.[5] Perhaps one of the most remarkable digital Outlook Towers in China is in the city of Wuhan, which is really three cities in the heart of the Middle Kingdom formed at the intersection of the Han River with the Yangtze. The Yangtze runs from southwest to northeast at that point, with the area above the Han and on the west bank known as Hankou, one of the original treaty ports that the Western powers extracted from the Qing dynasty by way of economic and political concessions after the Opium Wars in the mid-nineteenth century. Below the Han lies Hanyang, the Chinese settlement serving Hankou, and on the east bank, there is Wuchang, the ancient Chinese town that is by far the oldest and which is rapidly becoming, after Hankou, another major business district. The city's population is now about twelve million, but as with all city definitions, it depends how far from the Han–Yangtze river intersection (the geometric and economic center) one draws the boundary.[6] It appears that Wuhan is about the size of Greater London plus its outer metropolitan area. Its rate of population growth is still phenomenal: 6.5

million in 2000, 7.9 million five years later, 10 million in 2010, and almost 12 million in 2023.

To comprehend this scale and indeed the plans for how the city is generating ever more growth and development, the rather quaintly titled city hall—the Wuhan Citizens Home—itself a new building occupying some three million square feet, contains a remarkable exhibition of city-planning prowess. This is a sort of one-stop shop for a mixture of citizens' advice and local-authority services. Some 30 percent of this enormous building is given over to its exhibition, which contains at least three traditional physical models of the city and its region, all augmented by different forms of information technology to aid those interacting with the exhibits. The entire history of the master plans for the city since the late 1940s is also laid out here. Wuhan is such a complicated place and its growth so massive that this kind of exhibition is almost essential for any newcomer to grasp the scale of development. The kind of surprise that I encountered on my last visit there was due to the coming together of all these new developments, which can only be appreciated with respect to future growth by examining the scale of what is being planned. Just topped out, for example, is one of the largest skyscrapers in China, the Greenland Center, and there are plenty more like this in the pipeline. From a situation some ten years ago when the scale of business activity was somewhat low key and mainly based in Hankou, several business districts are fast emerging from a strange mixture of bottom-up and top-down planning. Complemented by a series of industrial clusters, the city is rapidly becoming one of the most visible examples of planned polycentric development. A far cry from the standard model.

A major feature of Wuhan's exhibition space is something equivalent to the city dashboard that we introduced in chapter 8. This, like many, is essentially a portal based on web-based pages that link directly to raw data with some simple translation thereof captured from real-time data feeds, usually involving data that can be archived from public APIs (applications programming interfaces). In essence, the Wuhan dashboard, like many, is simply a display of data available in real time, often close to the original data, with little or no interpretative spin yet being put on the data by those who compile it in this form, although this is coming. The most effective visualizations of such data are those pertaining to traffic and movement, for they provide a sense of motion of the functioning of the city in real time.[7] Other feeds, such as the weather or what is trending on Twitter

(Weibo in China) are often of serendipitous value, for they cannot easily be used in city planning other than as background information. In the Wuhan city exhibition, there is a dashboard monitoring traffic as it is moving in real time in the city, where the state of congestion on the system is graphed as flow maps. Pictures of this from webcams mounted on the roadside and a simple analysis of the patterns of movement over the past twenty-four hours are shown. This data is, in fact, supplemented by GPS data in real time from seven thousand buses, fifteen thousand taxicabs, and the almost three million or so users of the metro system, whose journeys are recorded from their RFID payment cards. The exhibition reflects a strange mix of past, present, and future—a veritable digital outlook on what Wuhan intends to achieve.[8] There are no hang-ups here about the fact that all of this is control, planning, and management from the top down, while at the same time those affected are being informed through exhibitions such as this, which present the world from the bottom up.

THE 3D DIGITAL REPRESENTATIONS

There is an iconic photograph taken toward the end of World War II of the town planner Sir Patrick Abercrombie and the city engineer James Paton-Watson, the designers of a plan for the reconstruction of the English naval port of Plymouth.[9] There they are shown surrounded by a variety of visual images and artifacts that define the city's future. They are pointing at a 3D model of the reconstructed town, but around them are various graphics such as traffic-flow maps and statistical bar charts, which provide a comprehensive picture of the various functions of the town that underpin its form. There is nothing remotely digital about this picture, but as we have already described in previous chapters, the material shown in the charts and flow maps now constitute the essence of digital computation based on mathematical models and much of the new urban science that we introduced in previous chapters. The architectural models of urban form, however, have taken much longer to become computable. Michael Goodchild, in a review of the history of maps that form the 2D essence of these pictures, argues that the reason why it has taken so long for maps and architectural models to become digital is largely because it was never considered that maps involved calculation. Calculation was the rationale for the first computers, and it took many years for the idea that we could calculate or rather compute the geometry of artifacts such as cities to be realized.[10] As we have explained, it was the development of the visual paradigm through computer graphics that established the fact that computation is as much about pictures as it is about numbers, and this realization took time.

However, the lodestone of the digital city is the 3D model. It lies at the heart of how we communicate ideas about the city to any of the nonexpert and even expert publics who are the targets for plans that are to improve the quality of life in their cities. Architectural models provide considerably more impressionistic visualizations than maps, and from the earliest days of computing, there was a force to think about icons of cities in ways that could be more easily communicated than models made of traditional materials—wood, cork, plasticine, and so on.[11] In chapter 11, we noted some of the early forays into computer-aided design, in particular Ivan Sutherland's Sketchpad, built at the MIT, where he demonstrated in 1960 for the first time not only how one might sketch objects in 3D but also how the designer might interact with the computer in developing designs for objects yet to be invented.[12] During the 1960s, progress with developing computational realizations of building geometries was fairly slow, largely due to the media on which such designs were visualized. Storage tubes with frame buffers were in their infancy, and most designs were ultimately transferred to line plotters that had difficulty dealing with large blocks of color or gray tones. Nicholas Negroponte, who founded the Media Lab at the MIT, developed a variety of such visualizations as part of his Architecture Machine project, but the focus here was more on how AI might be used in architecture than on visualizations per se of buildings and cities in 3D.[13]

There are two key problems in programming a computer to produce a geometric object in 3D. First, there is the problem of hiding lines that lie behind the view in question with respect to the perspective as seen by the user or anyone else viewing the scene. Such lines could be easily identified and then removed, but this took considerable amounts of computer time through identifying and sorting, and the problem could only be solved satisfactorily once computers had reached the point where such hidden-line elimination could be done routinely.[14] This only became widespread once computer memories and speeds reached the point where such routine calculation was possible, and this was at the end of the 1970s. The other key problem was the display of 3D content itself. Primitive storage tubes had been available from the early 1950s, but there was a twenty-year gestation period until the development of the minicomputer in the 1970s where such devices could be easily attached to such machines. In fact, Nicholas Negroponte referred, somewhat disparagingly, to computer graphics and art in this early era as equivalent to a Calcomp contest after the company that manufactured most of the plotters and held a competition for the best graphics and art each year.[15] It was not really until the invention of the PC and frame buffers were incorporated into screen memories that the computation and display of 3D objects became routine, and this did not happen until the 1980s.

Geometric models of buildings gradually became more elaborate and reached the point where they could be replicated within one scene many times. Rudimentary complexes of buildings were plotted on more sophisticated storage tubes throughout the 1970s, and by the time the minicomputer became widely available, many powerful storage capabilities that enabled 3D scenes to be routinely plotted and visualized using new methods of hidden-line elimination, state of the art rendering, and fast manipulation to produce basic animations all came on stream. The 3D city visualization that took the world by storm was published in the early 1980s by the architectural firm Skidmore, Owings & Merrill (SOM), which had been responsible for two of the most iconic skyscrapers to date: the John Hancock Centre (the second-tallest building in the world when it was first opened) and the Sears Tower (now the Willis Tower) in downtown Chicago, which from 1973 was the world's largest for more than twenty years. The firm is responsible for Dubai's Burj Khalifa, at present the world's tallest. In 1982, they began work on computer animations of all the significant buildings in nine US cities (New York City, Boston, Chicago, Washington DC, Denver, Houston, Portland, Los Angeles, and San Francisco), producing some remarkable flythroughs at rapid speeds, where the user literally flies between the buildings. The animations are generated from multiple perspective views rendered in sequence along the fly paths but with no hidden lines eliminated, thus providing wonderful wire-frame visualizations that impress on the user the complexity of building-frame construction in all its fine detail. Many of these early visualizations are now available as Vimeo and YouTube movies.[16] The SOM animations have provided the momentum for a movement that, to date, has produced many 3D digital models for most of the world's largest cities. In fact, it is impossible to count the number of such applications, for standard software now exists to produce such flythroughs rapidly, aided by rapid advances in computer graphics, gaming, parallel processing, new methods of rendering, and new forms of spatial data storage, access, and retrieval.

There are a number of quick fixes that can be developed to build 3D models. A favorite strategy is to divide the picture into 2D cross-sections and order these so that they are plotted as the viewer would see them, that is, in the order from the back to the front of the viewing frame. This is referred to sometimes as the "painter's algorithm" after the idea that each background is painted on top of its predecessor in the scene until the front of the scene is reached.[17] Other methods are sometimes invoked to compress more distant detail, so that the computation of whatever is in the frame is minimized.[18] One of the great advantages of 3D digital models is that the digital world gives us unprecedented powers to manipulate the content of the system in ways that can add and/or subtract from any view of the data, or picture of the scene, thus enabling models of the past and the present to be easily transformed into models of the future. Unlike traditional media that is hard to alter, where maps and models have to be constructed afresh each time the scene or view changes, digital models can be easily altered, and thus new development and redevelopment across many scales can be handled easily and cheaply. Moreover, once a digital model exists, it can be improved as new data emerges, while the fact that it can be reproduced at almost no cost means that the model can be easily transferred to other users either in situ or remotely across the web. One of the most attractive features of such models is that they can act as a visual repository for data as long as it has a spatial referent. Thus, data can be visualized within such models in lots of innovative ways, software even running models in real time within such visualizations. An increasing use is in communicating development plans to agencies and municipalities involved in managing, controlling, and enforcing standards pertaining to physical planning, transportation, security, rent control, and so on, as well as enabling emergency services to function more efficiently.

Once the PC revolution began, the games industry dramatically pushed computers into graphics, with many rendering tools being widely used to create 3D visualizations as backcloths to various kinds of object animation.[19] The work at Xerox PARC, which we have referred to several times in previous chapters, also forced the development of graphical user interfaces incorporated into Apple's first Macintosh machines, which led the industry in visual applications from then on. AutoCAD, now Autodesk, was one of the many companies that began with packages devoted to rudimentary drafting and 3D design, but it was really in the rapid evolution from mini-computers to workstations that 3D modeling really took off. Both Sun and Silicon Graphics led the field, with developments in real-time rendering and graphics transformations that were embodied in their hardware, thus generating very elaborate 3D visualizations of cityscapes, while specialist companies such as Pixar led the march into computer movies.[20] By the time the web was in prospect in the early 1990s, the notion of graphics being embedded into web pages came onto the horizon and 3D VR came into sight.

This is another story that we will outline below, but by the millennium, the idea of digital computer models of entire cities had become a reality. In fact, surveys back then revealed no more than fifty applications worldwide, but the number was growing rapidly, and before the decade was out, it was hard to count the many significant applications that had been developed. The scaling down in software costs, the availability of digital maps at the land parcel and street level, and the existence of very fine-scale remotely sensed data from LIDAR provided the impetus for the construction of many 3D models in the manner that SOM had demonstrated two decades earlier. The hidden-line elimination problem was long gone, and the emergence of many different media other than the computer screen for 3D modeling began to dominate applications. We will introduce these below, for 3D models now simply provide the foundation for many new approaches. CAD is now dominated by new tools, new software such as 3ds Max, Rhino, Grasshopper, FreeCAD, Sketchup, and others involving parametric modeling and computer-aided manufacture. Many different kinds of computation are being linked and merged from GIS to mathematical simulations of city structures to online participation, with new ways of crowdsourcing as well as delivery in real time across the web.[21] These we will describe in what follows, for they are taking this field into new pastures, largely unrecognizable from the early days of graphics and modeling.

AN EXPLOSION OF VR SYSTEMS

Once we were able to generate 3D representations of cities quickly enough for users to be able to fly through them and of sufficient size and detail for those who were exposed to them to consider them representations of cities that looked realistic enough that they could identify with them, the field really began to take off. In the last twenty years, 3D digital city models have become the basis for a wide range of new technologies that enable users to interact with these models in diverse ways. Before we sketch the many applications that now exist, we need to organize them into various types, for what we are talking about here is a very wide range of visual renderings, multimedia, and ways of interacting with such models. These continue to proliferate, and the field is thus confusing and perhaps confused. Just as there are almost as many urban analytic methods and models as there are analysts and model builders, not to say users, then there is now as wide a range of 3D model types, which we will refer to, rather loosely, as different kinds of VR.

Starting from our 3D model of the blocks that make up the buildings of a city, the blocks themselves can be used as a geometric filing system to store the data that composes the city, the data being the attributes of each block defining location in terms of the three dimensions.[22] From this data, various geometric measurements can be made, which are used to define not only size but also how far one can see, areas associated with daylighting, and, of course, densities when associated with such attributes as to how these building blocks are populated. It is also possible to construct aggregates of these blocks and associate them with other data that is defined for administrative units such as census areas and tracts at higher levels of scale. We can populate these models with physical networks such as streets, subway lines, internet links, and so on and associate these with flows of traffic, email, and many other types of supply chain. Such models have also been used to simulate actual flows on these networks in real time, while the user is also able to rotate, zoom, and fly through the model in user time while traffic can be flowing in a simulated real time. This mashing of times can be disorientating, but it is like many examples of VRs in which different simulated realities are compared across different spaces, geographies, and temporal sequences.[23]

There are many ways of rendering these 3D models. Sometimes, what seems to be preferred is, in fact, the wire frame, without any hidden-line elimination, but more usually the blocks are rendered either in one color, possibly with shadows being cast, or in many colors, reflecting the materials that actually compose the buildings in the scene. Sometimes, photographs are associated with the blocks, and sometimes, they are used to render the block itself with photorealistic detail. Other links can be added to the block such as sensors that display not only flows but also continually changing attributes such as the weather, social media activity, and so on. These, in turn, can be linked to web pages, thus turning the 3D scenes into a portal that delivers even more information that makes up the visualization, other than that which can be easily associated and pictured as pertaining to the building. In short, such visualizations can be linked to other software, usually through the net if that software is remotely accessible or directly if the visualization is on the desktop.

The media through which digital cities and VRs can be portrayed is now diverse. At the very beginning, it was line plotters, line printers, and oscilloscopes that quickly evolved into frame buffers. These were embodied in the PC and workstation screen that enabled 3D scenes to be displayed and animated, and an obvious extension of this was to sets of screens arranged so that they might deliver the richest experience possible. Screens have turned into video walls of various kinds, and holographic displays have also been attempted. To an extent, these remain somewhat specialist, as, very often, VR rooms such as CAVES and theaters have to be constructed, and these tend to be idiosyncratic to the problems in hand. Interacting with such media through headsets such as the Oculus Rift is now quite common, but this sort of apparatus is again pretty specialist and clumsy to the extent that it is difficult to see such interaction becoming routine.[24] VR theaters exist where various media, particularly video walls and screens, are used to portray visualizations where many users can interact in a normal setting, thus aiding public participation, collective crowdsourcing, and a variety of interaction experiences that focus on problems to be discussed and jointly solved.[25] The Decision Theater at Arizona State University in Tempe, which first opened a decade or more ago, is as good an example as one can find,[26] but in many of the Chinese city exhibitions referred to above, such theaters often form part of the media used to communicate ideas about city histories and their futures.

Embedding ourselves and our 3D city models into one another through various VRs has rapidly become possible since the millennium. To an extent, these extensions are part of cyberspace, which have evolved into the metaverse, which we noted in our first chapter and which we will speculate on a little more in our last. Much of this has come from gaming and, more recently, internet gaming, and the highest profile for these are built around virtual worlds, software in which digital content of many kinds can be imported. For example, 3D city models can form the landscape of these worlds, while users appearing in the form of avatars can interact with the objects in such worlds in many different ways.[27] A user appearing as an avatar can navigate within such worlds, exchange information via text and/or speak with others who have a virtual presence as avatars in such worlds, while also pointing to websites and other internet sources. This illustrates the principle that once a networked world becomes visual, one can exist within it and engage in many, if not most, activities that can be performed in the real world, as long as such content can be made digital.[28] Of course, most of our behavior will remain non-digital, but everywhere, these kinds of augmented realties are being introduced as novel ways of interacting socially, engaging in economic exchange, and enabling structured problem solving and design.

We have noted many times the idea that when computers become small enough in physical terms, they can be embedded in almost anything and in any profusion. This lies at the basis of the IoT, where many objects can be connected and embedded at many different scales down to the finest, where microchips can be woven into physical materials, linked to the internet, thus offering connectivity in what appeared in the past as the most unlikely of locations. It is easy now to see how the universality of digital computation can flow through any devices that can be connected, at least in principle. In a world of limitless connectivity across any and every scale, life will be very different from the world of the past as well as the world we are living in now. Although devices are being rapidly connected, it will probably require a different type of networking from that which currently makes our connectivity possible, that is, fiber-optic wires and wireless routers, and new forms will be required to enable us to build a comprehensive skin of endless connectivity. What this future will be is impossible to predict, but we appear to be well on the way to a world where unlimited connectivity will be the norm. It will be a world of augmented reality, where the distinctions between the real and the virtual begin to dissolve as we build ever deeper layers of software into our natural and built environments—in short, in constructing the metaverse.

Augmented realities that mix the real and many different versions of the virtual in different combinations are becoming commonplace, although the media to access such environments is still quite primitive, consisting mainly of mixing screen technologies of different kinds. Appearing as avatars and making more inclusive virtual environments accessible, for example, through glasses, headsets, and such paraphernalia, is still quite cumbersome. Until it becomes more seamless with respect to ordinary behaviors, it will always remain strange, probably focused mainly on specialist usage and, of course, games. One prospect is that the real and the virtual no longer pertain to the present but rather to a version of the future, and already there are examples of augmented realities that mix future designs with the present. For example, new technologies illustrated in terms of human interactions in the movie *Minority Report* show the principal actor being bombarded by digital information pertaining to his preferences, which reflect different futures as well as different presents. This idea of blending the future into the present is something we will take up a little later in part IV of this book, but it is in the context of VR that such experiments are beginning.

From one perspective, all digital representation and simulation implies a VR in which the world is abstracted and then manipulated using various kinds of computation. Indeed, one could argue that any kind of abstraction based on our ability to stand back from the real world and make sense of it involves constructing a VR, although in terms of digital representation, the way we manipulate such a world is very different from anything we did prior to the introduction of computers in the middle of the last century.[29] Digital computers enable a generic mode of manipulation that involves producing changes to the virtual world that are impermanent and disappear once the devices are switched off, notwithstanding that such simulations can be captured and archived digitally. Strict VR systems, however, are somewhat narrower than digital representation and simulation. In general, for the virtual worlds we will introduce here, these are characterized by an intrinsic link between the computable world itself and the way we, as users of this world, interact directly with it. Unlike the wider spectrum of computer simulation and urban analytics, VR systems engage the user interactively in that the user changes the world, with the world also changing the behavior of the user. VR systems always involve this two-way traffic; they are thus said to be "immersive," meaning that users interact directly with the world itself and each other in this world. But as we will see, there are many different kinds of immersion across a continuum, from relatively passive interaction to complete immersion where there is little obvious difference between the user and the world itself.

Our implicit definition of VR is clearly illustrated by its short history. In the mid-1980s, Jaron Lanier, who is the accredited popularizer of the term, devised various artifacts that enabled users to interact directly with computable objects in which they could manipulate their own actions using devices such as the data glove and 3D goggles now called the "headset."[30] Right from the word go, this kind of interaction came to characterize VR, with the user seeing, pointing at, and touching the digital media in diverse ways, hence changing its form through what was quite literally hands-on computation. Just as games were integral to driving the PC revolution forward, they were also instrumental in pushing VR. By the mid-1990s, gloves and related devices were followed by the development of entire environments where the user could be immersed in a physical space such as a theater or CAVE whose walls were controlled by computable media and in which users could interact with this media in 3D through various modes of touch.

Most VR systems until then were based on physically connecting users to machines, but by the late 1990s, once computers had more or less converged with communications through the internet, VR came to be extended to networked environments. Online games emerged, while virtual worlds were constructed using network software producing environments that, although local, were controlled by users at remote locations. In particular, the emergence of virtual worlds represents a synthesis of networked communications, 3D virtual environments, and interactive gaming platforms where users can appear alongside each other as avatars. As we have already noted, these avatars can interact with other computable objects, some controlled by other users acting autonomously with respect to other objects in the scene but with some objects simply behaving, being programmed from outside the immediate VR scene. Such worlds enable many different kinds of feedback between humans and computable objects that generate realities that are very different from the real world but which enrich the real world through such interactions as well as providing environments to explore future worlds yet unrealized.

Currently, we can distinguish between several types of VR environment that link users to computable representations and simulations in a two-way fashion. First, there are stand-alone personal environments, which represent the lineage from Lanier's data glove to contemporary headsets such as Samsung's Galaxy Gear VR. These are primarily used for gaming but are now completely affordable and represent the wave of the future in personal

VR. High-end versions such as those produced by Oculus VR might also be networked, and it now goes without saying that all the systems available can exist on the desktop in networked form.[31] Second, there are purpose-built environments, such as VR theaters and CAVEs that enable large numbers of users to participate in a virtual scene (usually, but not always, without themselves being virtual), where human–human as well as human–computer interaction is important. These systems are increasingly used for professional and scientific purposes, although they can be used for gaming. Third, there are virtual worlds that exist on the desktop of networked environments in which human users can appear as avatars alongside computable objects with which they interact.[32] The Unity platform provides such a virtual world, with the focus on many uses from gaming to professional digital design.[33] Fourth, there are mixed environments, which consist of a meshing of human interactions, global network access, gaming environments, and general-purpose access to the digital world, often using various human–computer interactive tools such as those based on point and click, or even on sensory inputs. These are hard to classify, but they do represent the most widely used of all VR environments, largely because they consist of stitching very different forms of human–computer interaction together. Last but not least, there are hybrid analog environments, where digital representation is projected back onto the real world and where human interactions take place with both analog and digital representations of the same phenomena. Sand tables, physical data tables, even touch tables, and all kinds of surface access represent the contemporary realization of these developments.

In our descriptions that follow, we will begin with traditional stand-alone desktop environments, which traditionally have been associated with single users but which are rapidly being ported to the networked world. Our focus will be on examples of 3D representations of cities that extend the idea of the digital map and immediately embrace the key idea in VR as the user moves through the scene. Motion is central to VR, usually through the way users interact with the media but also in more functionally based virtual geographic environments through processes that are embodied in the scene.[34] We will then move to networked environments, which give equal importance to the user and the scene within virtual worlds, either on the desktop but increasingly across web-based portals that define the ways in which such worlds are configured. We will not spend very much time dealing with purpose-built hardware–software environments such as VR theaters, for these have become commonplace. In many professional and now retail environments, large-screen displays with some multimedia are widely available, and even TV technologies are embracing VR in the home. The cutting edge is now much more focused on usage and the processes of engagement with the media than on the technology, and in the examples that we will introduce, we will focus on these features. We will then discuss mixed-media realities, which are often called "virtual geographic environments" (VGEs), and illustrate how we can integrate different kinds of media in desktop and networked form. Last but not least, we will move back from the virtual to the real world, illustrating how we might think of VRs as analogs of the real world where we mix VR and ordinary human engagement. As we have implied, VR is now everywhere in contemporary life, accessible from the most common of our devices, our TVs, and our smartphones. What we will provide here is a snapshot of this world as it applies particularly but not exclusively to the geography of the city.

VIRTUAL WORLDS: MIXING HUMANS, COMPUTABLE AGENTS, AND GEOGRAPHIC MOTION

By the early 2000s, the relevant hardware and software made the construction of 3D environments routine. GIS systems such as ERSI's ArcGIS had by then incorporated 3D content and the functionality to display it within their software, while companies such as Autodesk had extended their CAD software into systems such as 3ds Max, which could quickly generate city-wide scenes through which the user might fly.[35] Quite large-scale 3D city models with thousands of blocks could be constructed in a matter of hours if terrain, building polygons, and height data from LIDAR were readily available.[36] Google Earth released its 3D globe in 2004 and opened it up so that users could begin populating this with 3D building blocks. Models built for London by our own group are typical in that they use the detail of the built environment quite sparingly: out of about fifty thousand building blocks in any viewing frame for this model, only some 1 percent are rendered in high detail. The purpose of this kind of VR is to provide a 3D version of the 2D map—to populate the 3D content with attribute data pertaining to the building blocks while also enabling users to layer different surfaces across the scene associating the underlying geometry with other key geographic layers such as pollution, flooding, and so on. Such models in Google Earth also enable us to use all the functionality of that reality (such as Street View) as well as layers of data pertaining to points of interest, traffic flows, and so on, which inform users about what is in the scene.[37]

The long-standing idea of using these 3D models to visualize the aesthetics of the city is not the main focus for this kind of modeling, as the interface is designed as a working link to the kinds of data the user can generate on the fly. As we noted earlier, the focus is more on using these models as portals to retrieve information from data, as well as analytics that are useful for interpreting the city in ways that enable a deeper understanding than purely emphasizing the city's aesthetics. Various analytics are accessible through such interfaces, while the model can also be populated by importing movement streams generated from data and predictions from agent-based models of pedestrian and traffic flow. This is one of the features that the Virtual London (ViLo) metaverse developed for the center and inner area of the city enables. Another feature that has emerged in these kinds of 3D block model involves populating them with data that is not usually associated with 3D representations. In the development of real-time streaming of data from computers and sensors embedded in the built environment or used in a mobile context to capture media through smartphones, such data can provide a visual animation of how geographic spaces and their attributes are continually changing.[38]

Rather than providing an animated map—and such maps are also part of the wider domain of VR—the 3D model can be used to tag data that changes in real time with respect to its building blocks. If we then associate the numeric values of the data to the building heights—in effect, a mapping of real-time locational data to the locations associated with the building blocks—we can vary the building heights as some transformation of the real-time data.[39] ESRI's City Engine software can be used in this way, where the heights of the buildings are proportional to the number of people located at those buildings, these numbers being based on many different varieties of occupancy from, say, the rents that people pay to the number of Tweets being sent, tagged to the geographical coordinates of the building blocks in question.[40] Although there are plenty of illustrations of how social media data such as Tweets vary spatially across the city, for example, we have not seen anyone link any of the global databases of Twitter data to 3D visualization in real time, but this is now eminently feasible.

Our examples so far have mixed various computer technologies in diverse ways and do not show any of the purist focus of VR characterizing earlier developments. In our extensions to 3D technology on the desktop, networked systems enable agents and real users to be mixed, while real-time streaming of data associated with the emergence of the smart city is now providing a strong direction to such virtual worlds. Earlier, we noted that in city exhibitions before the Digital Age, city models were built of traditional materials and were used in participatory discussions between the many stakeholders and the experts involved. We pointed to the picture of the town planner, Abercrombie, and the engineer, Paton-Watson, examining the model for reconstructing the naval base of Plymouth after it had been destroyed during the World War II.[41] Much of this detail can now be replaced by digital models, CAD models for the city model, mathematical models for simulating traffic, statistical models examining the employment structure of the town, and so on. But the whole scene can also now be easily imported into a virtual world, with Abercrombie and Paton-Watson appearing as avatars, controlled, of course, by themselves or even by others, mixing the real with the virtual, blending these conceptions in many different ways. This we consider is fast becoming the norm in cutting-edge VR systems.

There are many types of virtual world in which these realities can be constructed, such as Second Life, which gained enormous momentum during the early 2000s.[42] But continual innovation almost guarantees a fast turnover of software systems that make use of new media, and one of the platforms that is now widely employed is Unity 3D. Working prototypes of this platform exist where the media is the virtual 3D city through which a user can navigate and upload and download attribute data associated with any location that has been populated with such data. Unity is not strictly a virtual world in the sense of Second Life, but this is simply a matter of taste, for Unity is essentially a games engine that has multiuser capabilities through the usual kinds of networked systems that enable many users to interact in a virtual environment. In fact, VR systems such as Unity tend to be more restrictive than massive multiplayer online worlds, but much depends on the extent to which the user requires rapid motion in navigation.[43] In the somewhat more applied worlds that the digital technologies associated with the computable city pertain to, shoot-'em-up capabilities, central to many game engines, are not required, and thus there is an increasingly wide choice of platform.

Many virtual worlds do not focus on navigation per se but rather on how we can capture and import information about the experience of the real users and how this can be incorporated in virtual movement through the scene in question. Imagine a user or a participant walking through a scene and being equipped with sensing devices that pick up noise and pollution that is experienced as the user encounters different sensory experiences relating to what they see and hear and what they are exposed to. This kind of data can be captured via mobile sensing devices, which are portable, linked to smartphones that enable the sensory data that is captured to be imported continually into an archive of data on the web that is thence picked up by the apps on the smartphone. This then controls the user as an avatar, providing immediate feedback as to the way the environment is perceived. It seeks to encode the avatar's behavior into a form that is influenced by the sensed data.[44] In short, the virtual world and its avatars behave according to the real data that is imported into the world, and users are thus able to condition the behaviors of the virtual agents. The prospect is in sight of generating much better behavioral data from this kind of experience, which, in turn, produces much more realistic navigation in virtual worlds.

All we can illustrate here is some sense of how such virtual environments are built as an impressionistic collage of devices. With the emergence of low-power and inexpensive sensing and wireless communication devices, crowdsourcing and mobile sensing has the ability to monitor a wide variety of environmental change and human activity at the city scale. Using geo-sensor and mobile data collection, massive and timely environmental data can be acquired for simulation and analysis through a virtual world. To achieve this, portable kits based on a sensor, mobile applications, and an external network server collect environmental information from specific sensors and communicate via Bluetooth, transferring the measured data (e.g., air quality, temperature, humidity, noise, etc.) to the user's phone in real time. The mobile sensor, along with a handheld sensor system, make up the basic environmental data-collection station. As an integrated platform for data management, analysis, and representation, sharable cognitive spaces can be built that help in enhancing our understanding of environmental protection, sustainability issues, and spatial awareness.

VGEs are essentially digital environments that integrate diverse VR technologies but with strong graphic imageability. They also link input data to process simulations of many kinds.[45] They are beyond representation per se, although they invoke all the main features of computer cartography, GIS, and 3D visualization that we have described in previous chapters in this part of the book. The key to VGEs is that they simulate processes in the geographic environment as well as engaging multiple users in analysis. Unlike VR game environments that may be networked to involve many users, VGEs are more analytic in their usage and enable professional and research users to engage in analysis in a coordinated but active way. They thus represent a form of crowdsourcing but without the frenetic activity that characterizes many games, while they also require users to be expert or at least to be exposed to learning and generating greater expertise about the projects and problems of their concern.

VGEs can extend models of air pollution to the regional scale, where such models show that environmental pollution has a wide sphere of influence, high occurrence, strong outburst events, and intense spatial and temporal incidence due to global climate change and related human activities.[46] Its impact on regional ecological security and sustainable development is substantial, and thus a collaborative system for sensing and measuring its impact is required, and this lies at the basis of collaborative VGE, where we are able to carry out distributed environmental simulations. These experiments not only help construct environments that are difficult to build but also save costs and resources in developing policies to mitigate the dis-benefits that such environmental pollution generates. The system is a convenient way to package information for experts and decision makers who are able to communicate through VGEs and collaborate in designing and testing the impact of environmental policies. Moving to national and global scales, VGEs tend to be less visually interactive but nevertheless embrace many more decision makers of different types, reflecting local elements in the global picture. Global change is both a complicated and comprehensive issue involving population, land use, the Earth system, and multiple agencies, reflecting many kinds of mission and politics. In this context, human activities are affecting the Earth system at an unprecedented rate.

Meanwhile, urbanization, the continuing but changing population explosion and its consequences for migration coupled with an excessive use of resources, has severely hindered sustainable development. Utilizing a multidimensional representation of the geospatial system, VGEs let us visualize the spatiotemporal variational processes of Earth system mechanisms, which help define the impact of human activities on the global system. VGEs that encapsulate global change are being built as open platforms for collaborative simulations and evaluations focused on model integration, reuse, optimization, and collaboration. Model integration considers how scales that are mismatched can be fused, how different VGE models developed by different teams with different development styles can be integrated and reused, while flexible patterns of interaction and collaboration for different modes of collaborative modeling and optimization can be developed. These systems are rapidly being developed at the present time and promise to extend VR technologies way beyond their traditional game-like usage, merging more generally with a massive range of new technologies, from real-time sensing and the streaming of big data to global decision making. These involve many different renditions of digital Earth informed by models and data that are essentially being developed in remote locations, accessible through cloud computing. This again extends all these technologies into a form of metaverse, which we might define as the Earth's emergent digital skin or even a new digital atmosphere for the planet.

THE REAL AND THE VIRTUAL: MIXED, AUGMENTED, AND BLENDED ENVIRONMENTS

Earlier in this chapter, we illustrated traditional iconic models and how these had become digital forming the essence of 3D virtual environments such as those that characterize the virtual city. In the rapid move toward the digitalization of society, there is now a feeling that something is being lost by separating the virtual from the real, and there are a variety of initiatives in VR that are moving these technologies back to the material world. Many have found that it is extremely difficult to engage users to their full extent in a purely virtual world where they appear as avatars, even if their presence is controlled by themselves from remote locations. In short, users are much more receptive to discussion of the problems that virtual worlds seek to represent and motivate if they are able to do this in the context of the real world. In this context, the VR theater is important. To this end, there are various ways in which physical projections of the virtual world onto the material might be used to structure this kind of engagement. For example, touch tables of various kinds, projecting data and simulations onto the physical medium of the screen as a table, the table as formed by sculptural surfaces such as sand or by harder physical media such as wood or plastic, are being used in ingenious ways. Many groups have produced analogs that involve digitally generated content that can be projected back onto a real scene manufactured from conventional materials—a physical map or model that engages people in discussion, analytical thinking and design in ways that are much richer than solely within a virtual world. We referenced the sand tables, which are quite widely used in the city exhibitions in China. One of the most elaborate is that developed by the Redfish Group in Santa Fe, NM, to simulate the hazardous diffusion of wildfire and associated phenomena, where a CA model simulates the spread while the user activates various policies to stop the spread by pointing at the displayed visual functions in the media on the surface of the sand.[47]

There are many other blends of the virtual with the material. A person can fly through a virtual world easily built from Google Earth, for example, in which various other media exist, navigating by performing various actions—essentially moving through the scenes as a bird might fly. Facilities such as this can be built using Microsoft's 2010 Kinect—the motion sensor for the Xbox 360 gaming console—which provides an intuitively obvious medium for navigation without any obvious controller other than the human user.[48] This is a kind of augmented reality, but for the same scenes, more immersive forms can be developed where the user wears the Oculus headset, whose goggles portray the 3D experience of flying through a virtual city. As we noted above, these technologies are now quite affordable as, for example, in the Samsung Gear VR, and there are now countless ways of enabling many users to coordinate and share experiences using such technologies. Touch tables, despite the somewhat difficult problems of controlling for the touch, are improving and, combined with material objects, are very useful to show animations of an underlying digital map or geometry onto which objects can be placed. A good example of such an augmented reality is in using material objects to activate and control a simulation of movements across a city region. There is such a model of the London riots in 2012 using agent-based simulation, whose media is projected onto such a table. To manage the riots, there was a considerable police presence, and it was possible to test out policies experimentally by repositioning the police agents as icons on the table. The agents' positioning above the table and their activation through pressure/touch enabled the riots to be contained, as well as the diffusion of the other agents to be manipulated. To this was added a cost–benefit calculation associated with minimizing the damage that was done by the rioters. In this sense, it is easy to see how users of the system might configure policies that optimize the role of policing, with such VRs also having the potential for many other purposes.

We are still in the midst of a digital revolution that is working itself out with respect to the merging of the virtual and the real in the metaverse, while in the next part of this book, we will explore how important the organization of technologies are to the development of a digital world where the real blends with the virtual. The concepts of the cloud, the platform, the convergence of the real and virtual, the material and the ethereal, and hard-ware and software in its widest sense will dominate our discussion as we seek to provide speculative insights into what digital society will become as the momentum for the smart city continues to work itself out. As part of this transition, VRs now dominate usage to the point where the traditional applications that were largely immersive are now being augmented by all kinds of media juxtaposed in diverse and often unusual ways. What we have not explored here that is almost bound to be significant in the next decade is the spreading out of computers into ourselves. The spaces we inhabit are being rapidly computerized, but our own bodies will be the next frontier as modern medicine continues to embrace computation. Medicine as software will continue in the treatment of illness, but the notion that we will embed computer hardware into ourselves is likely to generate all kinds of unprece-dented and strange applications that will move this field on once again as we find new ways to communicate with one another. The moral and ethical considerations will be massive, of course, and are likely to be highly disrup-tive. We will stop short of this here, but it is always in prospect in the discus-sions that we will initiate in part IV of this book.

IV PLANNING AND ORGANIZATION

Our story so far has told us that the evolution of information technology has been dominated by successive miniaturization, exponentially increasing communications bandwidth, and ever-increasing speeds of access. This has led inexorably to computation becoming central to all kinds of organized activities in the arts and the sciences and in government, entertainment, and business. During the 1990s, the idea of computation linking hardware, software, and data across networks was consolidated around the World Wide Web, but in the early 2000s, another sea change was thrust upon us. As soon as computable devices became individualized and personal (as, for example, in smartphones), this became one of the dominant ways in which information could be captured, generated, and communicated. These technologies have led to a multitude of platforms that offer a wide variety of information services from social media, online commerce, education, health, and retailing, with business practices everywhere beginning to fuse with information technology.

In the last part of this book, we sketch the genesis of these platforms, first noting the convergence of computation across networks that released a technology surge in the early twenty-first century. In chapter 14, we introduce the new building blocks that now form the foundations of the wired society, while in chapter 15, we explore the emergence of platform economies that are powered by universal computation and increasingly accessed using small pieces of software—apps—that are reconfigured incessantly. The deluge of data that we are now beset by has sharpened many age-old problems involving privacy, confidentiality, and surveillance, and although our focus in this book is not specifically on these questions, they do act as a foil to our wider discussion of the digital transformation.

Our thesis that the world in general and technologies in particular are intrinsically unpredictable has been graphically demonstrated by the recent pandemic that has rocked the world since early 2020. Only in late 2022, did it show signs of becoming permanently endemic, but it has led to many changes in the way the digital transformation is working itself out. The dominant forces based on the extent to which populations want to live near or far from their places of work and which are contained in the standard model that has dominated the way we have thought about cities for the last hundred years is under severe scrutiny. We will draw these issues together in chapter 16, where we will explore the contours of what idealized cities might look like, particularly in a post-pandemic, computable world. In chapter 17, we reinforce the themes that have focused on the unpredictability of what comes next, thence pointing a path to the future.

14 THE TECHNOLOGICAL CONVERGENCE

In an extreme view, the world can be seen as only connections, nothing else.
—Tim Berners-Lee (2000)

THE WORLD WIDE WEB

When Tim Berners-Lee made the final touches to his protocols for enabling us to hyperlink between remote computers, thus inventing the web, and Marc Andreessen and his colleagues at Urbana-Champaign invented the first of many graphical user interfaces that provided user-friendly access, the stage was set for dramatic growth in connectivity. This not only embraced science but also extended to entertainment and eventually to interactions of many kinds, evolving rapidly from email and various social media that the web quickly spawned. The decade before the millennium involved a multitude of initiatives, which evolved the requisite degrees of interoperability between networks and machines so that, by the year 2000, most people who had some form of email could send messages and eventually files to one another. Connectivity was the name of the game. Like the growth of the railways in nineteenth-century Britain noted briefly in chapters 1 and 7, the great weight of investment in broadband and related network systems that emerged meant that a boom and then bust of immense proportions beset the world. The dot-com boom, as it was called, led to many new companies being fashioned around the internet, already by this time a vast free resource, providing the world with fiber optics, which, in fact, laid the groundwork for a massive surge in new technologies once the millennium began.

To an extent, you can see the history of the last eighty years since 1937 as falling into two distinctly different periods. Until the invention of the web, the entire period was dominated by the development of hardware and then software, both eclipsing network telecommunications a little, but with a very strong focus on making it possible to compute remotely across distance. In short, the elements of how we compute across many scales, how we link computers, how we use them to communicate data and messages, as well as how we organize storage and how we index all this so we can search efficiently—these were the key elements that a networked society requires, and they were being rapidly put in place during this earlier era.[1] In contrast, since the early 1990s, hardware has advanced rather differently, becoming ever more personal, while the memory, capacity, and speed of computing have continued apace, as enshrined in Moore's law. The last thirty years have been dominated by networked computing across many spatial and temporal scales, while platforms have emerged for many styles of large-scale computing. These consist of plugging many computers together, leveraging the power of the crowd or even the power of the net itself to develop new and focused applications with important ramifications for everything from ecommerce to social services, transportation guidance, and online financial services. This list seems endless.[2] Hardware developments have simply continued the miniaturization of computer power and related chip-based technology as well as the ever-faster convergence of computers with communications. These developments, in turn, are increasingly involved with embedding computable devices into objects that are able to communicate with one another and with us. In the 1990s, the focus moved toward the IoT, but computation was fast becoming so all-pervasive that choice was often no longer an issue with respect to what kind of computer or network would be needed.[3] It is now the relative convenience of access on grounds other than hardware and software that has become significant.

We need to stand back a little at this point and consider the kind of information infrastructure that had been built by the time the web was invented. Essentially, the model that evolved was based on single but always multiple-purpose machines—universal machines—that slowly and surely first embraced their users by providing direct lines to the terminals that users employed to interact with the machine. In the 1970s, at Xerox

PARC in particular, users became connected to one another, not within their own machines, for this had always been possible since the earliest days, but rather with their individual machines being networked together. At the same time, as this local interaction emerged, so too did global interaction with the beginning of the ARPANET, and it is no accident that the most obvious of applications soon emerged: electronic mail. Users could email anyone who was connected to the network in question, anywhere and at any time. Networks then became connected to one another at the same scale or at different levels above and below one another, and by the early 1990s, at least in principle if not in practice, everybody on a computer could be connected to everyone else, for diverse purposes, but specifically for email.[4] All one needed was a computer, a telephone line, and a modem that enabled you to couple and then uncouple your machine to the net.

To some extent, email was frowned upon by the established orthodoxy, but it was the killer application that gave the internet its most basic rationale. In essence, this is simply the realization of Metcalfe's law, and it is no exaggeration to say that we are rapidly approaching the point where everyone is connected to everyone else. With some eight billion people on the planet, this implies more than sixty trillion possible links or pathways. This is the great entanglement of networks that was unleashed about thirty years ago, and all subsequent activities and initiatives involving the internet are based on this super exponential growth in connectivity.

The reason why we have been at pains to spell out that the internet has such profound network effects is based on the obvious fact that any one node is reachable from any other, although the connectivity is now so dense that this is hard to measure.[5] As we will note later, this total connectivity makes it possible to influence passively if not actively what happens at any node from any other. There are many things that are defined in this way, ranging from social networks to the way one searches a network whose nodes contain information. In fact, when the internet first emerged, the more sophisticated bulletin boards such as Online America, Compuserve, Prodigy, and so on, which were almost portals to the internet in their own right, were not able to grasp the potential that this emergent phenomenon had for the provision of a multitude of information services. Because they advertised themselves as the on-ramps to the information highway, these internet service providers were pretty much walled off from the wider net, and only slowly did it dawn on these companies that with their millions of subscribers, they could quickly propel themselves into the rapidly expanding global domain.[6] The downside at the time was that many of their subscribers wanted to send email to others who were on different networks, and this kind of interoperability took a while to arrive.

The internet, of course, raised the stakes with respect to search. Scientific computing, which drove the original need for computation quite quickly, gave way to the manipulation and storage of data, where computation became largely associated with data processing. In fact, many now see computation as mainly associated with data from which information is generated. For many years, it had been clear that the most powerful computers tended to generate the greatest volumes of data, and as soon as the internet left the halls of academia in the late 1970s, there was clearly a focus on building systems to facilitate searching increasingly large databases. To an extent, the first browsers were primarily search engines, although their basic role was to visualize the inputs and outputs from such search. Mosaic and its reincarnation as Netscape were primarily focused on these applications, notwithstanding that in those early days, there was little focus on how a good search engine might be designed. Many such engines basically started up to great flourish but bombed quite quickly, as there was no real business model to support them, or rather the typical model that gave the engine away free to small users and then specialized in marketing and sales to larger seemingly cash-rich corporations was never a money spinner. The only one to emerge during the late 1990s that almost lasted is Yahoo, which was started about the same time as Netscape in 1994 as an index or directory of websites, ultimately graduating into search as well as a whole range of other web-based services. Yahoo, through mixture of good management, good luck, and a shifting focus, outlasted most of the competition, and only now at the time of writing has it has entered its twilight years.[7]

The message, of course, from the early web that it might constitute a new medium for making money is somewhat salutary. Suddenly, it would seem, in 1993, as soon as Tim Berners-Lee got permission from CERN to make the basic software free,[8] all and sundry assumed that here was medium where, if you could reach everyone, you should be able to get rich quick. Of course, all this was a long way from what actually happened. As we will see, even though the history of the last thirty years has been very different from the preceding fifty, with the focus being less on technical scientific development and much more on how one might build good organization to exploit the infrastructure of the wired society in the best possible way, the times have been particularly turbulent. How one might grow a sector to get the balance of all these different types of infrastructure correctly positioned was no mean feat, and the 1990s, up to the dot-com boom, were perilous times for many companies that sought to exploit the power of the web. Not only was there the question of how the internet should be made computable if one was to use it to search for something, market a product, or simply buy a service, but there was also the issue of what the net should be best positioned to achieve.

In the early days, Tim Berners-Lee was adamant that it should be used exclusively to share scientific work, but this was simply disregarded by those who saw it as a great money-making machine, as we will detail in our subsequent discussion of how the web now lies at the core of the wired society. There are many avenues we could follow, but at this point, it is instructive to look at two of the originators of web-based activities: Google, first founded in 1998, as an example of search, and Amazon, set up four years earlier in 1994, as an example of using the net to sell. These are two of the biggest publicly traded companies worldwide today, following an astonishing rise in business power entirely due to the technologies we have been describing in this book so far. Indeed, the top companies worldwide at the time of writing in terms of market capitalization are all those that define the Digital Age—namely, Apple, Microsoft, Alphabet (Google), Amazon, Tesla, and Meta (Facebook) in that order. These and the hierarchy below them are the basic platforms that power the wired society, at least in the industrialized West, notwithstanding the rise of similar corporations and companies such as Tencent, Baidu, and so on in China and Asia-Pacific.[9]

SEARCH: THE INTERNET'S ULTIMATE APPLICATION

One of the implicit conditions that Tim Berners-Lee first prescribed for any web page was that it would not be just a repository of data or information but rather an index or catalogue to other web pages.[10] To this end, the web page would usually point to other relevant web pages, with their relevance being implied by the choice of web links made by the person who designed the page in the first instance. In this sense, one might define a web page as a set of pointers—itself included—to other web pages where the addresses in question are URLs. It is rare now for web pages not to have such pointers, although the web in its most popular form first and foremost can be seen as a repository for information rather than a pure index to other web pages, which is a somewhat nihilistic idea in any case. Thus, the first web pages were sources of information with links to other web pages, and from any one page, the tree that could be built up by tracing links in successive order provided some measure of the complexity of the website itself, as defined from its source or home page.

As soon as computers reached the point when very large volumes of quantitative and textual data could be stored in memory and hence manipulated, the need for systematic search became essential. By the time the web started, the internet itself contained pointers to many remote sources of data that could be accessed using pre-web technologies, particularly using various file transfer protocols. It was hardly a surprise when the first applications were designed to automate search. The first search engines were primitive in the logic they used, and if you wished to find a website relating to some piece of information, you would clearly begin by finding all those websites that contained that information. You would then take each of these websites and, in turn, search the tree of websites emanating from that site where the information also appeared. At some point, you might find that you would revisit the information on the source site, and in this way, you would go round and round the web, extracting and sorting the locations for the relevant information.

Where to stop the search would be an issue, of course, and how you would figure out where to direct the user to the website that was most important with respect to the information under scrutiny were problems that the first search engines simply ducked. Engines such as Excite and Altavista, which were popular during the so-called browser wars when Netscape was pitted against Microsoft Explorer, gave respectable but not particularly stellar results, and in general, their output was not very stable. It often depended on where one began one's search.[11] In short, the structure of the web was not taken into account whatsoever. In a world where the relevance of information was a positional problem, any algorithm that did not take account of the positioning was destined to fail or, at best, produce poor results. A website and its links may be key to some piece of information, but if the websites that were generated directly and indirectly from the source in question were important, then some way was needed to combine all this information into a form that might order or rank the websites in terms of their relevance to the user.

Onto this scene burst Google in 1998. The origins of the company began in Stanford University's computer science department a couple of years earlier.[12] Its two founders, Sergey Brin and Larry Page, were worrying about the reliability of search, and they proposed to themselves and their advisor, Terry Winograd, that a good topic for their respective dissertations would be an exploration of how one might develop the logic and algorithms that would make a good search engine for the web. Their essential starting point was that the web needed to be treated as a total network, and any ranking of a website should take account of the relationships of not only that website to all others but also all others to that website. This is the problem of simultaneity in graphs, where the number of relevant links *from* a node needs to be explored with respect to the number of relevant links *to* a node. In this context, then, it is a problem that is as old as the hills. It involves extracting the relative importance of nodes based on what is linked to what and this needs to be done for the entire graph. It is a basic problem in linear algebra involved in finding a unique solution to the system of equations that defines the stability of a graph. However, when a graph has a million nodes (which was the case for the web in 1998 when Google first became a significant search engine), this is a nontrivial problem, and it requires very fast algorithms to effect solutions. When you put filters or constraints on the nature of the information such as making sure that only those websites with x and y attributes are selected, the problems become trickier. When it is really blown up with all variety of constraints, rather special algorithms have to be invented to cope with such search. Moreover, in solving this kind of problem, one has to have access to the entire network, the entire web. This was the problem that Brin and Page posed to Terry Winograd in 1996 so that he could requisition on their behalf sufficient computer capacity and processing time that they could begin to catalogue, that is, index the web with respect to all its pages.

Out of this in 1997 came the essential algorithm that lies at the basis of Google's search: PageRank.[13] At first, Google did not have any kind of business model in mind, but the fact that the software was distributed freely and the fact that the search engine was so good and elegant saw the number of users grow super-exponentially, first from ten thousand searches a day in September 1998 to three and a half million a year later in 1999 and thence eighteen million in 2000. It more or less doubled every two years until 2010, after which the increase began to slow, but as noted earlier in chapter 5, it now accommodates more than two trillion searches a year.[14] This sort of growth could not be accommodated by Stanford, even though the web engine was simplicity itself. It did not load images and was based on clean simple text, which was parsimonious in the extreme, all designed for fast web access. However, through some judicious advice, the company spun off from the university in 1998, still without a business model but with some rather elite investment from venture capital that came in from the wings. The sheer scale of growth and the fact that the avowed mission of the company was to index the world's information provided enough rationale for it to be given some breathing space. Moreover, although the embryonic company was founded in the slipstream of the dot-com boom, its early reluctance to go hell for leather for growth gave it a singular advantage to build up strong foundations. These were resilient and virtually untouched by the nuclear winter that followed the end of the browser wars and the dot-com bust. The company remained in the throes of academia for longer than most spin-offs, and this gave it time to devise a strategy for how it might make money,[15] which would, of course, be through advertising on the very web it was searching.

The problem with advertising on the web—that is, labeling websites with banners and icons that point to places and other websites where users might purchase a product—is that the process of labeling can be somewhat indiscriminate. In the early web, it was not possible for anything more sophisticated than a user clicking on a website and an advertisement being displayed for the advertiser to be charged. It did not matter whether the user actually acted on the information or even took the information in. This was the same somewhat indiscriminate blanket model that had been used for decades by the industry. The problem with the web was that very often a user would visit a page many times, but the advertiser would get no sales from this, as the overriding nature of the website itself was of interest, not the advertisement. However, if the advertiser made a payment only if the advertisement was activated—that is, if the user clicked on the advert to bring up the advertiser's website or whatever related information, then this was a possible basis for charging. You might visit a page a thousand times and see an advert each time but never click on it, but if you visited a page three times and each time you clicked on it, you would secure three pay-ments for the owner of the website. In this way, the site's advertisements could be highly focused and this pay-per-click model quickly became the industry standard in terms of advertising.[16] Google did not invent this, but it refined it remorselessly, and now it is an essential element in the whole process of search and hence the source of much of the company's wealth.

When you visit a city or a country, at some point, you need a map. But the links that you traverse using a search engine also provide a map of the internet. This is a set of links that defines the topology, not the conven-tional Euclidean geometry of the net. It is not Google Maps, although one could speculate that a more conventional map of the net rather than just a very large network graph might exist. I am unaware of a version that is solely geographic, but I am sure one could easily be produced as a submap of Google Maps and linked geographically to more conventional graphic forms.[17] Here, for the first time, we do see some order in the material that Google is able to pull up for filtering to enable it to be represented spatially. To an extent, if the user places his or her own geographic filters on the search, what will emerge will contain locations. In fact, this is possible for certain Google products such as Google Maps itself, which can be searched for niche geographic attributes such as the locations of buildings, restau-rants, government offices, and so on. We noted this before in our discus-sion of mapping in chapter 12.

EXCHANGING, BUYING, AND SELLING: THE ONLINE ECONOMY

There is one last application that exploits the full power of the web that we need to describe before we shift gears and examine how the internet is being used to replace traditional markets. This involves another peer-to-peer or rather node-to-node application that so far has been the fastest-growing web service ever: Napster.[18] This platform reached eighty million users from a standing start in mid-1999 only to last no longer than twenty-five months in all from its inception to its demise. When you have a network where everything can be connected to everything else, then it is possible to think of an exchange of some sort between any node in the network and any other. If you want to swap something or buy something with some-one at another node, then you have to find out who they are and what they are selling. Then you need in place a mechanism for enabling the sale or purchase to be achieved, but the fact that the network exists at all gives you this possibility. You could charge for the product or exchange it for free. If you had already paid for it and it was easy to copy at almost zero cost, you may well make it for free. If it were a digital product, then it would be all the easier to ship it across the net using file transfer to the recipient, and in this way, a thriving modus of exchange might be established.

To an extent, it depends on what the object in question is, a good example being music. If you own some tracks that are on a hard disk, in principle, you could gift these to another user, who, in turn, could give you their own tracks, and the trade would be beneficial if both received the tracks they wanted but did not already have. This sounds like heaven, for, in principle, you could receive all the tracks ever produced and so could everyone else. The catch, of course, is copyright. Even if you were to own the tracks, having purchased them, this is only a copy of the original, whose tracks are not your copyright but rather the copyright of the author of the tracks or the record company in the first place, whoever—but they are not your copyright, as your purchase of them is only a license to play and hear them. Napster was set up to facilitate this kind of file sharing. It went viral in late 1999, particularly among college students, who all seemed to have access to virtually unlimited broadband and enormous hard disks for storage.[19] It led to some college authorities restricting Napster because up to 85 percent of all college computational capacity was being used in some places. At its high point, in mid-2000, it had reached almost one quarter of all college students in the US.

Of course, the music industry did not see it coming, but once it arrived, there were strenuous efforts to shut it down, which, in fact, were largely successful, but by then, the damage was done. The floodgates were open: if you have digital goods, then the market for them is changed beyond all recognition with respect to traditional goods, for if those goods are available on the web as pieces of information, then they can be shipped by anyone with access to that information, anywhere, at any time. The process is virtually impossible to police, as is currently witnessed by the vast amount of unsolicited digital information that anyone with an internet presence is now being bombarded with. The most important issue is that, in many cases, the digital goods in question do not have monetary value, their ownership is problematic, and their transfer is frictionless. This is beginning to change the rules where digital products rely on extensive exchange, especially in publishing, where the rules governing book purchase in the old economy are gradually being rewritten.[20] What form any book will eventually be in, whether digital, print, or audio, is a choice that, once made, will determine the marketplace it may reach, and this will very much depend on the media that it is available in.

There is nothing very sophisticated about setting up a website where products are listed for sale and where potential buyers can initiate a trade using various means where their sale can be effected digitally. How easy or hard it is depends to some extent on what is being traded. Products that are relatively compact, come in multiple runs, and can be easily differentiated from one another in bulk tend to be good bets when it comes to online selling. It is thus no accident that one of the first online websites was Amazon, which was first set up in mid-1994 by Jeff Bezos, who decided to specialize in selling books. To an extent, there is a massive markup on new books, and thus the economics of online book selling always looked good from the word go. Moreover, books are homogeneous products. For a given title, each book is invariably the same, and thus you are unlikely to discriminate between two separate books on the basis of their content, color, or size. There is also a degree of unambiguity about a book. The distribution of books too, in terms of their retail market, was also controlled by a small number of massive suppliers, and thus it was relatively easy to identify where one could source a book from. The fact that increasingly books were becoming digital did not escape Bezos's attention either, and currently, the proportion of sales of books in digital format on Amazon is well over half the world's total. Bezos's early entry into the field and the fact that his original company became one of the biggest in the world is probably more due to the single-minded nature of the business and of Bezos himself. The company's extension into selling consumer products of many different kinds and its recent subsidiary businesses in Amazon web services have been particularly motivated by the company's need for massive storage, both digitally for the platform itself and in terms of the very large-scale physical warehousing that is required for physical shipment of its non-digital products.[21]

The site that Bezos set up had some clever features, of that there is little doubt. It loaded quickly, like Google—with minimal graphics at first—and each page represented essentially a book whose information directed the reader or user to other pages where similar books were located. In time,

other features of the book—the author for example—were used to tag various web pages and to provide summaries of an author's works. Early on,
the notion of reviews based on readers' comments on the quality of a book
were recorded, and on this basis, good works were likely to receive more
positive feedback and vice versa. But the key statistic was the relative popularity of a book, with many different numerical measures being computed
and displayed on the web. Sales are the obvious one, but page views are
another. Although the website may be uncomplicated, it was only possible because of software that ensured good security, which was developed
primarily by Netscape in the early dot-com boom. Indeed, every other website where digital credit cards are used to enable transactions to be made
uses this technology. The key innovations in Amazon really lie behind the
scenes in the remarkable organization that developed to enable books to
be sourced, delivered to Amazon's now massive warehousing locations, and
then shipped onward to customers using a variety of traditional transportation services.

What, of course, this led to almost immediately was the demise of the
high-street bookshop, not completely but in sufficiently large numbers to
produce a gap in the geographical distribution of retailing.[22] In many countries, the bookshop was one of the clusters of community facilities that
represented an anchor point in the town or village, and its loss has very
definitely left a hole in the urban fabric. Barnes & Noble, with its book
superstores, managed to hold out against Amazon using a mix of online
offerings that copied Amazon, and its early association with AOL helped,
but even the various management-restructuring activities over many years
have not kept the group in profitability. At the time of writing, the company is still vulnerable, although it appears to be on a path to recovery,
but throughout the bookseller industry in general, the power of online
sales involves products that one does not need to handle before purchase.
There has been substantial reinvention of the bookshop during these years,
with new functions appearing within them. But in general, the lesson is
that when it comes to online sales, there are now many products that lend
themselves to new forms of organization associated with the internet that
collapse the sales cycle, thus reducing costs across many activities and dis-
rupting the traditional structure and locations that have previously served
to ground these activities within the urban fabric of the town and city. For
much of what we have said here about bookshops, the same might be said

for retailing in general. Entirely new ways of marketing what, in the last analysis, are physical products are emerging all the time.

SOCIAL MEDIA: SOCIAL NETWORKING

As soon as you have the rudiments of a network, which means potential users at locations—nodes—who have the ability to pass and/or manipulate information with respect to other users located at nodes, which link to the user in question, you have the basis for social interaction, which is now generically called "social networking." In some contexts, social networks appeared many years ago within single computer installations—usually in universities and research centers—where users were able to send one another messages as a form of internal email, thus building up some basic community. As soon the ARPANET began to grow, the same kinds of network appeared, albeit in rudimentary form, and in the cluster of networking technologies that began with the invention of the PC, predominantly in Silicon Valley, a mixture of the hacker culture with the hippie lifestyle spawned the idea that people should communicate through their machines.[23]

One of the first online communities to develop was known as the WELL (Whole Earth 'Lectronic Link). The futurist Stewart Brand, author of the *Whole Earth Catalog*, first published in the late 1960s, was deeply involved in the culture of invention around the PC with respect to the notion of bringing information and computers to the people. He worked with various inspirational inventors in Silicon Valley such as Douglas Engelbart and Ted Nelson, who we came across in the previous chapters in association with the development of computer interfaces developed at Xerox PARC, and in 1985, he established one of the first online communities—the WELL.[24] Remember that, at the time, the Ethernet was still behind closed doors at Xerox PARC, and the protocols for the ARPANET were in their infancy. The idea of networks, of course, was in the air, with various newsgroups beginning to feature on the emergent internet and basic email systems such as BITnet being used within academia. The WELL was, in some respects, the first example of a virtual community, and in this sense, it had a pseudo-geographical referent, although this was never exploited. It began life as a dial-up bulletin board, of which there were hundreds in the late 1980s, but its focus on linking together Silicon Valley aficionados such as members of the band *The Grateful Dead* gave it real exposure. Arguably, this was not a social network per se because in social networks people tend to link to others whom they know or know of, whereas virtual communities are drawn from people who have similar interests.[25] But it does represent a focus on linking together electronically, which is one of many types of network that began to grow in parallel to the internet in the late 1980s and early 1990s.

It is worth noting that in the 1990s, alongside the development of the internet and the World Wide Web, there was a dramatic growth in the science of networks. The first social networks in formal terms were identified in the 1950s in political science and social psychology, and basic ideas in graph theory were invoked to explain things such as how central one actor was to another, as well as the potential for different actors to reach one another through a minimum number of links. The small-world problem originated in this community—the notion that in any network of any size, you could reach everyone within six links—the so-called six degrees of separation—due to the way networks developed organically as well as randomly.[26] The big obstacle, however, to the development of social network theory was data. Only very small communities could be sampled with respect to figuring out who communicated with whom and for what purpose, and thus most empirical studies were confined to very focused networks such as those involving niche groups of politicians, those seeking jobs, those involved in community action, and so on.[27] It really took the development of the network of networks—the internet—to blow all this wide open and to enable scientists, for the first time, to begin to mine the deeper structures of very large networks. In parallel, network science, which essentially was based on the statistical study of networks, emerged, and networks began to be classified into different types with different structures.[28]

All this, of course, was history and background with respect to the development of online networks spinning off from the web. What constituted a social network or indeed social media was and still is a moveable feast. We cannot catalogue all the networks that have emerged, evolved, and died since the web began, but most began after the dot-com boom once the web had moved from its passive phase, which was called Web 1.0, to its interactive form, Web 2.0.[29] The most complete social networking site or platform is Facebook (now called Meta), which, unlike many other sites such as Myspace, evolved in quite a careful sort of way, although its growth was astounding. Facebook emerged as a site where college students in quite a cloistered environment—beginning at Harvard where it started—could reach out to one another and find information pertaining to their classes and indeed their interests based on the courses that they were studying or about the college dorms in which they lived. Mark Zuckerberg, its founder if not the originator, pursued a fairly single-minded goal that was to replicate a virtual community with close ties to the real community, and in some respects, the original context was geographical. It was started in early 2004, and within a few months, it was spawned to other Ivy League colleges. The growth was massive: it reached a million monthly users by the end of 2004, and once it was opened to the world at the end of 2006, it had twelve million. It grew to half a billion by 2010, and by the end of 2012, it was a billion. It doubled again to two billion by mid-2017 and reached 2.9 billion in early 2022, and although its growth is rapidly decelerating, Zuckerberg's continued vision that this would be a site that would reach all the world's population has almost but not quite been borne out.[30]

What exactly does this platform do that has attracted so many so quickly? At one level, all it is a website with many pages that are customized to its users, covering just about everything that relates to social and economic life. One of the key drivers of the site is that its users need to link to other users whom they know. You may meet new people on Facebook, but its roaring success has been due to the fact that its core relates to users who know one another and who are able to extend and enrich their virtual network in ways that bind them together. As it has grown, it has differentiated itself into nondomestic and nonsocial activities, particularly business. Such a massive platform, of course, attracts all kinds of media, and it is now a source for news, even a maker of news (and fake news) in that other media link to Facebook and use its content. For a long time after it had started, Zuckerberg resisted taking the company public, and in some senses, apart from its acquisition of Instagram and WhatsApp, it has built on its own base around its own software. Where it is headed is unclear, as indeed is the case with all social networking sites. At the time of writing, the markets are not looking favorably at this sort of Big Tech. Doubtless, it will continue to evolve, but it is possible that a sea change in how internet users see their acquisition of news about their friends or anyone or anything else will change its nature. The history of this kind of software and organization has always been somewhat volatile, and so it is extremely difficult to get any sense of where the internet is heading in terms of social media.[31] New social media platforms are emerging all the time, for example TikTok, the video-sharing site, which was barely heard of before the pandemic began in early 2020 but now boasts some one billion users.[32]

Before we steer our discussion in the next two chapters back to the infrastructure of the wired society, how this will affect the nature of the city, and how we will live within it, there is one last social media site that is worth noting, and that is Twitter. Twitter, as well as other similar sites, lies almost at the opposite extreme to Facebook in that it achieves its impact by allowing its users to distribute ("tweet") short text messages (SMS) of no more than 140 characters in length to their followers—those who are associated with the user in question. Followers can pass the message on to others, their own followers, by "retweeting" and can associate any message with a generic subject area that would start a link or relate to an existing thread of messages that pertain to that particular subject. One can see the messages of any of its users unless the user chooses to hide certain tweets, but in general, it is an open platform to everyone. You do not need to be part of the platform to look at the information that is tweeted. That is all there is to it. It might seem highly nihilistic in that, on the face of it, there is no prescribed content, and it is up to the user what is tweeted. But as a source of instant news, it is useful, and its media can be used in many different ways. Twitter was started in 2006 and grew rapidly at first, but its growth now appears to have tailed off.[33] In the first year, it grew from the initial tweet to a million tweets per day in 2007, a hundred million sometime in 2010, and by 2013, it had reached half a billion, but its growth appears to have stalled at this level, with no more than this in mid-2022.

It is hard to know what can be extracted from these kinds of media. The notion of being able to actually trace geographically social networks is problematic. In the case of Twitter, all one has is the name of someone you might notify when retweeting a tweet and the locations of your followers from their profiles. A very small number of tweets can be geo-located with respect to the fact that the user's GPS is occasionally switched on, but only about 1 percent of all retweets use this facility actively. Moreover, whether someone tweets is often dependent on where they are, and where they are does not necessarily determine what they tweet about. Constructing the social network in the manner of Facebook, where it is up-front and explicit, is almost impossible with messaging services such as Twitter, and thus the impact this has on the city is extremely hard to gauge.

There are many other internet services, of course, that form part of the broad embrace of social media, and those that are most important in terms of having more than half a billion users that we have not detailed are Whats-App, Tumblr, and Instagram. The Chinese sites, in particular WeChat, Tencent, Zone, Baidu, and Weibo, are very significant,[34] but the situation there appears unstable due to ongoing crackdowns on the Big Tech companies in China. Of course, Google itself has spawned many different media sites, ranging from storage and software through to sites that collect more personal information and news such as Google+. All of these continue to evolve, but there are couple of key conclusions from this more recent history. First, there is now a very strong disconnect between hardware–software and social applications of this kind, and it is almost as if hardware is no longer a problem. We will come back to this later because, in certain contexts, there are exceptions where hardware is still extremely significant. Second, many of the features that we have noted here in our discussion of new media will recur again and again in the rest of this book as we further fill in the detail of what the computable city is all about. Much more recent developments will find their place in those discussions.

15 THE TWENTY-FIRST-CENTURY TECHNOLOGY SURGE

Six decades into the computer revolution, four decades since the invention of the microprocessor, and two decades into the rise of the modern Internet, all of the technology required to transform industries through software finally works and can be widely delivered at global scale.

—Marc Andreessen (2011)

BUILDING THE WIRED SOCIETY

By the first decade of the new millennium, most of the basic technologies were in place to seriously develop the earlier visions of the wired society that futurists such as E. M. Forster to Arthur C. Clarke, James Martin, and Alvin Toffler, among others,[1] had speculated upon. Computers, their software, the focus on sorting, searching, storing, and manipulating data, and the ability to do all this at a distance provided a superstructure that could be quickly exploited to reap the advantages of technologies that promised to increase the wealth of our cities, their sustainability, and, more generally, the quality of life for all those who could embrace this new future. Computers had been miniaturized to the point where they were being embodied in the most personal of devices—in phones and digital notebooks—and in the first decade of the new century, the smartphone appeared, marked by the launch of Apple's iPhone in 2007. Manipulating data, creating information, and accessing the internet thus became something that one could do from a handheld device, and in this sense, a truly global society was about to emerge. Moreover, once cameras also merged with smartphones, the visual world merged with the numerical and the textual, just as it had done a generation or more before with the advent of the PC. The scene was set for a very different world from the one in which computers had first emerged.

So far, we have avoided any discussion of the dark side of these technological revolutions, and although these final chapters will hint at some of these, we consider these to be part of a parallel debate that anyone absorbing this material will need to consider. Questions of who are enfranchised by these digital technologies, who are disrupted by them, whose privacies are invaded, and a host of negative implications of the new technologies, as well as positive ones, pervade this debate. We will address some of these and, as a last gesture, leave pointers to where these debates may be joined in full. With the coming of this new world, none of the segregations and divisions in social structure that marked the old have disappeared. Gender, ethnicity, and wealth continue to dominate different degrees of access to the whole range of technologies that now dominate our working and social lives. At the time of writing, the global geopolitical situation has morphed to an energy war, where a very large percentage of the population in developed countries are rapidly being driven into poverty. The way technology divides young from old involves the differential abilities to employ new digital technologies, where the elderly, who have not grown up in world dominated by the keyboard and the ability to deftly download and piece together data and software now essential for everyday living, are being massively disadvantaged. Numerous information divides are being created. These are issues that we will briefly return to below, but they require a much broader canvas than this one on which to understand their wider social implications. This is not meant as an apology but simply as a signpost to another debate that cannot be joined here but which is nevertheless an essential theme in the discussion of how unpredictable the future will continue to be.

Marc Andreessen's quotation, reproduced above from his article "Why Software is Eating the World,"[2] suggests that a sea change in the nature computation began in the early twenty-first century—a change that is almost as profound as the emergence of the digital computer itself. This is largely due to the fact that Moore's law has finally taken us to the point not only where everyone has a computer[3] but also where computation is or soon will be literally everywhere, embedded in our own senses and sensors all around is, in our mobility, in our social interactions, in our work and play, in quite literally most tangible objects. The organization of computation is becoming ever more important to the point where computers are penetrating every fabric of society. The emergence of a new style of personal computing involving many small pieces of software—apps—which we now use routinely is changing the way we interact with machines and with each other.[4] We will begin by noting how organization at various levels, but particularly in business and in government, is being influenced by computation and the way in which automation is closely related to such organization, and this will take us into contemporary conceptions of computing as platforms that are quickly taking over the functions of work-related organizations. In parallel fashion, new software has quickly emerged, enabling us to recreate business and social meetings built of individual windows on the world for each of us who connect to each other through the net. The pandemic has dramatically accelerated our capabilities for meeting virtually while dramatically pushing us to remote working, shopping, education, health care, and a variety of online activities. All of this is likely to change the form but most likely the functions defining both the low- and high-frequency city. We will continue to speculate here, as well as setting the wider context for trying to anticipate the world that will follow this one.

TECHNOLOGIES WITHIN TECHNOLOGIES

The difficulty of figuring out the meaning of the computer revolution is that each technology is firmly but often opaquely built on existing but preceding ones. You cannot really explain the evolution of any hardware or software without knowing the details of those that proceeded it, as earlier techniques and tools combine and recombine with one other and with new.[5] Moreover, information technologies are continually scaling both up and down as the digital skin that now dominates pervasive computing[6] begins to combine the many components that we have described in previous chapters and which are identified above by Andreessen.[7] At the largest scale, software and data are combined into platforms that are global in their impact, as defined by McAfee and Brynjolfsson in their book *Machine, Platform, Crowd*, which charts the massive impact of these all-pervasive technologies.[8]

The rationale for defining this as sets of technologies that nest into one another in complex and convoluted ways is based on the fact that once the computer was invented as a consequence of the mechanical and electrical industrial revolutions, new technologies have come thick and fast.

The digital technologies that emerged in the middle of the last century marked a relatively clean break with the earlier industrial revolutions that were dominated by steam and electricity. The prior era was one of tabulation, where the physical motion of machines such as the devices made by Hollerith dominated transactions processing.[9] The subsequent move from atoms to bits represented not only an evolution of the technologies of the industrial revolution but also a transition from what in the past was a largely physical world to one that was becoming virtual. As we will see, in the new technology surge that is just beginning, there is a slight sense of a move back from bits to atoms as software is being embedded in everything from artifacts to animals, from man-made to natural objects, from built to natural environments.

What has been happening since the millennium is a massive convergence of all the technologies that have been invented over the last eighty years or so, driven by miniaturization and recombination, which have enabled layer upon layer of hardware and software to be compressed into forms of computation that are now barely understood by most of their users. This, you may say, had always been the case with mechanical and electrical technologies of the first and second industrial revolutions. You do not need to know the workings of the internal combustion engine to drive a car, but although sometimes we can operate quite easily all-pervasive technologies, sometimes this is not so easy, as the rate of change of this compression can be daunting. Almost as soon as one learns how to handle a new technology in the form of new software or a new device, another one appears, knocking the first out of the water. The evolution of software is now so fast and so all-pervasive in terms of apps and operating systems that virtually every time you switch on your smartphone, something new has been uploaded, usually without your permission. The rate of change is astonishing, and it continually takes us by surprise.

Thomas Friedman refers to this transition as being dominated by the speed at which complex software and hardware is being compressed and how it is all being connected. He argues that by connecting things together, the product cycle has massively shortened, and the friction of solving problems and designing new artifacts is being reduced all the time to the point where markets themselves are becoming frictionless. The machines that we use to access and manipulate such markets now involve compressing a myriad of technologies into devices that one can access very easily, and it is this compression of systems that makes the resultant complexity much easier for us to use. He says, "When everything and everyone becomes connected, and complexity is free and innovation is both dirt-cheap and can come from anywhere, the world of work changes." This, he argues, marks the transition within the transition, and it is as much an organizational one as a product of hardware, software, and networks.[10]

There are many ways of defining the threshold that divides the first part of the digital revolution from what is now going on in our contemporary world. For our purposes, a very convenient way of defining this transition is from a world where computing was largely seen as something that was separate from the world at large to a world where computing is deeply embedded within it. It took Moore's law a good fifty years from the invention of the transistor to reach the point where computers were small enough to provide universal access at extremely low cost. Once they were completely personalized, small enough to be mobile, wireless, and connected to anything with an internet locator, the world suddenly changed. To all intents and purposes, this can be dated to between 2007 when the iPhone was introduced and the years that followed when other mobile operators copied what Apple and, before that, Blackberry had introduced.[11] Communication and computation became mobile. Into this nexus came the idea that small computable devices and software could be embedded into many different kinds of object to the point where the line between the new hardware and its software and all the other man-made or natural materials used began to blur.

From around 2010, software began to enter the fabric of our material world in such a way that nobody could remain aloof from it. So many tasks suddenly became reachable to digital organization and hence computable, and for many functions, the use of computers became essential. New objects, practices, and procedures supported by software and dependent for their very existence on this software were becoming routine.[12] Prior to this time, most software, and its concomitant hardware, was gradually being rooted in the material world, meaning that the material world could no longer be designed, changed, manipulated, or even operated without such software. In fact, when computers were first invented, the use of computing machines in business was well advanced, and there was a growing industry devoted to transactions processing based on mechanical devices for computation that had originated in the late nineteenth century in the US from Herman Hollerith's development of punched-card technology for organizing the 1890 US population census. In fact, Hollerith's original company—the Tabulating Machine Company, acquired and renamed in 1924 as International Business Machines (IBM)—seemed almost ready and waiting for the advent of the digital computer, and unsurprisingly, commercial applications began quite quickly.[13]

In this manner, computers began to spread into organizations for routine transactions processing, which were initially based on payroll and other financial procedures that related to the operation of any business in which wages and profits were the dominant driving force. At the same time, by the mid-1950s, mainframe computers were being packaged into smart boxes that took up considerable space within any organization but which represented a very tangible asset that required their own manpower to operate, program, and manage basic routine functions and to link these to the key operations of the organizations in question. Clearly, the size of the organization was reflected in the scale of the computer system required, and the mainframe vendors such as IBM, Univac, CDC, DEC, Data General, and the host of such companies that flourished from the 1950s onward until the advent of the PC varied the machines they produced from the smallest—hardly desktop but with very limited core memories—to the biggest, which were essentially the first supercomputers. Profit-making operations, where speed and efficiency of transactions processing was essential,[14] were the key organizational applications creating these demands. But government too, with its nonprofit welfare mission, also set up substantial demands for such machines, in particular for processing social security, population, tax, immigration, and related welfare payments, as well as their own routine operations involving payroll.

By 1960, many organizations had computer centers that formed an important part of their administration, and if they had scientific functions, there was sometimes a pairing of computers between the administration and the scientific side of the organization. Although computers were becoming part of the organization of such firms and agencies, they remained largely separate in terms of their management. There was no sense in which computation was a central function for the whole organization, and the idea that computation could ultimately become the structure of an entire organization was a million miles away. This was despite the prescient speculations of Vannevar Bush, Alan Turing, and the original pioneers, who argued that computers would eventually scale down, would probably be able to communicate with one another, and would decentralize their office functions to the work practices of individual employees.

The big hardware companies such as IBM also had a monopoly on the creation of software, and most routine business operations were run using very standard software packages. But even in some firms and local government agencies, there was experimentation with writing computer programs designed to address specific tasks, which were often idiosyncratic to the particular organization involved. What did emerge during the 1960s and 1970s were large-scale data centers devoted to transactions processing, which became an increasingly important part of those organizations that were large enough to afford and sustain them. In particular, in government, municipal data centers began to develop information systems that focused on using widely available data pertaining to local taxation, land use, even traffic, as well as the entire budgeting operation of the organizations involved.[15] These systems tended to be particular to the locations and cities where they were developed and, in this sense, were prior to the development of more generic software such as GIS. Indeed, in the 1960s, what emerged were associations of professionals involved not only in basic transactions processing in organizations but often in scientific usage where large databases were critical, these forming the cutting edge for the diffusion of computation within many organizations.

Only when computers scaled down to the point where individuals could dedicate them to their own pursuits and only when they could begin to communicate with other computers and devices that enabled them to input data and generate outputs did organization become as important as hardware and software. From the invention of the microprocessor in 1971, this did not take long, and by the early 1980s, Apple, IBM, and a host of other start-ups had begun to manufacture what were then called microcomputers but which more generally became known as PCs (after the IBM PC). In the 1970s, as we have already noted, LANs were invented at Xerox PARC, and the rudiments of software for many office tasks emerged. Quite suddenly, in the early 1980s, the notion that even the simplest of office tasks involving typing letters, calculating budgets, even keeping files in digital format on peripheral storage devices such as floppy disks all came together, and a revolution in office organization began.[16] The core of this were the word processor and the spreadsheet—two essential pieces of software that dominate many routine office tasks to this day and have now spread out everywhere in such a way that we are all now involved in office organization, be it our own or part of the wider organization where we work.

It is hard to provide a sense of how quickly and how thoroughly this revolution took place, but within ten years, by 1990, it was almost complete in that most PCs were networked locally, external storage on servers that could contain both data and software was under development, and computing for such tasks as well as for scientific activities was now available on portable desktops—laptops—thus introducing the idea of computing on the move. At this point, it is fair to say that the organizational implications of hardware and software had penetrated most organizations to the point where they were slowly being considered as part of the organization itself. This was not quite the organization in its entirety—this would have to wait for even more scaling down and networking—but computation and communications at least had now become an essential part of most large organizations. In chapter 14, we sketched the way the internet emerged in the last decade of the twentieth century, along with wireless technologies that enabled many office environments to connect geographically across diverse locations. This occurred in such a way that many organizations that were spread out geographically could communicate quickly and efficiently, share data and information, and centralize computation in the form of processing power and software in such a way that they could generate economies of scale through networking. The client–server architecture rapidly emerged in fully-fledged form, and increasingly, organizations began to weave computation and networking into many aspects of their operation.[17]

During this period, large software companies specializing in databases and networking technologies such as Oracle grew, and although they never acquired the status of the new internet companies such as Google, Amazon, Facebook, Apple, and Microsoft, they drove the industry very strongly with respect to the way companies began to organize themselves around software platforms. The same might be said of the networking companies such as Cisco and many that piggybacked on the successive revolutions that threw computation to the wider edges of the information society. The stage was thus set for the emergence of some very different kinds of organization, now referred to as "platforms,"[18] where the business model is often very different from those in most companies and agencies of the past. The economics of these platforms reflect a very different economy too, as we illustrated in the last chapter. These reflect a networked economy that depends intrinsically on the products of a digital world, which are fast merging with the traditional physical products of our material world.

THE RISE OF THE PLATFORM

To discuss the emergence of the most significant organizations that have grown on the back of the digital revolution, we need to revisit the principles of how the digital economy is beginning to influence the production and distribution of products in the sort of economy we outlined in the last chapter. The first products that were produced and exchanged in such an economy were essentially informed by digital technologies. To an extent, the early internet economy was based on using new information technologies to generate products that, in themselves, mainly involved information—bits to bits—whereas the dominant modes of digital exchange now involve processes linking bits to atoms, with the prospect of the future information economy being based on using digital technologies to link any material (measured in atoms) or virtual product (measured in bits) to any other. The key issues in thinking of information products are, of course, that they are much more flexible in every sense than material ones. Information in the form of software and data, once created, can be infinitely subdivided and duplicated instantly, and in this sense, their marginal cost is near zero.[19] Moreover, the digital goods being traded can also be produced with complete accuracy. In this sense, perfect copies can be produced instantly in the case of software or data. McAfee and Brynjolfsson define these kinds of goods that are now being produced in the digital economy as being "free, perfect and instant."[20] This threefold distinction is the basis of what is fast turning out to be an entirely new kind of economy—one that is beginning to develop rapidly and which will come to dominate contemporary society for the rest of this century. In some senses, although we are now comfortable with the idea of the computer being a universal machine, there is still a sense of wonder and fiction—of magic even—in the continuing rapid transition from the material to the virtual.

The term that is increasingly popular to describe these systems is a "platform," which consists of networked software that enables us to produce new forms of information, with the potential to use that information to generate better material and information products.[21] In this sense, a platform is an environment in which users who are linked to it are able to communicate and disseminate information that can involve everything from business to academia to government to social interaction and entertainment and much else, where this is enabled through digital networking. In this sense, then, a platform is a software environment. There is a generic principle at work here that underlies many of the platforms that have emerged since the recent technology surge began about a decade or so ago. This is based on the fact that a platform links two or more networks, which, in their simplest form, are systems linking producers to consumers—suppliers to those who initiate demand. Both groups have their own networks that enable producers to reach consumers and consumers producers comprehensively, efficiently, and almost instantly. If you have a digital good that you wish to sell, then you can use the power of the network to reach consumers who, in turn, can buy this product if the price is right. Producers and consumers can do this in an entirely liquid way, implying frictionless markets, at least in principle, as long as there are minimal regulations that might distort the market. In this way, it appears that the market will find its own level quite quickly and efficiently, with demand equating to supply.

In the case of producers generating physical or material goods but across digital networks and consumers getting access to these goods using their own networks, then the products supplied are much less easy to supply quickly and often show characteristics that restrict the rapid process of increasing demand and supply for pure digital goods. In the case of these pure digital goods, supply can then be increased very easily, as it is costless, while demand can react just as quickly. The products are perfect in the sense that they can be reproduced identically. In the case of products that take time to produce, there are still positive feedback effects because the products are being supplied across networks. This increases demand, and there are more pitfalls in ensuring demand and supply are balanced in these cases. In fact, the economics of these markets are not well worked out at the present time, but nevertheless, the network effects involving free, perfect, and instant products are generating very new types of organization.

Many of our institutions that have dealt traditionally with information goods such as libraries, where the hard-copy book, journal, or magazine is their stock in trade, are moving in part to organizations that are largely virtual. Although it does not appear currently that hard copy is likely to disappear, much has become digital and can be distributed across networks, where perfect reproduction is possible and where access tends to be immediate. There are still quite dramatic distortions in these markets in terms of control, ownership, copyright, and access, but if one considers the process of production of a book, say, then once the author has spent their labor producing the work and a modest production and distribution cost has been fixed, such a product can be put on a platform, and consumers can acquire it digitally or can even simply read it online, where the cost declines in proportion to the number of accesses, reaching near zero very quickly. The cost, in fact, is controlled by the platform, and if we are talking about a product that can be produced perfectly and instantly millions of times, as bestseller books usually are, then effectively most of the profit is attributed to whoever owns and develops the platform. Amazon, which we described in the last chapter, built its business in this way.

The same kind of structure relates to other information goods such as software itself, which now is bought and distributed online. The massive profits made by the largest information technology companies such as Microsoft, Apple, Oracle, and so on relate to the fact that it is software that is their core business, not hardware, which is simply the means by which access to their products takes place. The hardware might be packaged in such a way as to draw consumers in with respect to design and distribution—the Apple stores and the company's attractively designed devices are obvious examples—but ultimately, it is the information that these products make accessible that is their core business. Music, of course, is another similar domain. In fact, what is beginning to happen is that such software companies are marketing their products in the form of a service—"software as a service" (SaaS) is the typical lexicon now used.[22] Such companies are using their platforms to license their products on a subscription basis, the products being centrally located, with the consumers who subscribe being assured of all updates and complete maintenance, where the software host takes on many operations that the consumers themselves require in adapting the software to their specific needs.

The platforms that have emerged around social media such as Facebook (Meta) and, to a lesser extent, Instagram, Twitter, WeChat, and so on draw consumers into what are essentially free services that enable them to interact with others for many different purposes. Initially, these are often for social interaction between friends, but as these platforms have grown, they have supported this growth essentially through advertisements, with the large scale and low cost of such advertising leading to massive profits. The average revenue per employee in these new kinds of company is enormous compared to traditional businesses that produce material goods; all the costs of production of the platform itself are tiny in comparison to the revenue gained from advertising. Their size in terms of their reach across the population is now in the low billions of world population, largely a function of the fact that these networks piggyback on the internet, which is essentially still free. The information goods that they sell can be produced at a marginal cost near zero, while their distribution can be almost instant. There are still limits on such platforms, as the world is only just waking up to the fact that when the internet is free, there are all kinds of undesirable effects emerging that range from extreme criminality to unsocial behaviors that destroy personalities and communal activities. In short, such platforms can generate disruptions that destroy well-balanced systems that have grown more slowly, incrementally, and adaptively over many years. This is nowhere clearer than in the case where information technologies are now being used to deliver traditional products of a material kind such as transport and housing. In fact, it is in cities that significant disruptions are clearly visible.[23]

In terms of our focus here on cities, we stand on a threshold where these new forms of digital organization are changing the form and the way we function in cities. To speculate on what is happening due to the recent technology surge, we will say a little more about the two platforms Uber and Airbnb, which we have noted in previous chapters. It is worth cataloguing here the way these platforms have grown and the extent to which they are changing the nature of personal travel and the supply of housing in big cities.[24] Uber was started in 2008 when its founders were in Paris unable to get a cab and decided the way forward was to solicit such services over the internet. Essentially, the platform is based on the notion that if you own a vehicle and wish to carry passengers for a fee as regular taxi services do, then if you make your position in any location known to the platform, the platform will put you in touch with a potential rider (or fare), and you and the rider can select to engage in the transaction or not. In short, the platform links its network of drivers, who all operate privately, with the network of clients, also private, through an app on the web, where the client can see the position of the available vehicle nearby and select one that they wish to travel in. You may think that this is a dangerous situation in that the private vehicle and driver do not have to meet the strict regulations that taxicab drivers need to meet in terms of their licenses, and there is no way of ensuring a safe ride. But in fact, the rider, as the customer, is usually mandated to rate the driver and vice versa, and this reputational system appears to work well enough so far in terms of ensuring safety. The problems with the system all relate to regulation, but it is easy to see how the system is almost costless and can generate enormous profits as the platform creams off a slice of the fare simply for providing the way in which driver and rider are able to communicate.

This is much the same with any of the new platforms. As Tom Goodwin said, "Uber, the world's largest taxi company owns no vehicles, Facebook the world's most popular media owner creates no content, Alibaba, the most valuable retailer has no inventory and Airbnb the world's largest accommodation provider owns no real estate. Something interesting is happening."[25] Moreover, Uber does not pay drivers but simply acts as the middleman to ensure that payment takes place, it does not own any road space, it does not own any internet, it does not really regulate—no wonder, in its present form, it is valued at many billions of dollars and is enormously profitable. But there are problems, and these relate to the fact that traditional transport services, not only taxis, which are usually regulated by municipalities, but also bus and even train services, are being affected by the emergence of these unregulated services. In fact, there is now substantial evidence of the decline of taxi services in places where Uber is largely unregulated. There are many other undesirable effects relating to serious disruptive effects on public bus services that are provided in areas where subsidies are needed to ensure any kind of service. Last but not least, insurance for drivers, the actual fee paid relative to what the market can bear especially when new entrants are coming to the market, as well as hours worked all raise problems that detract from the ubiquity and convenience of these kinds of services.[26] The economics of services such as Uber appear to be increasingly vulnerable.

Similar disruption appears to be taking place in the case of the Airbnb platform, which provides a service that links those who wish to rent their accommodation for short stays to potential customers wishing to visit places for similar short periods. These traditionally are marketed to tourists. The Airbnb platform, like Uber, puts renters in touch with those who wish to rent, thus linking the network of those providing accommodation with the network of those who require it. There is some regulation that has been introduced relating to length of stay, but in general, what appears to have been happening, particularly in housing markets where there are big restrictions on supply and particularly in world cities such as London, is that Airbnb has led to rises in housing prices in the rental market, as well as reductions in supply of housing to potential tenants who need accommodation over the long term. This is pitting tourism against housing, and again, the disruption tends to occur in terms of second and third order, not direct effects—that is, their impacts are unanticipated before such services come to be established.[27]

As these platforms depend on networking so that large numbers of producers and consumers can be linked together, they depend on advertising in one form or another. They also depend on highly decentralized patterns of ownership and local management. But most importantly, they depend on fast and immediate access to massive digital storage facilities. In short, they depend on the cloud as being one of the ways in which computing has spread out globally. Free storage for customers has to be paid for by the platform providers, but individuals are now gaining access to many cloud services, which can be quite profitable as more and more data that we wish to store is generated. The classic example involves digital photographs that we now produce routinely.[28] The number was estimated from ad hoc surveys as being about 1.7 trillion worldwide at the time of writing (2023), and this is growing at some 20 percent per year. The storage demands for this are enormous, and if one wishes to keep even a little of this data, personal cloud services such as Dropbox are required. In fact, Apple market a cloud service specifically for photos captured on their iPhone that prompts you to buy such services when the number of photos on your phone hits a certain threshold. In many of the examples we have recounted here, such services are now almost mandatory for the simple reason that, without them, the device or software on the device that you are using fails to work. These are fast becoming central to the future of the digital world.

A MULTITUDE OF SMALL PIECES: SOFTWARE APPS

At the other extreme from storage on the cloud, individual users store data and programs on their handheld devices, usually smartphones, using software now referred to generically as "apps." The term came from the first generation of software for the first microcomputers, which were referred to as "applications," or rather, in the case of new and innovative applications such as the first spreadsheet, VisiCalc, written for the Apple II, as killer applications. The association between these kinds of software and the fact that many of the first were developed for Apple was not lost on the company and its founder, and during the 1980s, Steve Jobs kept referring to any software application on the Mac as an "app."[29] By the time mobile devices emerged, the term had become uniquely associated with *App*le and the iPhone. The way the platform for mobile devices was developed, particularly by Apple, meant that it was closed to other vendors with different operating systems.[30] Thus, apps were written for one platform or another but were exclusively made freely available or chargeable for a small fee from the particular stores that emerged to distribute apps for different devices. In the case of Apple, this was the App Store, while for the other widely used device based on the Android operating system developed by Google, this was Google Play. Other platforms exist, of course, but Apple and Google are the main ones, with currently about 1.9 million apps distributed by Apple and about 2.5 million by Google.[31]

We are used to seeing these sorts of numbers in the millions for users involved in various platforms and the number of computers available, but it is somewhat remarkable that there are this number of pieces of individual software available as apps. In fact, in 2019, it was estimated that there were about two hundred billion downloads of apps from the various platforms and that this number was growing at about 10 percent a year. Apps generalize to every piece of software on a mobile device, including, of course, the basic applications that are used for accessing email, the web, and any function that links the device to the outside world. But apps can be divided into those that contain such links and those that are self-contained once they are downloaded to the device. Such self-contained apps are the easiest to describe, as they do not seek any kind of interactions with the outside world through the internet and act in the same way that desktop software worked on the PC prior to the time when such computers were connected over LANs or to the internet. The best examples of these apps are games that have simple functionality, good graphics, and do not rely on large databases. Most applications, however, cover a very wide range of functions that rely on access to the outside world through the internet so that they can link to other applications but more particularly to databases. The range of apps is now so great that the majority of users use apps to access the web from smartphones, and estimates in 2020 suggested that about 53 percent of all web users accessed the web in this form.[32] This proportion is currently growing at about 4 percent a year, but this is slowing quite quickly and looks set to converge to a steady state of about 60 percent of all web users by 2025.

The other feature related to downloading apps onto smart devices involves the issue of giving unfettered access to the user's data on their devices by the software companies who produce the media in question. It now appears that many apps contain facilities for uploading new data to a device or capturing other data from the device, particularly geographical location and other means of individual identification. Often, permission is not even requested from the individuals whose device and data are being accessed, and increasingly the user is not even aware that such data is being collected, despite assurances by the platform owners that such collection is not possible. For most platforms, it is possible to hack into the devices that we use to access the data stored on or related to the platform in question, although this does not mean compromising the platform itself. It is the device that is hackable, and this is a very gray area in terms of what is possible. It relates strongly to questions of security, privacy, confidentiality, and access to personal information. It is an enormous field that cuts right across many of the computational functions and procedures that form the core of the smartcity movement. In the next section and in the last chapter, we will return to these issues, for they are rapidly growing in importance as the century moves on, and they are so significant that they could change the direction of how the digital revolution might work itself out in the decades ahead.

The range of functions is now staggering. The number of apps for the Android phone is growing at around six thousand each day, while the app that is currently most used is Uber followed closely by Instagram and then by Airbnb. In fact, the various platforms that we introduced in previous chapters such as Facebook, Amazon, and so on define their own apps, as well as containing and linking to a great variety of more individualistic apps associated with the platform app in question. The average number of apps on a smartphone is about seventy, and a typical user in the US clocks up about 275 minutes screen time while launching ten distinct apps each day.[33] It is easy to figure out that insofar as one can define an average person, then that person spends around four hours a day manipulating information with an app. This begs the question as to what a typical user was doing before the smartphone and apps were invented, but the answer is probably watching TV or engaging with print media or simply talking. It remains an open question as to whether apps in general have as much educational value as the functions of the past that they have replaced. We will probably never know.

The biggest generic growth area with respect to apps involves monetary transactions, just as one of the largest routine applications on PCs and laptops is still spreadsheets for accounting. We will discuss these questions of payment and banking below, but the existence of apps for such transactions is now beginning to distort and change the whole way in which cash and credit are being handled in the modern economy. In 2018, I was introduced to WeChat, the Chinese messaging service, which combines social media with online payment, started as recently as 2011 by the company Tencent. In six years, one billion users have signed up, mainly in China, which is 70 percent of its current population.[34] Such explosive growth is unprecedented for all the information technologies we have introduced so far, but social media has never been simply an exclusive medium for casual conversation, as its consequences are much further reaching. My colleague recently tried to buy some everyday items in a Chinese supermarket in Wuhan and found that they did not take cash or credit card, only payment using WeChat, and I myself recently found my credit card would not work in the high-speed train station in Shenzhenbei, where my host also had to pay for my ticket using WeChat. Business cards, so long the darling of the Chinese businessman and conference goer, have all but been abandoned in face of the ubiquitous QR (quick response) codes on WeChat that enable you to connect with anyone who wishes to share their data and profile with you. I never cease to marvel at the breakneck speed at which the Chinese seem to engage in using this kind of social media, for it appears that whenever you send a message, the answer is almost immediate. It is hard to know how such dexterity in the use of this messaging software is acquired, but the entire population of China now seems to be connected in a way that is hard to fathom and even more difficult to anticipate with respect to what happens next.

At some point in the not-too-distant future, you will be unable to pay using cash or credit card as we all succumb to the handheld device, which no longer acts as a phone by any measure. Just as the British Post Office has almost disappeared, voice, which for a long time has been the dominant means of analog and digital communication, is rapidly declining in favor of other forms such as text and the whole cornucopia of social media. This, however, is likely to be a blip, for there is every prospect of voice returning as the dominant mode of interacting with machines as translation dramatically improves as new interfaces to our devices are developed. In fact, all these technologies depend on telephone (land) lines at some stage of the transmission, but the traditional telephone is declining in importance. As we pointed out in the early chapters of this book, there is very little left of the once-powerful government telecommunications agency in the UK that had the monopoly on everything involving voice communications. When a business is near virtual as telecommunications, its physical presence becomes less and less important. There are many ways to chart this change, but the decline in the number of minutes of telephone call usage from around 103 billion to 54 billion between 2012 and 2017 in Britain is dramatically greater than the increase in minutes spent making mobile-phone calls, which increased from 132 billion to 148 billion over the same period.[35] This suggests that phone calls on landlines in general are decreasing at about 10 percent per annum in terms of volume, and this is in the face of rapidly falling costs. By the end of this century, telephone calls from landlines are likely to have disappeared, while it is impossible to say what the growth or decline in mobile-phone calls will be. But what is clear is that the telephone will be gone, and in its place will be . . . whatever the computable city throws up—we simply do not know.

This history of money is based on the transformation from tokens that began to replace the pure barter of goods in preclassic times to the development of credit notes—"promises to pay the bearer"—and thence currency, usually organized by the nation-state. However, cash, as we still call this medium, is fast disappearing.[36] Yet, this is not a new phenomenon, for the first credit cards were introduced in the US in the late 1950s. These cards are fast migrating onto smartphones, which are increasingly being used for activating payments in multiple domains. There is little doubt that the decline of cash, as in coins and notes, will continue apace as entire currencies move to the net. In the UK, use of cash in circulation has sunk from about £25 billion to some £7 billion in the last ten years, and it is easy to see that, like landline telephone calls, it will probably all be gone by the middle of this century, notwithstanding the usual stubborn resistance that means a residual is likely to remain.

Banking has now largely moved online, and while currencies rooted in digital technologies such as Bitcoin are the subject of some wild experiments, these suggest that the need for the nation-state as the underpinning guarantor will be necessary for a while yet.[37] What all this has to do with the smart city other than the fact that it represents one of the many facets of information that takes place by and large in cities is hard to figure out, largely because it represents another critical layer of invisibility in networking and information flows that are near impossible to gain access to. Even if we were able to gain access and produce a synthesis of the relevant networks and flows, it remains hard to measure. In fact, in real-estate markets, where land and buildings defined by cash flows are now largely virtual, as well as in the broader economy in general, which is dominated by online transactions,[38] it is hard to see how these flows are affecting the physical structure of the city. But this issue has remained and continues to be our starting point for our examination of the computable city throughout this book. That such markets are being turned inside out by the emergence of digital money and digital transactions is already clear, but we have little idea as yet how to develop an urban analytics that makes sense of this complexity.

There is one last feature of the recent surge in technology that began in the early millennium years that we still need to consider, and this is the development of apps that pertain to what traditionally was called the "public sector." Often, these apps are used by individuals and even created by individuals through various forms of crowdsourcing, but they are encouraged and motivated by public, nonprofit agencies such as local municipalities and other public-sector organizations. Many city governments, for example, have seized on the idea of apps as being ways of delivering services to their citizens, residents, whichever groups they have a responsibility for, while at the same time soliciting and eliciting responses from their citizenry that will help them deliver better services.[39] Information technologies originally emerged from departments of government responsible for transactions processing, but these began to spread out within local government organizations that morphed into what were called a generation ago "research and intelligence departments." More recently, these have become more explicitly related to the information services role of authorities, where, in a big city such as Boston for example, it is now called the "Office of Urban Mechanics."

These still fulfil the same sorts of functions in terms of service delivery and information and simulation for plan making and policy analysis as they did of old, but now the use of online web resources and apps and the delivery of open data dominate their operations. If you log onto the Boston page, what you see is a fairly slick set of categories that reflect these functions, and if you drill down into them, you will find various web resources that you can utilize, as well as apps that enable you to see what is happening in the agency in real time.[40] Reporting problems such as noise, litter, crime, building hazards, and so on through these apps or through the web is now fairly routine, while related public (as well as some private) agencies dealing with city-wide functions such as property, transport, jobs and so on are creating apps that enable intelligent interactions with many municipal services. These enable the public at large to query transportation systems for timetable information and disruption, explore house prices, investigate the labor market, and access a host of other functions that, in principle, make life easier and improve the quality of the urban environment. Much of this information is now available through the kinds of dashboards that we introduced and described earlier in chapter 13.

We could detail all these functions at length and illustrate how they vary across the world's cities, but most of them reflect the traditional role of local governments, which involves services mandated by central governments.[41] It is likely, however, that such governments will increasingly reflect the organizational characteristics of the kinds of platform we have hinted at here. But, as in much of our thinking about a future digital society, the extent to which this organizational change will take place is fuzzy, confused, and uncertain. Public agencies dealing with agreed social objectives cannot avail themselves of the same kinds of revenue streams that characterize the platforms that deal exclusively with information products such as Google itself, Facebook, and the other giants of the industry. Local governments deal with products that are both material and virtual but depend for their functions on local taxation, which is very different from the elastic markets that have grown up around free access supported by advertising. In our remaining chapters, we will continue to explore the kinds of disruptions that appear to have become generic to the digital world we are evolving toward.

But before we do so, we should at last open the door to the many social issues that emerge from what we have so far defined as being the key elements of the computable city. We cannot do these justice here, but the deluge of data that is now streaming almost uncontrolled from our various devices, the question of who owns this data, our ability to develop software and make it available in a form that others can access and change, the abilities of different individuals and groups to engage with these new platforms and services, and the increasingly obvious divides that are taking place where segregation along the lines of age, ethnicity, and gender—all of these pose major issues that we need to reflect upon in our speculations of what will come next and are part of the intrinsic unpredictability of our future.

DATA, PRIVACY, CONFIDENTIALITY, AND DIGITAL DIVIDES

As we have sought to show as explicitly as possible, the digital revolution began with scientific applications that were as much to demonstrate how complex numerical problems could be solved rather than simply confronting theory with data. But by the early 1960s, the dominant applications were for data processing, and as the revolution quickened, databases came to dominate most applications. In this, as we outlined in chapter 14, search and storage were key. Once devices had been scaled to the point where they could be used to sense human and physical activity associated with the emergent information infrastructure, the focus began to shift to data science, big data, and real-time streaming. As devices to detect all this became personal, the age-old question of who owns the data that is and has been collected moved to center stage.

This question of ownership was much less complex when the majority of the data in question was mandated and regulated by government and related public agencies, but once data associated with individuals began to be generated, particularly from activities where the data in question pertaining to individual profiles was not intrinsic to the transactions involved, the question of its legitimacy and legality rose to the fore. In the last decade, the emergence of social media sites where users are enticed but rarely forced to hand over their data or where providing personal data is a condition of participation in the site has sharpened the focus on what data is stored by those who control the media sites in question. In the public context and in the halfway house of what is public and private, particularly in matters of health, confidentiality is to the fore. So far, the Hippocratic Oath, which lies at the core of medical practice, appears to have held up, although data pertaining to non-health matters but linked to health remains a problematic area. The problem of surveillance surrounds these questions, and there

is now considerable disquiet about the way in which the largest media platforms and Big Tech collect such personal data, selling it on and using it in
countless ways that imply degrees of surveillance that are politically unac-
ceptable but hard to identify.[42]

The question of whether the data has value to the individual providing
this, especially if the data is captured unwittingly, raises serious problems of
ownership although there are now many media forums that solicit personal
information and provide some fee for services rendered. The most problematic areas are where partial snapshots of data are captured in an ad hoc way
and then put together—synthesized—to provide a much richer data set from
the value added. In this manner, countless agencies, companies, and platforms are constructing large data banks, whose synthetic data is only in
part confidential, with the resulting products remaining in a large gray area
where ownership, controls on privacy, and confidentiality are blurred. Such
issues are of some importance in developing tools to explore and design
cities in their high-frequency mode, and increasingly, data that is simply
a product of the technology being used—mobile-phone calls and other
similar data that is transmitted in real time across networks—fall into this
category. Even map services where the software can be downloaded for free
often contain elements that track the user's use of the software, and this
is particularly the case where such users are unwittingly tracked moving
physically in real time.

There is little doubt that when we examine the access to information
services, data and related software, and the ability to use the tools provided
to interpret it and employ it for many different social and professional activities, user populations are segregated in many different ways due to the obvious distinctions based on class, income, culture, education, and so on. This
has implications for the online economy, while the data itself that is gathered online is itself only representative of the individuals and groups who
engage in such exchanges. In building up a picture of the computable city,
the data that is collected in this way is unlikely to be representative, and
there are many limits to the data that is becoming available in those ways.
For example, it is obvious that those who do not partake in the online economy do not generate data about these activities, and it becomes problematic to extrapolate to include entire populations. This involves synthesizing
online data with conventionally collected data by direct observation or questionnaire. Such data sets introduce many new complexities into the data we

have to use and this makes the digital transformation reflected in data more and more difficult to disentangle from change in general.

There is much more we might say about the software, data, networks, protocols, and the constraints that are being imposed on the digital world by those who control the various elements that Andreessen[43] noted in the quotation that begins this chapter. All of these have important implications for our future society and the way we continue to fashion and develop new information technologies. To take these arguments further involves a shift in focus from the lens of technology to that of society, from a physicalist perspective on cities and their design to a social one. But before we pull these ideas together and point to the future, we will have one last look at the ways various individuals and groups have begun to articulate ideal cities in the Information Age, smart cities akin to ideal cities. This this will set the scene for our conclusions, which will impress once again the unpredictability of future technologies in the context of the unpredictable future city.

16 ORGANIZING SMART CITIES

No automatic system can be intelligently run by automatons or by people who dare not assert human intuition, human autonomy, human purpose.
—Lewis Mumford (1970)

COMPUTABLE FUTURES: SMART CITIES OR SMART CITIZENS

The picture of the computable city that we have painted so far is largely concerned with the ways in which new information technologies are changing the form and functions of the existing city as articulated in the standard model and its successors presented from chapter 6 onward. But as the digital transformation of society has evolved, the notion that there might be idealized variants of future cities, where such cities are built around new technologies, has become significant, and there are now a number of distinct proposals for such idealizations, some of which are being implemented. Based on differing technologies, we can now broadly divide contemporary cities into three types: cities that are absorbing new information technologies into their fabric from the bottom up, cities that are resistant to such changes, and cities that are being planned afresh using the principles of the information society to evolve what we might call "idealized smart cities" or sometimes "smart new towns."

In fact, there are two other themes that cut across these types. First, the notion that smart cities are more about smart citizens and the way they use these technologies, and second, the subtle but significant changes that have occurred since early 2020 when the global pandemic began and led immediately to dramatic changes in the way people adapted their behaviors to new movement patterns and locations in cities. In this sense, our three types need to be considered with respect to how new technologies are used intelligently by their citizens, and how the pandemic has affected their form and function. These appear to vary substantially between different geographical locations and different cultures.

As we will explore in our next and final chapter, the ways in which we have changed our different modes of meeting and interacting with each other since the pandemic began continue to have important and probably long-lasting impacts on the structure of cities. Part of the unpredictability of our urban future is directly attributable to the unprecedented impacts that new norms for social distancing, new preferences for low- versus high-density living, and changing preferences for using different modes of transport are having on our cities. As well as summarizing different designs for idealized smart cities, we will factor these changes into our discussion of what the future city will look like, and we will explore the extent to which the rapid strides toward a computable future are being influenced by changes in behavior due to the pandemic alongside the introduction of many new technologies that we presented in the previous two chapters.

Our quest to live in cities is deeply ingrained in continually improving our social environment in search of a better life. Some call this our progressive instinct, some think of it as our natural greed, others are more generous in thinking of it as the motivation to improve the welfare of those whom we relate to.[1] Whatever and wherever the origins of this quest lie, if we look at cities since they first emerged more than five thousand years ago, we have spent as much time thinking about their ideal form as about the problems that such ideal forms are designed to solve, resolve, and alleviate. Go back to nomadic times more than ten thousand years ago, and the earliest expressions of art and the written record in cave paintings appear to reflect our desires to improve our immediate environment, if only to hunt better and to secure something more than mere subsistence.

Part of our signature on the planet is our ability to seek out a better life, and it is this, to a certain extent, that defines our rationality, indeed our humanity.[2] To chart this in terms of cities, it was the Greeks who first gave substance to these ideas, and from the writings of Plato onward, the ideal city has been largely represented as a geometric construction whose form reflects the embodiment of various social and economic functions that define the way in which its physical parts combine to produce a coherent form. Geometry has been the watchword in elaborating the ideal city, probably because humankind's first and most obvious articulation of form was from the heavens, which, in turn, inspired religions that place God, Earth, man, and very often the city at the center of the universe.[3]

When people come together in cities, they cluster at one place to engage in exchange. The notion that the ideal form of the city is arranged around some core, often historically the place where those who governed the city located the pinnacle of their economic and political power, has been writ large in our imagining cities throughout history. This centralization is entirely consistent with control of territory around such focal points. Access to the ideal city has often been articulated as movement to a focal core, arranged in symmetric radial fashion, star-like in every sense, thus reflecting our long-standing image of the heavens. To some extent, these ideas are implicit and thus reflected in our standard model that we have taken as a default for the form of cities throughout this book.[4]

This fascination with form has dominated the search for the ideal city since prehistory, and it still does. Ideal cities, in whatever era they have been proposed, are usually conceived as pure geometries and symmetries, but another feature of these utopias, ever since cities first emerged in ancient Sumeria, has been a concern for their ideal size. For thousands of years, the largest cities fell well short of one million people, for the technologies to go beyond this were simply not available, as we have pointed several times already in this book. In fact, estimates of the largest cities slowly rose from the first recorded settlements—the city of Jericho[5] had about two thousand people as far back as 5000 BCE—and over the next five thousand years, the largest city populations increased gradually. The biggest seems to have been Babylon, which reached a hundred thousand in 1200 BCE, with Carthage and Alexandria reaching half a million or more between 300 and 200 BCE. Rome touched one million for a long period between 150 BCE and about 300 CE when the empire collapsed, and it began to fall back to half a million, which it reached one hundred years later in about 400 CE.[6] From then on, cities rarely reached one million in Europe (although they did so in China) until the industrial revolution began, with London growing from one to six million during the nineteenth century. The largest cities such as Tokyo then grew to twenty million or more by the end of the last century, and the biggest agglomerations are now upward of sixty million, with the Greater Bay Area (Hong Kong, Shenzhen, Guangzhou, Zuhai, and Macao) being the example par excellence.[7]

THE IDEAL CITY

What is clear from this history, which is echoed, of course, in the growth of the world population, is the fact that our view of the best or ideal size of a city has changed as we have become more familiar with ever-bigger cities. Although there is little consensus about the first speculations concerning the ideal size for a city, they probably go back to the time when cities first emerged. But it was a little later in classical times—in the Athens of Socrates and his philosopher disciples, Plato and then Aristotle—that the first clear statements emerged. Although Athens existed in quite a highly urbanized landscape of some three hundred thousand people in the wider Attica region, the city-state or polis was considerably smaller in the order of thirty thousand. It was in this context that Plato articulated his ideal city, which consisted of some 5,040 citizens.[8] The number is not completely arbitrary in that Plato argued that this was a number that could be subdivided many times into a large number of groups, which could provide a hierarchical organization that was ideal for a stable and balanced population, but it also implied that the best city size might be smaller than what was then the average size of the typical city-state. In fact, throughout history there have been two strong forces: the first arguing that the biggest contemporary cities are too big and thus any ideal should be smaller than this and, in contrast, the second, which is a much more recent view, arguing that the contemporary city is not big enough and needs to be bigger to realize its true economies of scale.

During the Renaissance in Europe, many ideal town plans were drawn up, and most of these implied relatively small populations in the order of thousands, but this was largely because the focus was based on replicable geometric neighborhoods, which became unwieldy and unrealistic once they were scaled up. Leonardo da Vinci's urban idealizations assumed a city size that was typical of the Italian towns of that period—something in the order of no more than thirty thousand people, although the largest towns such as Milan were more like a hundred thousand in population size.[9] However, during the late nineteenth century, there were some who argued that the ideal city should be hundreds of thousands, thus going with the flow as new technologies enabled cities to grow ever bigger than the old limit of about one million people. However, this argument was not generally adopted, for the evils of the industrial city dominated the debate about city growth. Much of urban planning legislation that was beginning to be put in place was designed to lower densities and encourage a back-to-the-country ideology, but the advantages of living in large cities did not escape the notice of many commentators and early urbanists.

Yet, the quest to think of the ideal city as being small, intimate, accessible, and a mix of town and country rather than big, congested, lonely, remote, and with a hard urban edge continued to reassert itself. In particular, the great wave of enlightened industrialists, among them Robert Owen, William Lever, and Joseph Cadbury in Britain, proposed many new settlements from the mid-nineteenth century onward, but these too tended to be small, more like factory villages, with populations in the thousands if not the hundreds.[10] These idealizations found their best expression in the garden city movement associated with Ebenezer Howard, who argued that the evils of the industrial city should be replaced by a pattern of satellite cities—what came to be called "new towns"—based on a mix of town and country. This, he argued, was the best of both worlds, a return to the past without abandoning a future based on the newest technologies—the railways and telecommunications—still being the glue that held industrial society together. Howard argued that the ideal new town should be around thirty thousand in size, a number not dissimilar from the average size of town in Britain in those days anyway, but configured in such a way that a distinct hierarchy of forms and different town sizes would produce a network connecting small to big.[11] Howard did not imply that big cities were passé, but he did argue that the urban world should be configured with much smaller settlement sizes than the rapidly growing cities that had been spawned since the beginnings of the industrial revolution.[12] In this way, he implied that pollution from steam power that would continue to dominate heating and movement could be minimized.

All this changed with the development of the modernist movement in the arts and architecture, and it was technology once again that played its part in fostering bigger cities. This time, however, it was growth in the vertical dimension rather than the horizontal. It was the architect Le Corbusier who struck the first blow in his *City of Tomorrow*, echoing but diametrically opposed to Howard's proposal in his 1901 book *Garden Cities of To-Morrow*. Corbusier's utopia was published in 1929 in which he suggested a city composed of sixty-story tower blocks in wide-open parkland surrounded by residential blocks some six stories high. This would generate a city of three million people. Three years later, Frank Lloyd Wright proposed almost the diametric opposite in his plan for Broadacre City, which was composed of acre plots, one for each household, thus producing a city of some five thousand people, strangely reminiscent of Plato's magical number 5,040. But Corbusier's dream of massive utopian cities, as portrayed in his plan for Paris, remained unfulfilled, as the twentieth century was still dominated by the notion that the most livable cities should be small towns.[13] Worldwide, the quest for low density became the watchword for city planning as cities grew through the advance of suburbia powered, of course, by the automobile. Only when Asia-Pacific began to open up in the last twenty-five years of the century did the quest for bigness come back onto the agenda, and it thence emerged in a somewhat different way from the dominant model of the ideal city that hitherto had been largely physical in emphasis.

It was Alfred Marshall, as we noted in chapter 6, who first articulated the notion that as cities grew bigger, they captured an increasing proportion of production (and consumption), which he defined as "economies of scale." As the scale (or size) of a town increases, there are more than proportionate returns to scale, meaning that the town becomes more than proportionately richer.[14] In general, big towns seem to be richer than small towns on a per capita basis. As cities developed through the twentieth century, the stigma of their industrial origins declined as they became cleaner and less polluting and as the health of their populations improved. By the millennium, larger towns were being considered "cool" places to live, and there was a selective return by younger populations to live in the central city. Moreover, as the economy became more and more digital, large cities attracted those most able to innovate in these domains, and their scale economies associated with jobs in high tech and financial services became more and more part of their attraction. In short, big became beautiful in quiet contrast to Ernst Schumacher's "small is beautiful" mantra, which had dominated thinking about cities in the second half of the previous century.[15]

To an extent, this idea that large cities generate more than proportionate returns to scale is not new, despite it being loosely accredited to Marshall.

It has been self-evident for many centuries, ever since the first cities began. Indeed, it is the underlying logic to the formation of any city, where clusters of like-minded people come together to pool their labor and to specialize in the production of goods that can only take place as concentrations of population as cities get larger and denser. Although economies of scale in cities have been assumed for many years, only in the last two decades has there been a sustained effort to measure these effects, and even now, the evidence for such scale economies is mixed. As the scientist Geoffrey West demonstrates in *Scale*, and Luis Bettencourt in *Introduction to Urban Science*, the average wealth of cities tends to increase more than proportionately with city size.[16] If a city were to double in size, then its wealth per head of population could increase by as much as 15 percent, although different cities in different cultures and in different historical eras manifest different rates of increase. Most, however, tend to show an equal or more than proportionate increase, that is, if a city grows in population by 1 percent, its wealth grows by anything from 1 percent to 1.5 percent.

There is little doubt, however, that cities reveal such economies, but the crucial question is not whether the city is more than proportionately richer on average but rather how the city organizes its wealth among its different social groups. If a large proportion of the city's population experiences a fall in income as the city grows in size due to more intense competition for jobs, but a much smaller proportion of its rich population continues to accrue income more than proportionately with respect to their existing wealth, then it is quite possible for the total wealth of the city to increase per capita but for the numbers of poor people to also continue to grow and get more than proportionately poorer. This occurs because the rich accrue wealth that wipes out the decrease in wealth of the poor. There are many considerations like this, such as the cost of living, that need to be factored in for different income groups, and thus the notion that bigger cities are always more attractive places in which to accrue wealth is by no means clear once distribution and costs of living are taken into account. Nevertheless, we continue to live through a time when bigger is better, despite this flying in the face of the strong tendency for urban society to be more and more segregated with respect to the distribution of the wealth that is generated. Even Plato, in *The Republic* in about 360 BCE, said, "any city, however small, is in fact divided into two, one the city of the poor, the other of the rich; these are at war with one another; and in either there are many smaller divisions, and you would be altogether beside the mark if you treated them all as a single State."[17]

Later in this chapter, when we introduce a sample of the digital new towns that the smart-cities movement has thrown up, we will continue with our description of how these utopias are formed. This will focus on their geometry as well as their scale, which still tends to dominate the way many continue to think about what such a smart city might be. In fact, as we shall see, digital new towns are still largely fashioned around the architect's dream sketch, with all the bling and razzmatazz that the high-tech future can conjure up. But before we explore these examples, we will dwell on two rather different conceptions of the high-tech future with respect to our images of the smart city. Almost as soon as the glue to link us all together through the internet was invented and our presence on the net was formed through web pages, people began to speculate that the smart city could indeed be fashioned as a set of web pages. Indeed, web pages and their hyperlinking provided one of the first forms of virtual city that were suggested, and although much of this hype has passed, it is still worth noting the elements of this future. Our second preliminary set of ideas involves the development of the virtual city in more physical terms, through computer-aided design and 3D representations. Once we were able to describe the physical city fabric in digital terms, then great poetic license emerged to fashion it differently. Many examples of idealistic 3D cities continue to be generated, and these will provide the foil to our later discussion with respect to how real, in contrast to how virtual, smart cities are being evolved.

INTELLIGENT CITIES: NETWORKED CITIES BEFORE THE INTERNET AGE

Thirty years ago, were you to jet into Changi, which was then Singapore's sparkling new airport, you would have been greeted by posters announcing that you were now entering the intelligent island. Ever since the airport opened in 1981, Singapore's government had pursued a plan to automate every facet of the island.[18] Its government had long ago realized that a small city-state with no natural resources to speak of would have to embrace a future where its citizens would have to live on their wits, and this would mean developing and applying new technologies, particularly communications technologies, consistent with its role as a key port at the crossroads of Southeast Asia. In that year, the National Computer Board (NCB) was set up to evolve such a plan,[19] which would see the number of computer installations/centers in government grow from a handful to more than a hundred in ten years. The project focused on developing a range of integrated software projects that would automate everything from the way trade was conducted to the way in which services were delivered to its populations and the way transportation could stitch the various parts of the city-state together with each other and with the rest of the world.

The focus on building a networked city began in Singapore as well as several other places in Asia-Pacific before the Internet really took hold. Throughout the 1980s, the prospect of fast networks using fiber optics grabbed the attention of several entrepreneurial cities, which many argued would be crucial for their future economic development. This was an act of faith at the time, and even as late as 1989, the notion that the world would become wired in the way it has done since was by no means a certain prospect. Singapore was at the forefront of establishing what was then called "ISDN"—an Integrated Services Digital Network—which was a digital network associated with public telephone lines where both voice and data transmission were merged.[20] Like many such technologies, it was well ahead of its time, but it was only one of various related networking technologies that Singapore experimented with. As part of the NCB's plan, several different digital networks were developed before the internet itself began to dominate—networks such as Tradenet for automating the port, various delivery networks for information services in housing, education, taxation, and so on, all of which, by the mid-1990s, had produced the world's first online society. Although Singapore, then and now, was hardly planned as a utopia—even though its politicians and civil servants considered it to be a model for the future—it certainly continues to aspire to be one built around networking and communications. Although it has none of the superficial glitz that the smart new towns we will describe below have, this is a city where information technologies are being built from the bottom up in the time-honored fashion of how cities evolve naturally. Singapore is still a model of how to do it, as indeed are a number of other developments in the same region that are also worth noting.

The idea that one should build the city of the future around the wired society was an idea that grew rapidly in popularity through the 1980s as new fiber-optic technologies came on stream and as computers were scaled down to the personal level. We have recounted all this in considerable detail in previous chapters, but it established a context and an environment in which the idea of building infrastructure to enable rapid communications in business and industry really took off. The notion of providing high-tech clusters in science parks also grew in popularity alongside the military–industrial complex in the 1960s and 1970s, which became key focal points for economic development.[21] But at this stage, there was little sense that new information infrastructure was needed for the city itself to support new industries. Silicon Valley, Route 128 in Massachusetts, and Research Triangle Park in North Carolina were examples not of embedded communications infrastructure but rather of entire high-tech districts where the development of hardware and software and the labor market to generate this were located in close proximity. These were places that produced information technologies rather than heavily used such technologies in their production, although in time, of course, such areas would become heavily wired. The internet and the kinds of sensing technologies we now have to engender this would eventually underpin the infrastructures needed to support the production of high tech.

During the 1980s, the idea of whole cities associated with new information technologies emerged in the form of technopoles, particularly in places such as Japan. The inspiration, even back then, was the development of what the Japanese called the "Fifth Generation Computer Systems" project—a long-term effort to develop mass AI.[22] In fact, this project faltered, as Japan itself did economically in the late 1980s. and in any case, it was based around the concepts of AI in its strong form, not on the pattern recognition focus that the current field of AI—weak AI—now has. Nevertheless, during this period, there were a number of high-profile projects relating to wiring entire districts of cities. In Malaysia, the area from Kuala Lumpur west to the coast in the Klang Valley became the basis for a multimedia super corridor based on economic incentives to attract the world's high-tech companies such as IBM, Microsoft, Cisco, and so on.[23] It was designed to catapult Malaysian economic development into the twenty-first century. In contrast, in Japan, there were many projects designed to renew areas such as the one in Kawasaki, an older industrial suburb sandwiched between Yokoyama and Tokyo where the installation of the latest networking technologies linking high-tech hubs was designed to provide new jobs and new industries that would depend on rapid digital communications.

As we have remarked, all this took place before the internet became established for multiple activities well beyond the narrower scientific purposes it was originally developed for. In 1984, the original ARPANET was split into its academic and military components, but by 1989, a new network, NSF-Net, had been developed by the National Science Foundation (NSF) to take over nonmilitary research in academia and to link the six supercomputer centers that NSF had established.[24] Of course, when we say new networks, much of this infrastructure was based on existing networks at regional and metropolitan level as well as communications hardware that was a mix of the public and private sectors. By 1995, NSFnet had quite quickly transitioned to the commercial internet, and although this extension and deregulation was not without controversy, by the end of the century, the internet as we conceive it now was more or less in place. This was aided enormously, of course, by the invention of the web, in particular by various graphical user interfaces—browsers with the capability of hyperlinking from one computer to another—thus establishing the kind of connectivity that we now take for granted. Singapore, Malaysia, Japan, and many other similar networking initiatives around the world merged effortlessly into this wider context.

THE FIRST VIRTUAL CITIES

In exploring the future form and function of the smart city as an idealized utopia, several directions emerged as soon as the internet became widely available. None of these represent the kinds of comprehensive utopia that are encoded in digital cities that are designed and planned from scratch, but all of them portray some aspects of their form and functioning that is ideal and virtual, if not quite real. This is often only a way of communicating information about services as well as enabling their distribution, over the web of course, if they can be consumed as digital products. Such sites can contain all the information about what a town can deliver through its municipal government, and one such site that was in the vanguard of these ideas some twenty-five years ago—the website for the Italian city of Bologna—displayed a perspective map with 3D icons representing hot spots where the user could click and be directed to more detailed websites that contained the information services in question.[25] These kinds of web pages were a cross between the activity of virtual tourism and a way of delivering better services—a more coherent way of transmitting information. In this sense, the information was about the real city, but displaying it digitally and activating links to it represented a better way of making that information known.

There were plenty of web pages constructed in this manner, where towns and city governments displayed their wares, most of them displaying cartoon-like links to existing information but some taking on a semifictional character in the quest to communicate effectively. In chapter 13, we noted examples of digital cities, where the power of graphics was used to portray the physical character of the city, particularly in 3D, but early representations of physical form were produced as flat web pages, where maps and diagrams of various kinds were used to communicate information about various services associated with the places in question. These kinds of display, based on icons representing different functions, which provided entries to other web pages through hyperlinking, are still the dominant way of communicating a very wide variety of information on the web. In fact, the vast majority of web pages used to communicate rely on this kind of linking, and if everything of significance in terms of the city can be connected, then this represents one kind of virtual environment that is close to the real thing.

Before digital models of buildings and streets in a city became routinely represented in 3D models, there were many kinds of virtual environment that depended not on such geometries but rather on the power of hyperlinking. In particular, virtual worlds emerged in which users remote from one another could meet in a virtual space defined by some software shareable across the net. An early example of such a virtual world was Alphaworld, renamed Active Worlds in the mid-1990s, where users could quite literally construct their own environments, their own dwelling spaces, chat in cyberspace, and generally act out whatever existence they might imagine.[26] The environment itself encouraged building and development, and what emerged from this and other similar worlds were configurations of physical buildings and streets in this virtual space that appeared town like. In Alphaworld, locations of thousands of users could be extracted from the virtual places where they acquired their real estate. Perhaps unsurprisingly, the city of Alphaville within this world was strongly dominated by radial routes around its clear city center. The first virtual settlers located near one another around the town's central point. Settlers coming later had to locate on the periphery of the town, and as the town grew, a much stronger radial pattern than one might see in the growth of a real town became established.

This was largely due to the fact that in virtual space, coordinates were used to locate and navigate, and it appeared that users—settlers—used regularly spaced coordinates on a grid more frequently than irregularly spaced coordinates. In fact, the growth of the virtual town provided its own dynamics, which at least in principle provided some understanding of the way a real town might evolve. What is in some ways remarkable but always explicable in hindsight is the fact that there was no centralized collective mandate. There was no government in this virtual world. Yet, the form and geometry of a classical town emerged from many bottom-up actions, mediated, of course, by various basic rules of behavior applied at the individual level— yet another example of the way the standard model from chapter 6 recurs in different guises.

We need to note one last set of examples of virtual cities that have some utopian flavor, and this is no more or less than the use of 3D digital models to portray fictional futures—planned cities that only exist in the computer model rather than in any kind of reality. To an extent, the kinds of blue-sky thinking and speculation that characterize the schemes of the great visionaries from Leonardo da Vinci to Le Corbusier onward are now being fashioned in countless ways within 3D computer models. These are much faster to manipulate into futuristic forms than their non-digital equivalents produced using pencil and paper, although there is no guarantee that any of them will surpass the "imagineering" of the great reformers, philanthropists, and architects of the past. In fact, one of the problems of the Digital Age, with all its consequent wealth of opportunity that we have yet barely got to grips with, is the fact that there is now an order of magnitude increase in the numbers who can avail themselves of digital technologies and produce great work. Among this growing band of experts, there are bound to be important visionary statements of the future that those with a facility for digital design will be able to generate. How to even identify these from the great mass of digital content is a real challenge for society in general, and there is an increasing number of significant works that will never reach a wider audience, trapped by the sheer volume of content that no one can easily, perhaps ever, absorb. This has already been a major constraint on the development of the very best science during the last fifty years or more, and although incidental to our story here, it is a major challenge, along with confidentiality, privacy, archiving, and curation, as well as what we are able to share with each other in the Digital Age.

We may well ask where such virtual cities will end up as we continue to represent them as 3D digital models accessible from the web on any and every device. Soon, we are likely to reach a point where remotely sensed data at scales of one meter or less will be produced routinely on a daily basis. Such data will enable us to measure detailed physical changes in the 3D urban environment, and we will thus have continually changing but up-to-date models of the virtual city that we can use routinely for a vast array of functions, particularly measurements associated with such geometries. These models are increasingly being populated with socioeconomic data as we link their geometry to the uses and functions that take place in the individual building blocks, open spaces between the buildings, streets, and all the other physical detail that defines the virtual city. To see where we are headed, we will zoom back to Singapore, for as you might expect of a place that prides itself on being a model at the cutting edge of the technological future, here we discover the current state of the art when it comes to 3D models of the virtual city. Virtual Singapore takes pride of place in that it is being built by the National Research Foundation, which is part of the Prime Minister's Office.[27] This is located in the CREATE building—the Campus for Research Excellence and Technological Enterprise—at the National University of Singapore, where a cluster of world-class universities researching the future of Singapore are based. Virtual Singapore is the nearest thing so far to a digital twin running alongside the real city, where users of the model, who range from scientists and policy makers with a professional interest in the future of the city to the informed public and the public at large, can explore the detailed attributes of the city, building by building, neighborhood by neighborhood.[28] The model lets a user take everything from measurements of the size and volume of what they encounter to estimates of the amount of traffic congestion, crowding, and density of those who inhabit the city. The picture that the model offers you is essentially the kind of information that is contained in the best 2D map you can imagine but in 3D, where the city actually looks like the real thing, much more realistic in feel from any traditional map. You can quickly imagine that this is the kind of reality that will become routine in future navigation. It may be available in one's eyeglass or on one's phone as it merges seamlessly into the real thing through an augmented reality that is close enough to the real thing to be regarded as an actual reality, no longer a virtual one. This, then, is the future, where the real is mixed with the virtual and where the ideal is mixed with the real, where fact becomes fiction, and where the digital world begins to spawn many different variants of the virtual, tailored to a myriad of different purposes and functions. This is the world of digital twins that we will pick up again in our last chapter.

THE EMERGENCE OF SMART NEW TOWNS

At many junctures in this book, we have argued that cities are not planned from the top down but largely emerge from processes that are driven by individuals and groups of individuals who act from the bottom up without any overall coordination. Although we do see many highly regular geometric patterns in cities, where we do, these are usually the result of actions that are generated using simple spatial structures that manifest themselves in regular patterns at the smallest of scales. If they sometimes show up at larger scales, these are often the result of repeated applications of the same basic processes at ever higher levels. This is expressed in the notion of cities being fractal-like objects in terms of their geometry—ideas that have gained ground in thinking about the form of cities over the last thirty years.[29] The idea that cities evolve from the bottom up is entirely consistent with the observation that at the city-wide scale, they appear unplanned, the product of random growth whose dynamic is decentralization around some core or pole where growth begins, usually around the CBD.

To take this notion of evolution from the bottom up seriously, it is immediately obvious that the idea of a centrally planned smart city is as rare as similar proposals for planned cities prior to the Digital Age. Information technologies are implemented from the bottom up, as the history of computing that drove our initial explanations for the emergence of the smart city illustrated in part I. The fact that many computers can now be associated with a single individual and in many contexts—most of those who are able to operate such devices have access to at least one—implies that we are all smart and that an entire city can be smart. The onus then falls on everyone within a city to use such devices in as smart a way as possible, and the focus then becomes the smart citizen rather than the smart city. The answer to the question of what or where the smartest city is—or the variant of whether your city is smart—has the obvious response that this is a meaningless question. When everyone has a device that enables them to be smart, they have to use that device intelligently to be considered smart.

This is not to deny that there are good and bad ways of installing these new technologies in cities but that the predominant way in which cities become smart is from the bottom up, with individuals dominating the way the city's population acts in an intelligent manner.

In many respects, the current conception of the smart city is one of simply automating everything that can be automated, controlling it through state of the art technologies, making sure that the city works efficiently in that its temporal and spatial functions are optimized. This is a very narrow conception of smart in that it does not in any way relate to what the city feels like in terms of living and workspace—in short, in terms of the quality of urban life. It was Jane Jacobs, the most strident critic of contemporary planning in North America in the early 1960s primarily through her book *The Death and Life of Great American Cities*, who exhorted the public at large as well as the urban-planning profession to focus on the quality of small-scale environments. Her clarion call was to fight "big government" in its adoption of plans that would turn our cities into low-density, barren landscapes of concrete, where the local community was effectively wiped out, thus nonexistent.[30] Her image of the city beautiful was one where the diversity of the community on the street in terms of the support that individuals in close-knit communities would offer one another is not something that can easily be squared with the introduction of more automated technologies.[31] Insofar as she advocated us to build smart cities, although she never used the word "smart," her definition of smartness was light years away from that of the large-platform information technology companies and the rest who use the word "smart" to signify more efficiency through automation. Her notion of community and diversity is something that cannot be reduced to the quest for ever more efficiency. In this sense, technology cannot be easily related to ways in which urban areas can be improved other than through the installation of yet more computer and communications technologies, which play a rather narrow role in questions concerning urban sustainability and quality of life.

Once ideas about information technologies being embedded into public spaces emerged and were thence associated with the idea of the smart city, the idea of building communities afresh to link many components in the city that were subject to automation of various kinds quickly emerged. In some senses, one can think of these ideas akin to the early- and mid-twentieth-century examples of new towns, designed then to contain sprawl, to remove the evils of the industrial city from urban life, and to demonstrate how clean, efficient, and socially balanced communities could reveal the way forward. In general, the success of such new towns is mixed, and many do little more than simply contain urban sprawl but rarely reduce congestion or pollution very much, nor do they necessarily improve social balance. They, like every other town, are subject to the dictates of the forces determining the new urban economy, which is a mirror of the national economy.[32] Nevertheless, several smart new towns are under construction, and many more have been proposed. These tend to be a mixture of new technologies, with an emphasis on online consumption and production, intelligent energy systems to control the home and work, automated transport of various kinds, and the control of microclimate and pollution. The hardware for these components usually mirrors some idealized geometric constructions that lay out these features in regular patterns that do not usually bear much relationship to the underlying functions of the town. This was the case in the English new towns established after World War II, and it tends to dominate most proposals for towns, automated or not but built and planned from scratch.

In fact, in most of these smart new towns, form hardly follows function. Like earlier generations of such new towns around the world, form now only superficially relates to the activities that define the functioning of the town. The standard model is usually adopted in that there is a core to the town, traditionally the CBD, and although it is arguable that with the kinds of technologies that we now have at our disposal, such a core, could be dispensed with as activities might be carried out anywhere, the physical realization of these ideas still adopts previous conceptions of urban structure. This also goes for neighborhoods, and often a rigid educational and retail hierarchy is assumed, despite the fact that such structures really relate to the past rather than the present. To an extent then, smart new towns use an outdated model that does not connect form to function, for the functions of such towns in the Digital Age no longer require the past geometries of the non–Digital Age. In many of these new towns, as soon as the regime of control—the government—that is put in place to implement the plan is relaxed, the town begins to acquire the characteristic of those towns that grow organically from the bottom up, and soon, the patterns best match the behavioral desires of the residents. Brasilia, the planned capital of Brazil, is the example par excellence.

Idealized towns and the new generation of those that badge themselves as smart are as much a part of this quest for the ideal city as earlier ones and tend to follow three types of geometry. First, the most obvious, which has inherent characteristics of simplicity, order, efficiency, and rapidity in construction, is the grid. The Romans laid out their encampments according to such a regular grid pattern, reserving different parts of the grid for different purposes and uses. Many American cities are laid out in this way after the example set by the commissioners in New York City in 1813, where they pronounced that, henceforth, settlements should be divided into square plots tied to a grid that produced an orderly organization.[33] In the case of Roman camps, which were regarded as temporary, where they did become established into the fabric of urban life once the Roman Empire fell, the pattern of settlements remained grid like.

Towns, of course, need to capture their populations into a form that enables them to interact with their dominant center, assuming that this is the way they developed. This is our standard model introduced in chapter 6, where the population lives at lower densities at increasing distances from the core. The form that best fits this relation is radial in structure, like the spokes of wheel radiating from the hub, star like, as we noted earlier. This form emerged more slowly, naturally almost in comparison to the grid, but combinations of these patterns can be seen from prehistory, particularly in the flurry of idealized towns proposed during the Renaissance in Italy.[34] In fact, there is a wonderful example in Ohio, where, during the nineteenth century, a village was planned in an agricultural landscape where the center of the market and the routes to this core were splayed out in radial fashion, with the place being named "Circleville." Over the course of some forty years, the townspeople gradually replaced the circular radial structure with a grid, and by the end of the nineteenth century, it no longer represented a Renaissance-like geometry but rather one that mirrored the dominant grid pattern of North America. The historian of towns, John Reps, referred to this process, somewhat tongue in cheek, as the "Squaring of Circleville' in his book *The Making of Urban America*.[35]

These, then, are the dominant patterns around which towns planned from the top down reveal, and this goes as much for recent speculations about smart new towns as for towns prior to the Digital Age. The first such towns were proposed over a decade or more ago and were badged as not only smart but also sustainable in their design. Masdar, adjacent to Abu-Dhabi airport in the United Arab Emirates, is a case in point.[36] Designed as a low-carbon future, where solar arrays power many features of the town from electric vehicles to energy in the home, the town is designed to minimize the desert heat by bringing many new design principles involving streets and buildings to bear on the way heat is distributed and its morphology cooled. The town is laid out on a grid, which is consistent with the delivery of energy to each part evenly, and appropriate building materials are used to lower the overall heat of the town to something like 10°C less than the surrounding desert. The town was designed ultimately to reach about ninety thousand people, but this has been compromised by many changes foisted on the implementation of the plan due to the volatility of the local economy and by changes to the design to ensure that a feasible, working built environment can be constructed. Many of the features of the new town are experimental, and although the design is based on state of the art technologies, the concentration only on public transport has been relaxed due to the fact that the town is likely to be occupied by those with little appetite for traveling in this manner. In this sense, Masdar is for the wealthier groups in the local community, whose concern for sustainability is of little importance.

Then, there is Songdo on the western edge of Seoul in South Korea, again a new town first proposed around the time that Masdar was, more than a decade ago.[37] It too is located adjacent to the capital's international airport, Incheon, and in this context, it is more like a high-tech science park, attracting many information technology companies such as Cisco, whose products are being used to build this high-tech future. It is in a free trade zone, and this undoubtedly provides the impetus for its economic base. Songdo too is a showpiece, where waste removal is automated and where the town is wired for efficient energy use and access to global information. In fact, its urban design is more conventional than Masdar's, although it is also laid out on a grid. From one perspective, it bears an uncanny resemblance to Le Corbusier's *City of Tomorrow* as a series of tall towers set in a relatively spacious green setting, where biking and walking seem to dominate its transport.[38] The town is part of South Korea's Ubiquitous "U-city" program, and it is designed to be much bigger than Masdar, with up to a quarter of a million inhabitants when it is completed.[39] In some respects, it is much better integrated into the wider development of its home metropolis than Masdar is, but already, the wave of interest in U-city has passed, largely because all cities now have the phenomenon of ubiquity due to the fact that most populations who want personal devices that let them connect up to the world's information now have them. By the time you read this book, more than two thirds of the world's population will have smartphones.[40]

These towns—and there are several such developments in train—are statements of intent rather than serious attempts at building sustainable, smart communities. This is not because those who have proposed them are not serious about their mission but rather because they tend to be showpiece developments—designed more to show the intentions of their advocates, largely for experimental purposes, rather than robust and workable templates for the future city. In this sense, they are beauty contests with respect to what information technologies might achieve in the smart city. Many of these efforts become strangely dated almost as soon as they are rolled out, for the future of information technology, as we have seen throughout this book, is impossible to predict. The new towns proposed in the last decade could barely have anticipated the rise in social media and the connected online just-in-time society that has rapidly arrived almost everywhere. We will chart some elements of what this future might be like in the last chapter, but for the moment, these smart new towns are more demonstrators for how garbage might be collected invisibly, how low-carbon heating and energy use can be implemented across a wide area, how public transport can be automated in ways that make it easy to use, and how electric vehicles might substitute for conventional ones based on power from fossil fuels. Almost all of these new towns were not able to anticipate the current hype relating to autonomous vehicles, while how human behavior is being altered in the city with respect to the rapid rise of social media and instant communications has rarely been factored into their operation. The mismatch between their physical form and the human behaviors that enable their functioning makes the operation of such towns opaque, if not largely unpredictable.

URBAN OPERATING SYSTEMS AND SPATIAL DATA INFRASTRUCTURES

Much, if not most, of the commentary on smart cities is media hype, with little focus on the underlying meaning of the impact of new technologies on either the form or function of the city. But more particularly, neither is there discussion of the changes in human behavior that are taking place, nor are the impacts on how we design cities for an unpredictable digital future being spelt out. We will discuss this in our last chapter, but here, it is worth sketching out how the smart city is conceived with respect to the ways in which the large vendors of information technology see the city as their marketplace. In fact, as soon as the big computer companies dealing in hardware, software, data, and networks became aware of the city as a particular focus, there was a rapid realization that the problems of making a smart city were not related to the products that were their prime business but rather to much more intangible social problems. We will not make an exhaustive catalogue of these here, but they focus on overall accessibility to different services and communities, problems of segregation and polarization, and problems of congestion and density, which are endemic to all cities where capacities of infrastructure channels have been reached. These are made even more acute by problems of housing markets and prices, problems of privacy and surveillance with respect to personal protection of the individual from intrusive monitoring of behavior, and problems of crime, policing, and emergency services. Most information technology companies are not able to do very much about these problems, for they are the domain of government and individuals, notwithstanding that their solution may well involve many partnerships with the private sector. Such issues, therefore, are usually tacked onto the end of and not well integrated into the smart-city solutions being sold as new technologies for cities.

All the largest tech companies such as IBM, Cisco, Siemens, and so on have divisions that purport to deal with smart cities, but these tend to be adjuncts to their marketing operations so that the wider impact of such technologies can be considered and publicized. The platform companies—Google, Facebook, and Amazon, with Apple, and Microsoft in the west and their near equivalents in China such as Alibaba, Baidu, Tencent, Weibo, and so on—seem to steer a little wider of the mark on the smart-cities phenomena, perhaps because their sphere of interest and operations is in some senses not about the city per se but rather about society in general. Nevertheless, all these companies and many more seem comfortable with the somewhat autocratic, top-down concept of an urban operating system, a platform for cities that essentially integrates the many hardware, software, and data components that describe every facet of the city.[41] To some extent, this notion is counter to the idea that cities evolve from the bottom up—the product of millions of individual and group decisions. Many of these companies have avoided proposing such bottom-up integrated operating systems per se, for their focus on embedding technologies into the urban fabric and linking them through the IoT has provided tacit support for such centralized applications.[42]

The most visible attempt at building an urban operating system, which is defined loosely as a platform to integrate all the sensors and raw data that they generate across a wide variety of functions in an extended area such as a city, is the proposal by Living PlanIT for a smart city built around such a platform in the Portuguese town of Parades. This proposal was built around two related drivers: first, the idea that the town would incorporate all the components of a low-carbon future with extensive internetworking relating sensors and their data to a variety of human functions, and second, a business model that was developed in parallel but involving bank finance that required investors to buy into the scheme. Unlike Masdar and Songdo, where the business model was part and parcel of central government financing, this was a private initiative requiring very hard decisions involving finance. This private consortium would manage the town as soon as construction began before it became part of normal routine urban management. The town was expected to grow like Songdo to upward of a quarter of a million, with very dense networks of sensors suggesting that there would be up to forty sensors per person[43] associated with different functions. The proposal again was first conceived more than a decade ago in an era almost prior to the smartphone and where the possibility that everyone everywhere could be connected through an integrated set of networks would form the platform for a city-wide operating system. As we have noted, networks, like everything else in the digital world, grow and evolve very largely from the bottom up, and thus an urban operating system such as this is more a conceptual thought experiment in that many functions can be rapidly integrated but without any central coordination that such an operating system conceived from the top down requires. This is not to say that such coordination is not required—it may well be but not necessarily in a completely comprehensive manner. Smart cities, as we argued earlier, build themselves from the bottom up, as information technology largely defines the way we use information as individuals.

In the 1960s, as part of the systems approaches that were being advocated for cities and their planning in top-down fashion, the analogy between cybernetics proposed by Norbert Wiener in 1948 as "the scientific study of control and communication in the animal and the machine" was almost taken literally. For example, in his book *Control and Urban Planning*, Brian Mcloughlin proposed that the analogy could be elevated to designing structures for local government that would enable plans to be implemented more effectively than the muddling through procedures that then and now tend to dominate such management.[44] As we recounted in chapter 10, this systems approach was extensively critiqued by those who saw the city as much more complex, indeed often assuming that such complexity lay beyond science itself. The rise of complexity theory took its place, where the focus was more on explaining how cities grew from the bottom up rather than being controlled from the top down. The problem with the smart-cities movement is that is has no intrinsic memory of these past experiences, being formed by well-meaning individuals and firms who have no knowledge of this history and the perils that have beset our attempts during the last century to design cities that produce a better quality of life, more socially equitable environments, and more efficient means of functioning than anything we have been able to do hitherto. The idea of the urban operating system is thus another attempt at control from the top down, building structures that imply some centralized control, where it is clear that all the information technologies developed so far imply decentralization and individual action rather than comprehensive planning.

The Living PlanIT concept of the urban operating system, although purporting to be comprehensive, is largely focused on the high-frequency city, as indeed are most of the smart new town initiatives that are currently being proposed and implemented. In fact, as we have noted many times before, the traditional focus on cities has been low-frequency events that articulate our understanding and planning of the city over much longer time periods. In this sense, an urban operating system manages data and controls sensors that measure the city in real time at high frequency, and most of them do not appear to consider what that data means over terms longer than days or weeks, which would imply some concern for the low-frequency city and its long-term planning.[45] Closely related to such operating systems are the portals and gateways that package real-time data into a form through dashboards and such like, which we outlined in chapters 12 and 13. Information systems that increasingly relate to longer-term events are also being constructed as part of making government and management more effective, thus extending the real-time high-frequency data to longer-term views of what such data means for the city. These kinds of information systems are closely related not only to spatial data as represented through GIS but also to notions about building spatial data infrastructure that relates networking and operating systems to the collection and collation of data at different frequencies, as well as the software that is essential for managing such data and extracting relevant information. The major characteristic of such systems is that they imply many connections between sensors, real-time streamed data, and the software used to process such data and thence to visualize it, aggregate it, and apply it to various kinds of policy making and urban management functions. To an extent, this is also related to the embedding of such software and its infrastructure into the built environment, activities, and actions that we dealt with earlier and which we will pick up again in the last chapter as we begin to sketch out what the future of the smart city in particular, and the twenty-first century city in general, might be.

In this book, we have not emphasized the detailed hardware that is being put in place as physical plant and structures that automate the many functions that are carried out in public places. Moving sidewalks, electric vehicles, smart energy systems in the home, drones that complement the traditional delivery services, transit on demand, and many other features at the level of smart building and smart streets occupy much of the concern for the smart city. But the smart city goes well beyond this. The smart city is not really about technology but rather about how we think about ourselves and the future of humanity in cities, how cities will function as places where we interact socially and do business, and how the global marketplace and global governance impacts the city at every level from the local upward. As the digital revolution continues to demonstrate, we simply cannot predict the future, and most plans for smart cities, particularly those in the form of smart new towns, represent visions that are a mix of reality and fiction.[46] Most do not go as far as the middle-term dystopias captured in science fiction and portrayed in movies such as *Bladerunner*, *The Matrix*, and so on, but many do suggest that the city of the future will be built around new kinds of material, automated transit, and energy systems that reduce costs dramatically. Sustainability is also the watchword that dominates some of these visions such as the new towns we have portrayed in Masdar and Songdo and the now abandoned plan for the Toronto waterfront fronted up by Sidewalk Labs, Google's reach into the domain of smart cities.[47]

To an extent, all of these visions find it hard to embrace the traditional concerns of cities and, of course, usually steer clear of the key urban problems involving segregation, poverty, unemployment, housing affordability, and such like. In fact, several proposals have run into extremely vibrant criticisms from community groups wrestling with these problems and seeing little by way of change from the status quo in these visions that seek to develop a better life though the medium of the smart city.[48] The concern over whose data is being captured by such schemes is an ever-present worry, with many suggesting, as Shoshana Zuboff has in her critique of Sidewalk Labs' Toronto project, that it represents "the frontline of an historic contest between surveillance capitalism and democracy."[49] In our final chapter, we will paint a picture of where all this is going, illustrating that the continued diffusion, decentralization, and miniaturization of information technologies is likely to lead to a future that is even more uncertain than the present. We consider that living in cities will be dominated by successive and continual disruptions as the cycle from one innovation to the next continues to shorten. This might suggest that the singularity so often referred to could well manifest itself in extreme polarization and segregation as an increasingly small segment of the population who are able to grapple with ever-faster technologies come to dominate the way our cities are structured and the way our society is governed.

17   THE UNPREDICTABLE CITY

Once a new technology rolls over you, if you're not part of the steamroller, you're part of the road.
—Stewart Brand (1988)

THE POST-PANDEMIC CITY

For many of us, the two years from early 2020 to mid-2022 represented as radical a disruption in our patterns of behavior as anything that we have experienced in our entire lifetimes. When COVID-19 hit in February 2020, it quickly became apparent that this would be a global pandemic from which no one could escape. Most major governments in developed countries invoked regulations that confined more than half their populations to work from home, to observe social distancing that advised us to keep more than a meter apart, to restrict social gatherings, to wear masks in public spaces as well as work environments, and to engage in as many pursuits as we could that traditionally involved physical interactions but in the newly formed virtual worlds of online interaction that we were able to access.

The impact on cities has been devastating. Cities are places where people come together to interact. Their rationale is based on being near one another. Glaeser and Cutler articulated the major problem we face when they said, "human proximity that enables contagion is the defining characteristic of the city."[1] In the urban world that we have built, which requires us to be near each other, infections of the kind associated with COVID-19 necessarily keep us apart, and therein lies the dilemma. Unless we conquer such infections, although with COVID-19, after nearly four years, we seem to be well on the way to doing so with vaccines, the future of the city is in doubt. The fact, however, that cities have remained resilient to various diseases since the time they were first formed does suggest that we have an uncanny knack for resisting the plague. This time, it is likely to be no different, but it is entirely possible that some of the disruptions caused by the pandemic, which are now reflected in our changed behaviors, are here to stay. In short, when the dust settles and it is taking a while, there may well be a new normal.

For most of the two years when the pandemic was at its height, there were dramatic falls in long-distance travel. My own experience sums up the situation quite well. I did not use the subway or buses for most of this time, I walked everywhere I went in London, masked up, and only two or three times did I take a train to the West Country. Towards the end of the pandemic, which began in the summer of 2022, when I did eventually plan a trip to Singapore, I went to the airport in record time using a brand new high-speed subway line that had been finished but which had lain idle during the pandemic and had only just opened. Once at the airport, the terminal I flew from was much as I remembered it apart from the fact that what was for sale in the shops had changed quite radically. What I did not expect was the rate at which new information technologies had been substituted for old during this time. In the lounge, there were no newspapers, only QR codes for you to get them online on your phone. The same with ordering food. Once on the plane, all the physical menus and magazines were gone. To enter this world, you had to be an aficionada with QR codes, and if you were at all anxious about the online world, then life would be even harder. Needless to say, everything about the regulations you needed to meet to travel could only be done online. In short, form no longer appeared to follow function[2]—a classic example of how the online world is changing the city. The form looks the same, but all the functions are different as our ability to change digital technologies is almost immediate, but a change in their physical form takes much longer to effect, as it always has done. To an extent, this is one of the central messages of this book: that in cities, computation is breaking the link between physical form and socioeconomic functions, and this has profound implications not only for understanding the future city but also for engaging in its planning.

THE NEW NORMAL

In the standard model, the location of places where people live and work and the patterns of movement from home to work and back are the key components around which cities have been organized since they first emerged in preclassic times. If you disrupt this pattern, a good deal of their functioning falls apart, as we have witnessed during the last two years as the majority of people have worked from home. As the pandemic moves to being endemic and as the population is slowly returning to work, the virtual world of meetings that grew dramatically during the pandemic has largely remained in place. When the pandemic began, it was software based on voice over internet protocol such as Skype that was most widely used, but this was quickly expanded using technologies for virtual meeting such as Microsoft Teams, Zoom, and Google Meet, which were based on better functionality for such virtual interactions.[3] The growth of these systems has been such that currently there are many millions of participants using this software each day—150 million for Teams, 300 million for Zoom, and about 50 million for Skype and Google Meet. These are very crude estimates, and even as the pandemic ends, their use and adoption are still increasing as their advantages for remote online meeting continue to outweigh conventional physical meetings in many non-pandemic contexts.

It is still somewhat remarkable that when the pandemic hit, we already had available technologies that enabled us to interact virtually, to substitute physical interactions with virtual. The information infrastructures that had been assembled for email and social media as well as the emergence of the online economy had, of course, been put in place over many years from the time when LANs, computer graphics, and the internet began to converge in the two decades before the millennium. In this sense, it is perhaps unsurprising that an online world capable of enabling quite effective substitutions to be made between physical and virtual interactions existed in the shadow of our normal behavior and practice. What is surprising, however, is how well this seemed to work, given that much of this was accomplished over relatively low-capacity broadband from home rather than work. The last mile, so often talked about as the great obstacle to a seamless online world, did not appear to be a major constraint any longer. This is perhaps a testament to the improvements in software associated with networking and graphical user interfaces that continue remorselessly, unobserved and unremarked upon.

To a large extent, this is the death of distance that Frances Cairncross, among others, argued was beginning to take place from the early 1990s, but it seemed to stall, as only a small proportion of workers embraced the possibilities of working from home.[4] In 2015, only 5 percent of the US workforce were working from home entirely. This dramatically increased during the pandemic to about 60 percent, but it is impossible to know what the new normal will be if and when the pandemic washes itself out completely.[5] A related statistic before the pandemic in 2018 suggests that about the same percentage of the workforce have occasionally worked from home from which we might imply that a much larger proportion of the population will continue to work predominantly from home once the new normal stabilizes. All of these, of course, are speculations couched in high uncertainty.

If you work from home as an alternative to traveling to your place of work, then there can be considerable disruption to the patterns of transportation that define the old normal. In fact, in the core of a world city such as London, in the financial quarter—the square mile—where there are some half a million jobs, various reports have shown that at the height of the pandemic when the country was in severe lockdown, the number traveling into the area for work dropped by 90 percent. In the rest of the country, this decrease has been some 70–80 percent, while there has also been an increase in car usage, particularly in the suburbs. The impact on public transportation has been significant in that the majority of such transportation has sustained enormous losses of income and will continue to be subsidized by the national government for some time yet.[6] It is still not clear what level of patronage these systems will return to, and so far, although the pandemic is officially deemed to have ended in many Western nations, the levels are still only 75 percent of their old normal. In fact, what is hard to factor in is the uncertainty relating to the pandemic itself. There is still need for caution, as additional variants of the disease are still evolving, and there is a residual fear factor among a significant proportion of the population with respect to contagion. Low-level behavioral change in terms of rules for social distancing will remain for some time, perhaps even becoming part of the new normal.

Working from home has also disrupted patterns of consumption involving retail goods, food, and entertainment. There has been growth in home delivery in these areas, and this has changed the pattern of trips involving freight of various sorts. The wider context involving supply chains is also critical to the new normal, which has repercussions through the many networks that connect urban functions in the city together and to the wider global context. Woven into all of this are factors pertaining to incomes and expenditures that are changing the relative profitability and sustainability of previously well-established urban services. Even the platform companies such as Uber and other ride-hailing services, whose focus on flexible individual mobility one might consider to be an asset during the time of the pandemic, have been disrupted in that the fear factor of traveling on transport that is not regulated in terms of health issues has suppressed these businesses quite dramatically.[7]

There are several other elements of the standard model that have been disrupted and distorted by the pandemic. The use of real-time streaming has increased massively with respect to entertainment, particularly sports, which have shifted online, while education, like work, has bumped up and down as the pandemic has gone through different phases. The young are much less likely to become seriously ill from the virus than the elderly, and in this sense, there is less concern for isolation, but nevertheless, education mixes age groups and also encourages interactions in classroom contexts, which are highly problematic for disease transmission. Health itself has tended to shift online, with many routine consultations that traditionally have been held face to face shifting to virtual consultations through computer-aided cameras to picture physical symptoms in diagnosis and treatment. The implications of all this are that ever-greater capacities are continually being sought for digital interactions, and the move to 5G should continue this massive expansion of networks that has continued to take place over the last half century.[8]

We are not able to trace the wider ramifications of the pandemic on the structure of cities, nor can we embrace some of the critical urban problems such as climate change, housing affordability, segregation, and polarization, although all of these are affected by new technologies, on the one hand, and the pandemic, on the other. But we need to note that the overall form of the city could change dramatically if more than half of all workers continue to work from home in the foreseeable future. There is already a sense in which cities are beginning to sprawl once again, with sprawl only somewhat arrested prior to the pandemic as populations began to weaken their love affair with the automobile. In fact, this trend has been turned on its head, with an increased demand for living further away from the core of the city, far from the highest-density locations, which prior to the pandemic were regarded particularly by the young as the most favorable locations for living and working. These decentralizing forces have begun to gather momentum, leading to a further decrease in the use of public transport that generally requires high-density population locations in the center city for it to be sustainable.

There are many other forces that are emerging that will change the future city, which are somewhat marginal to our concern here, but in the wider picture of exploring the future form of the city, they are critical. Inflation is one major issue that relates to contemporary geopolitics, where a sea change is taking place on a worldwide basis. Much of this pertains to the energy crisis and thence to climate change and the quest to build a world based on net-zero emissions. The extent to which digital technologies are buried within all of these issues is hard to unravel, but to set all this in context, we now need to restate once again what we consider to be significant about the digital world and the way these various challenges will continue to reinforce the unpredictable nature of this future.

PERSPECTIVES ON UNPREDICTABILITY

As we have noted several times so far, the industrial revolution that began in England in the middle of the eighteenth century has consisted of a sequence of revolutions, one following on from the other, each woven into the preceding revolution with increasing frequency. This rate of change has reached the point where it is becoming more and more difficult to absorb the technical challenges that each technology offers, with the consequent implications that these technologies are beginning to collide with one another. The environment that is being created shows increasing signs of being destructive in managing and using new technologies, with their speed of implementation threatening to separate different user groups from one another and removing any incentives to integrate these technologies together.

It has been impossible to predict every subsequent revolution, with the nature of computation changing at each stage. Clearly, the industrial revolution has been one where mechanical, electrical, and digital technologies have built on each other with a focus on hardware, software, and then data, with organization now being overlaid on this sequence. But for more than forty years, it is software that has been elaborated in different ways, with new systems being increasingly difficult to absorb. If we pose the question of what comes next, then we may be able to guess what is likely in the next five years, but for anything longer than this, it appears impossible. If we look back eighty years since our story began, another eighty years will take us to the end of this century, and who could hazard a guess at what computation will look like there and then.

Miniaturization, as embodied in Moore's law, has driven this process since the beginning, but the consequences of super-exponential growth in computer memories and speeds have led to developments that, in hindsight, are easily explicable but were not expected at the time.[9] The personal computer was entirely unexpected not because it was unpredictable in terms of its hardware—it was obvious that one could build these devices if one was able to fabricate computers at fine enough scales—but rather because it was its usage that was most unexpected. The notion of using such a computer in a personal capacity was not foreseen. Clearly, one could do science on such machines, but it was all the other applications that smaller computers fostered and generated that we did not predict—until they happened, that is. The same is true of networks, and arguably it was only the development of fiber optics that made massive connectivity involving machines possible at all.

Now that we are all equipped with smartphones, it is easy to predict that these will get faster and smarter, while more and more applications will migrate into these handheld environments. We may be able to predict that display devices will broaden out within the physical environment, with data being projected onto various sorts of screen in virtual worlds—from Google Glass–like technologies to various kinds of augmented reality or metaverse. The IoT will continue to diffuse to all sorts of physical artifacts, while different modes of vehicular transport will embrace new software applications. Autonomous cars, despite the hyperbole given to them over the last decade, do not appear to be very likely in their pure form, but these are all fairly predictable in that they simply take what we already have a little further forward. In fact, if you search the web for stories about what is likely in the twenty-first century, all you will find are predictions that build on technologies that we already have. Indeed, it is in the human domain that the real strides might be made—in the online economy, in housing markets, in Fintech, perhaps in computer-aided fabrication of various kinds, in health, of course, and possibly in transport. Yet, the future is wide open, and it may be that it is in other areas that the greatest change will come, building on the information technologies that we already have.

In parallel to new technologies, there are many other unpredictable features of the future of our cities, and the most obvious one is the pandemic, whose impact on the city we have been discussing. In chapter 1, we argued that unpredictable events can be classified into different varieties. If an event is genuinely something that no one has ever observed before, this is what Taleb calls a "black swan." This is usually a relatively simple event in contrast to the kinds of events that we characterize here as a new technology, which are complex and relate to other events prior to the event in question, occurring and with important implications for the events that follow.[10] In fact, the pandemic is not a black swan in these terms, for there is knowledge and indeed experience of similar events in the past. An equivalent event at the end of World War I from 1917 to 1919 was the Spanish Flu, equally as unpredictable as COVID-19 but an event that we could understand to an extent and one that was predictable in hindsight, although one that was entirely unpredictable in terms of when it appeared. And in this case, there are still a few people alive who remember the event. From prehistory, there have been a succession of pandemics that are ingrained on our consciousness. The original black-swan event, however—when black swans were discovered for the first time in Australia in the late seventeenth century—was unimagined and unpredictable when such swans were first observed. In this sense, then, our sequence of revolutions that are now colliding with each other and from which we can glimpse elements of the future are not black swans, but they are still intrinsically unpredictable.

To an extent, this says much about our inability to predict the future—defined as events that have not yet happened—as it says about the nature of computation. A little before the time when computers were first invented, in 1934, the philosopher Karl Popper published his seminal book *The Logic of Scientific Discovery* in which he argued that science could never predict the future, it could only ever falsify it.[11] In short, theories that generated predictions could never be proven. They could withstand the test of time and not be disproven, but at some point in the indeterminate future, situations could, in principle, emerge that might falsify any theory. Only in situations where the system of interest was closed from the rest of the world, shielded, if you like, from outside influences, could a theory be invented or discovered that is able to generate 'truthful' predictions and I am still uncertain enough to put this in parenthesis. This could only happen if all these outside influences could be accounted for, and, of course, in practice, this was impossible anyway. Popper's theory of the scientific method as one of falsification—searching for theory that can be falsified—is now largely the conventional wisdom,[12] although the recent growth in big data, machine learning, and weak AI has tended to cloud the picture somewhat, with those who still consider that the truth lies in the data flocking to this ideology. In this sense, although our argument is one of successive unpredictable events, their intrinsic unpredictability is never guaranteed. All that can be said is that all the technologies associated with the digital revolution have been unpredictable so far.

To complete our discussion of the changes that have dominated the last eighty years, although we have defined different periods within the evolution of information technologies from the invention of the computer onward (and we could have taken this back as far as the invention of printing, the invention of written language, and so on), the past history of invention and innovation has been characterized by waves of change since the modern era began in Europe. These waves associated with economic growth and innovation due to Kondratieff, Schumpeter, and like-minded economists and historians provide a tidy way of seeing how different technologies get invented, are disseminated, mature, and are then often destroyed or radically changed as new technologies follow in their wake.[13] These cycles appear to be speeding up if one examines the different technologies that have been invented over the last two hundred years. The telephone took sixty years before it reached 80 percent of the US population, the PC thirty years, but the smartphone only took six years, and some newer software technologies hit the majority of the population sometimes in a matter of months. This collision of technologies is what makes us consider we are heading toward a singularity.[14] Before I finished writing this book, ChatGPT was thrust upon us which gives the singularity a further twist.

THE TRANSFORMATION OF THE STANDARD MODEL

Early in this book, we introduced the "standard model," which in essence was a portrait of the map of a typical city in which its activities, defined by different land uses, were arrayed in concentric fashion around the city's center. This model is an ideal type, of course, and we have already noted how the pandemic has put the functioning of such an ideal type on hold, so to speak, as its functions, in terms of how people move and where they locate, become frozen. In fact, the standard model is an incredibly robust structure based on very obvious ways of locating different activities. The center is often the origin of settlement where traders first came to exchange their goods: the marketplace, which in classical Hellenic times was the agora and now in US cities is the CBD. This concentric pattern, where the rings relate to different land uses, is usually permeated with radial routes that channel the traffic associated with workers commuting to the CBD from areas of housing around the center, thus distorting the regular and perfect concentricity with wedge-like slices where accessibility to the center is a little higher than in the associated rings. The logic for the location of different land uses associated with these different concentric rings is based on a trade-off between the cost of travel and the price of space. In a perfect market, where everyone works in the center and receives the same income, you would pay more rent if you lived nearer the center, but you would pay less transport cost, and the city would be balanced in that everyone would incur the same total cost: rent plus transport cost would always be the same wherever one lived. The place where one actually lived would simply be an idiosyncrasy in preferences, for one would incur the same total cost anywhere. This, of course, was von Thünen's insight[15] that we elaborated on at length in chapter 6.

In fact, in reality, there is no perfect market. Where people live nearer the center, the densities are higher, and in many situations, people who can afford to spend more on transport would also be able to consume more space the further they lived from the city. In this sense, richer people would pay less per unit of space than poorer people who lived nearer the center. To some extent, this mirrors what has been called the "poverty trap," where the poor are forced to live at higher densities and pay more per unit cost for their space. In many cities, however, much depends on the quality of the space, and although the general rule in the typical nineteenth- to twentieth-century industrial city is that the richer groups tended to live in more peripheral locations and the poor nearer the center, this is sometimes reversed if there is a very high quality of space close to the center. This often occurs in larger cities with greater longevity, capital city functions, and a global positioning in terms of the world's major financial trading hubs. It is also bolstered by the fact that travel times in really big cities can be very long, and it thus becomes necessary for the very rich to move back or remain as close as possible to the center.

As cities grew from the start of the first industrial revolution, the focus on the CBD increased as the growth of new economic functions reinforced economies of scale—agglomeration economies—through a process of positive feedback. At some point, however, the center becomes congested, and diseconomies of scale kick in. Businesses as well as residential populations find the periphery of the city, or at least locations away from the city center, more economically desirable. This leads to the phenomenon of out-of-town centers and edge cities in the suburbs,[16] where newer manufacturing industries, services, retailing, and so on grow in a more specialized way than in the original CBD, thus leading to an urban area where the city is pockmarked by many such hubs of different size and scale.[17] We cannot develop this idea in any detail here, but it is worth noting that in all cities, the model of the city core and its periphery is actually one in which a hierarchy of urban functions of different sizes of subcenter actually dominates the urban landscape. As we have noted, a single center surrounded by undifferentiated suburbs is an ideal type—a convenient fiction that simply lets us start a discussion of urban form and function. This is the model that has been widely exploited by economists in their explanation of how typical agents who produce and consume generate demand and supply for land in cities while reconciling their competing preferences, balancing their budgets, and maximizing their utilities.

To an extent, the standard model is more than simply a force that concentrates all activity at the center of the city but also one in which centralization at the core and decentralization toward the periphery are held in constant tension. Variants of this balance lead to different forms of city, from the most highly concentrated or compacted around the CBD to the most diffuse and sprawling structures such as those that one finds in parts of the world where the automobile reigns supreme.[18] As we have noted many times during this book, transport cost is one major determinant of this tension or balance, and when transport cost changes radically, as it has done over the last two hundred years with the invention of new technologies for moving people, goods, and now information, cities usually respond accordingly. But before we elaborate on this thesis one last time, it is worth noting that as world population has grown exponentially through the nineteenth and twentieth centuries, the number of cities from the smallest to the largest has grown accordingly. As the location of cities is very highly constrained to coastal areas and their hinterlands around the planet, the amount of space for the best locations has always been limited.

Cities have increasingly begun to grow by fusing into one another, thus leading to a pattern of growth that is much more complex than the ideal city on the isolated plain where all these functions are internalized. Cities of this type may have a well-defined hierarchy of many centers and subcenters, but they are very different from those that grow into one another, fusing in ways that make their hierarchical separation at different scales and sizes of center much harder to unravel. The term "polycentric" is often used to define cities that fuse in this way, and to some extent, this is already an elaboration of the standard model.[19] In short, the standard model is simply replicated over and over again as cities fuse with one another, although the resultant metropolis is not one of simply overlaying these models next to one another, for this does not take account of the relationships between different components of each city and the way cities grow incrementally. The resultant structure is a good deal more complicated than a simple replication of one model many times.

From another perspective, the standard model will continue to reign intact as long physical transport incurs costs. Frances Cairncross's death of distance that we discussed in detail in chapter 7 is about how these costs are transformed as people get to travel at faster speeds and lower per-unit cost but where there are still substantial deterrent effects to traveling long distances, particularly for specific functions that require some form of face-to-face contact.[20] For example, limits on time tend to impose an upper bound on time spent traveling. Although this can lead to specialization, where individuals and agencies organize their activities into large chunks of time such as days in a week where, for people to engage, they are still able to live at greater distances but travel infrequently for longer periods, distance is still the great leveler when it comes to location. In fact, it is the emergence of other kinds of movement that may well be instrumental in the death of distance. Transactions processing began long before the computer was invented, with various mechanical aids being developed in the late nineteenth century. It was not the digital computer per se that led to routine information processing but rather its convergence with communications, the rise of LANs, and thence the internet and its many variants. Although the notion of living and working in an electronic cottage dating back to speculations by E. M. Forster, James Martin, and Alvin Toffler, the difficulties of building cities and society around isolated communities and groups has rarely formed the basis of any of the ideal cities that we discussed in the last chapter.[21] In fact, only now, when we are wired so extensively, can we begin to speculate once again that a very different form for where we might work and live is a prospect we should seriously consider as the century progresses.[22]

As new technologies provide us with many new choices for enriching our lives, the prospect of a world in which our cities are spread out at very low densities seems less of a distant prospect than ever before. The pandemic has polarized our thinking along these lines. If pandemics are to be a feature of contemporary life and because their timing is entirely unpredictable, given our existing controls on public health, that is, then the notion of a world where cities fuse into one another at very low densities is a distinct possibility, already existing in parts of the US where the automobile reigns supreme. Electric cars with a long battery life could well herald this as being the model of the future, and already when the growth of populations in cities over the rest of this century is considered, the seeds of megacities at relatively low densities are already being planted. In fact, in contrast to this, the structure of cities in terms of the standard model is still very robust, and it could well be that compactness and high density come back onto the agenda. We have explored how robust the form and function of the idealized standard model is through various thought experiments, and it is hard if not impossible to break the mold that is based on layers and layers of complexity reinforcing the radial symmetry in form that we still see in most cities.[23] We simply do not know whether this will continue to hold. The future form of our cities thus remains entirely unpredictable.

The standard model tends to focus on the low-frequency city rather than the twenty-four-hour city—the high-frequency city—which relates to the many new technologies that now characterize patterns of behavior determining the way our cities function. Our cities have become more complex as more technologies have been invented, as we have found many new ways of using them to enrich our various social interactions, and as an ever-increasing proportion of the population have become enfranchised to invent and experiment with their own use of these technologies. In chapter 9, we made the distinction between high- and low-frequency events that described the dynamics and evolution of cities, and in chapter 12, we noted new forms of urban analytics that we might use to improve the functioning of cities across these domains. As we have been able to embed more and more sensors into the urban fabric that help us in understanding the functioning of different elements in the city along with the volumes of data being generated by ourselves from our various devices—fixed and mobile computers and smartphones—the high frequency becomes low frequency if we have long enough runs of data that take activities that occur second by second and consider them over months, years, and eventually decades. There are now many sensors that have been active for decades, and notwithstanding the limits to the real-time streamed data posed by changes in technologies of their capture, for the first time we now have the prospect of exploring how cities function across many different spatial and temporal scales.

There is a sense here that as new information technologies are invented, they tend to converge on each other and, in this way, obfuscate the ways in which we are able to control and manage our cities. This is clearly the case in many contexts, particularly when patterns of interaction using different information infrastructures such as the Ethernet and Wi-Fi carry very different kinds of traffic that can confound their impact on how we interact with one another socially and commercially. In fact, the notion of all these new technologies merging with one another is only one perspective on the digital world that we have invented. A somewhat opposite view is that new technologies do not necessarily merge with one another but rather remain separate, simply extending the range of activities being automated through software and thus broadening the activities that can be translated into data and information. In short, as we invent new technologies, they simply spread their impact wider and wider, and in this way too, they can confuse our abilities to understand and predict the future form of the information society.

There are several themes in this book that we need to pinpoint before we conclude. The future is unpredictable in many dimensions—in terms of how new information technologies coalesce, diffuse, build on one another, and continually generate new uses, users, and applications. The future form of the city seems robust at one level, but contemporary cities are vulnerable to changes in patterns and modes of transportation and location, from the impact of the pandemic but also from diverse forces such as climate change, housing affordability, and demographic aging and migration. Cities are getting more complex as new technologies and behaviors change the way we function, and thus our theories of how they function are forever in doubt. But it is the role of distance and the physical geometry of the city that is most in doubt, and to impress this, let us turn back to where we began—to the Post Office, which laid the foundations for the digital revolution that we have attempted to explain here.

BACK TO THE POST OFFICE IN 2022

It has taken more forty years for the British Government to sell off the Post Office in an effort to disengage these functions from the state. As we described in chapter 2, the process began in the late 1970s as deindustrialization in Britain, and more generally in the Western world, began in earnest and as deregulation from the institutions of the centralist state became the focal point of government. It is easy now in hindsight to see the process as being strongly linked to emergent information technologies that would spread computation and communications everywhere, but one of the most enduring features of cities is that we can still see physical residues from past technologies long after they have lost their original functions. The biggest of the remaining functions at St. Paul's in the City of London, which was dominated by posts and telecommunications until the 1980s, is the BT headquarters on the site of the old Central Telegraph Office where Marconi sent his first wireless signal. The building was sold in 2019 just before the pandemic began, for telecommunications no longer requires that kind of space in such an expensive location.[24] In the recent past, many such buildings have been taken over by banks and related financial services, but although a hedge fund has purchased it, its future use is likely to be entertainment and shopping, some financial services, and upmarket restaurants, but even this is uncertain, as this transition in usage is passing as many of these activities themselves go virtual.

Next door to BT, the GPO building built in 1889 at 1 St. Martin-le-Grand, which was sold to Nomura Bank in 1986, has recently divested itself of its banking functions due to the demise of the firm as a banking giant.[25] If you were to ask what uses and functions the building now contains, you might be as surprised as I was to find out that some of the financial, human resources, and information services (what was once the computer center) in my own place of work, UCL, have leased much of this space, moving and expanding the administration of the college into the City of London from its main site in Bloomsbury. The word on the street, however, is that such a location for these kinds of services is a halfway house on the way to complete decentralization and possibly dissolution as these types of function become completely dispersed and automated in the new digital platforms that are emerging. Universities, you might expect, could become largely virtual due to the way information and knowledge can now be packaged and delivered, at least in principle. But it is an open question as to whether such support services that are quickly automating, as well as face-to-face interactions, physical lectures, libraries with physical as well as digital materials, laboratories, and all the traditional materials of education and research will continue to remain at the core of such institutions. It is functions and activities such as these that are now being located at St. Martin-le-Grand, and it is likely that in the near future, many of these will be automated—not just the back offices but also layers of not-so-far back offices, closer to the cutting edge, where many jobs will be lost as we travel headlong into the digital future.

It is somewhat ironic that in the course of this book, we have moved full circle from Marconi first demonstrating wireless technologies 120 years ago across the precinct of St. Paul's Cathedral to the new location of the information systems division in my own university, which is now adjacent to the heart of where the GPO began at St. Martin-le-Grand. The information mile has thus looped back from the University of London in Bloomsbury to St. Paul's. In a sense, it does not matter any longer where any of these functions are located, and herein lies one of the conundrums of the contemporary city. Form no longer follows function. What the buildings actually contain is no longer related to their original functions, and explaining this disconnect promises to be the dominant issue in thinking about how we might understand and plan cities and improve the quality of urban life in the near and far future.[26] You might think that this has always been the case, but you would be wrong. Many buildings contain elements of their original functions, but in the transition to the new digital world, many of the physical functions of the past that required particular locations are no longer significant, and the new ethereal functions of the future built around digital information are essentially rootless, placeless, decentralized, liquid, even diffuse. This is the true meaning of globalization for the future city.

Our journey from 1889 beginning at the GPO building traveling west on the information mile and north to Bloomsbury where the first internet connection was established in the computer center at UCL in 1973 now takes us back to the GPO more than two hundred years later where what currently passes for the UCL computer center now resides. This mirrors the massive transition in digital computing from a complete focus on its physical distribution to our contemporary perspective on digital organization. As we argued in earlier chapters, more than forty years ago, the notion arose that the dominant activity in computing was shifting from hardware to software, what was then called "orgware." This transition, which became clear as the computer revolution became all-pervasive, implied that hardware would continue to diminish in importance and cost, software would become ever more automated, and it would be the organization of computation that would preoccupy us and take the lion's share of the resources devoted to computing.[27]

To an extent, this has come to pass, and in a way, this almost defines our progress in charting the development of the computable city, where the first part of this book was devoted to hardware, the second to the physicality of the city, the third to software, and the fourth to this much wider foray into the way computation is organized and the way cities are being planned using new technologies. Insofar as we can glimpse the urban future, digital technologies will continue to weave themselves in and out of our daily lives to such an extent that after another eighty years has passed, by the end of this century, it will be almost impossible to reconstruct the origins that we have used to compose the story we have told here.

NOTES

CHAPTER 1

This is attributed to von Neumann in Paige Thompson's article "John von Neumann, the last great polymath," at https://www.sothebys.com/en/articles/john-von-neumann-the-last-great-polymath in 2018, but it is also accredited to a summary of von Neumann's life by Stanislaw Ulam in "John von Neumann, 1903–1957," *Bulletin of the American Mathematical Society*, 64, 1–49, 1958. There are a number of similar quotes by those close to the invention of technologies based on the computer, and these are introduced in the text, but some of these must be taken as unattributable sentiments and hyperbole.

1. Ananyo Bhattacharya, *The Man from the Future: The Visionary Life of John von Neumann*, Penguin Press, New York, 2021.

2. There is little doubt that Thomas Watson did not say these precise words, but similar phrases, often hyped out of all proportion to their impact, were made by many at the time. There is a good discussion in the Watson Wikipedia entry https://en.wikipedia.org/wiki/Thomas_J._Watson.

3. Klaus Schwab, *The Fourth Industrial Revolution*, World Economic Forum, Portfolio Penguin, Geneva, CH, 2016, https://www.weforum.org/about/the-fourth-industrial-revolution-by-klaus-schwab.

4. Alvin Toffler's *The Third Wave*, Bantam, New York, 1979, and Erik Brynjolfsson and Andrew McAfee's *Race Against the Machine*, Digital Frontier Press, Lexington, MA, 2016, provide incisive commentaries on the origins of the revolution in computation.

5. Schwab's successive industrial revolutions were preceded by Alvin Toffler's characterization of three waves first stated in his first major book *Future Shock*, Bantam, New York, 1970, where he defines the first wave as characterizing agricultural society, the

second as industrial, and the third, similar to Schwab's third revolution, as the post-industrial era marked by the rise of information technologies and the computer.

6. https://www.weforum.org/agenda/2016/01/the-fourth-industrial-revolution
-what-it-means-and-how-to-respond/.

7. Michael Batty, *Inventing Future Cities*, MIT Press, Cambridge, MA, 2018.

8. Patrick Vlaskovits, "Henry Ford, innovation, and that 'Faster Horse' quote," *Harvard Business Review*, August 29, 2011, https://hbr.org/2011/08/henry-ford-never-said
-the-fast. He may have said something similar in 1923. See https://quoteinvestigator
.com/2011/07/28/ford-faster-horse/.

9. Such as the technological singularity, https://en.wikipedia.org/wiki/Technological
_singularity. See also Ray Kurzweil, *The Singularity Is Near: When Humans Transcend Biology*, Penguin Books, New York, 2006; Nick Bostrom, *Superintelligence: Paths, Dangers, Strategies*, Oxford University Press, Oxford, 2014; and the science fiction writings of Philip K. Dick, William Gibson, Neal Stephenson, and others.

10. Thomas S. Kuhn, *The Structure of Scientific Revolutions*, University of Chicago Press, Chicago, IL, 1962.

11. Vannevar Bush, "As we may think," *The Atlantic Monthly*, *176*, No. 1, 101–108, 1945, https://www.ps.uni-saarland.de/~duchier/pub/vbush/vbush-all.shtml.

12. https://en.wikipedia.org/wiki/Turing_machine.

13. John von Neumann, *The Computer and the Brain*, Yale University Press, New Haven, CT, 1956.

14. Alan M. Turing, "Computing machinery and intelligence," *Mind: A Quarterly Review of Psychology and Philosophy*, *LIX*, *236*, October, 433–460, 1950.

15. Walter Isaacson, *Leonardo Da Vinci*, Simon & Schuster, New York, 2018; Jean Barone (Editor), *Leonardo da Vinci: A Mind in Motion*, British Library Press, London, 2019.

16. Newton's laws are explained in https://en.wikipedia.org/wiki/Newton's_laws_of
_motion.

17. James Cheshire and Michael Batty, "The era of the megalopolis: how the world's cities are merging," *The Conversation*, November 22, 2022, https://theconversation
.com/the-era-of-the-megalopolis-how-the-worlds-cities-are-merging-193424.

18. The idea that cities are transforming themselves into computers is a speculation that defines the emergence of the smart city. It is an extreme characterization, notwithstanding that it provides a powerful imageability for how cities are becoming computable. I wrote about this speculation in my article "The computable city," *International Planning Studies*, *2*, 155–173, 1997, but it is worth contrasting this with the robust and trenchant critique of this perspective by Shannon Mattern in her book *A City Is Not a Computer: Other Urban Intelligences*, Princeton University Press, Princeton, NJ, 2021.

19. Nicholas Negroponte, "Toward a theory of architecture machines," *Journal of Architectural Education*, *23*, No. 2, 9–12, 1969, and Stewart Brand, *The Media Lab: Inventing the Future at MIT*, Penguin Press, New York, 1989.

20. Klaus Schwab, *The Fourth Industrial Revolution*. https://www.weforum.org/pages /the-fourth-industrial-revolution-by-klaus-schwab.

21. https://www.weforum.org/about/the-fourth-industrial-revolution-by-klaus -schwab.

22. Sarah Barns, *Platform Urbanism: Negotiating Platform Ecosystems in Connected Cities*, Palgrave Macmillan, London, 2019.

23. Mark Weiser, "The computer for the 21st century," *Scientific American*, *265*, No. 3, 94–104, 1991.

24. A useful history of the early development of AI is presented in "Homage to John McCarthy, the father of artificial intelligence (AI)," *Artificial Solutions*, October 29, 2020, at https://www.artificial-solutions.com/blog/homage-to-john-mccarthy-the -father-of-artificial-intelligence.

25. George Dyson, *Turing's Cathedral: The Origins of the Digital Universe*, Pantheon Books, New York, 2012.

26. Michael Wooldridge, *The Road to Conscious Machines: The Story of AI*, Penguin Books, London, 2021.

27. Arthur C. Clarke's three laws, https://en.wikipedia.org/wiki/Clarke%27s_three _laws. There is an exploration of several interpretations of this law from scholars at the Cultural Research and Innovation Lab at https://lab.cccb.org/en/arthur-c-clarke -any-sufficiently-advanced-technology-is-indistinguishable-from-magic/.

28. Black swans are unpredictable events, named after early Dutch explorers to Western Australia, who discovered black swans, unseen and unheard of in their native land. For an excellent summary of the nature and conditions under which such events occur, see Nicholas Taleb, *The Black Swan: The Impact of the Highly Improbable*. Random House, New York, 2007.

29. Kate Crawford, *Atlas of AI*, Yale University Press, New Haven, CT, 2021.

30. Philip E. Tetlock and Dan Gardner, *Superforecasting: The Art and Science of Prediction*, Random House, New York, 2015.

CHAPTER 2

Quoted by Nalaka Gunawardene, "Humanity will survive information deluge," *OneWorld South Asia*, December 5, 2003, https://bazaarmodel.net/phorum/read.php ?1,462.

1. Of the several histories of the British Post Office, Duncan Campbell-Smith's *Masters of the Post: The Authorised History of the Royal Mail*, Allen Lane, Penguin Press,

London, 2011, is a tour de force, taking the story to the early 2000s. Much of the early section of this chapter relies on his descriptions and insights.

2. Anon, "The birth of pioneering electrical engineer Guglielmo Marconi," *The Daily Telegraph*, April 28, 2017, https://www.telegraph.co.uk/technology/connecting -britain/guglielmo-marconi-birth/.

3. The name change from "Post Office" to "St. Paul's" is described in the Wikipedia entry for Blackfriars tube station, https://en.wikipedia.org/wiki/Blackfriars_station, and in Alan Jackson's *London's Termini*, David & Charles, London, 1969, new revised edition, 1984.

4. Campbell-Smith, *Masters of the Post*.

5. The year 2019 marks the 190th anniversary of the move of the Post Office to St. Martin's Le Grand, https://www.postalmuseum.org/blog/190-years-of-londons -post-office-quarter/. It was a daily spectacle when the mail coaches left St. Martin's Le Grand in the late afternoon, racing each other along the post roads, which attracted large crowds to the colorful event. See British History online at https:// www.british-history.ac.uk/old-new-london/vol2/pp208-228.

6. If you visit the Post Office extension, which is now owned by Bank of America, there is statue of Roland Hill at the front, and if you point your smartphone at the plinth, focusing on the QR code, it will give you some history of the penny post (if it is working!).

7. Since classical times, commentators and scholars have held to the idea that cities are places where we come together to pool our ideas and exchange goods, thereby increasing our wealth through our division of labor and specialization, thus sharing the collective surplus. Jane Jacobs in her classic book *The Death and Life of Great American Cities*, Random House, New York, 1961, made this rationale the essence of her critique of modern urban planning, while more recently, Ed Glaeser in his *Triumph of the City*, Macmillan, London, 2013, has extended and deepened her argument.

8. Edison's project is described in Victor Keegan's *Lost London 8: Hidden History of Holborn Viaduct*, https://www.onlondon.co.uk/vic-keegans-lost-london-8-hidden-history -of-holborn-viaduct/, which also has reference to Amazon's new office in an adjacent location.

9. The location of Amazon's London head office, https://www.buildington.co.uk /london-ec1/60-holborn-viaduct/sixty-london/id/2862.

10. Matthew Green, "The fascinating history of Fleet Street," *The Daily Telegraph*, May 21, 2018, https://www.telegraph.co.uk/travel/destinations/europe/united-kingdom /england/london/articles/the-history-of-fleet-street-and-british-newspaper-industry/, and Ben Judge's summary of the founding of *The Daily Courant*, "11 March 1702—the world's first daily newspaper published," https://moneyweek.com/383504/11-march -1702-elizabeth-mallet-daily-courant-first-daily-newspaper/.

11. Harrison's Herculean efforts to crack this problem are admirably described by Dava Sobel in her magnificent book *Longitude: The True Story of a Lone Genius Who Solved the Greatest Scientific Problem of His Time*, Fourth Estate, London, 1995.

12. From Basil Mahon's *The Man Who Changed Everything: The Life of James Clerk Maxwell*, John Wiley, Chichester, UK, 2004.

13. James Clerk Maxwell, "On hills and dales," *The London, Edinburgh, and Dublin Philosophical Magazine and Journal of Science*, Series 4, *40*, 421–427, 1870.

14. Sara Rigby, "James Clerk Maxwell: The great scientist with a profound impact on modern physics," *Science Focus*, February 4, 2019, https://www.sciencefocus.com /science/james-clerk-maxwell-the-most-important-physicist-you-havent-heard-of/.

15. Richard Feynman, *The Feynman Lectures on Physics: Volume 2: Mainly Electromagnetism and Matter*, 1–6. Addison-Wesley, Reading, MA, 1964.

16. Kaiser Tarafdar comments on this in the Quora post at https://www.quora.com /What-was-the-relationship-between-James-Clerk-Maxwell-and-Michael-Faraday, where there are more links to their contacts.

17. Laurence Scales, "How television was invented in London," *Londonist*, 2017, https://londonist.com/2016/01/how-television-was-invented-in-london.

18. We will put the first internet connection outside the US in context in a later chapter, but its location at University College (now UCL) is described by Peter Kirstein in his post "Early experiences with the ARPANET and Internet in the UK," at http://nrg.cs.ucl.ac.uk/internet-history.html.

19. There are many biographies of Alan Turing, and these tell that he was born in 1912 in a nursing home at Warrington Lodge, 2 Warrington Crescent, Maida Vale, London. There is the usual plaque commemorating him at that address. and there is a good summary of his early years on English Heritage's Blue Plaques website, https://www.english-heritage.org.uk/visit/blue-plaques/alan-turing/.

20. An extensive survey of the development of the Colossus at Bletchley Park and the role of Tommy Flowers in the design of the first digital computer is given in Jack Copeland's edited volume *Colossus: The Secrets of Bletchley Park's Code Breaking Computers*, Oxford University Press, Oxford, 2006.

21. The best history of telecommunications, particularly the telegraph in the nineteenth century, is Tom Standage's *The Victorian Internet: The Remarkable Story of the Telegraph and the Nineteenth Century's On-Line Pioneers*, Weidenfeld and Nicolson, London, 1998.

22. The first passenger railways were opened from Stockton-on-Tees to Darlington in 1825 and from Liverpool to Manchester in 1829, and this heralded the start of a massive expansion of cities. For the first time in the history of the world, the urban population could live up to twenty miles from their work, if they could afford to

travel by rail of course. When the Liverpool to Manchester line first opened, it was hailed as a remarkable novelty, but it brought home to roost just how precarious the beginnings of the industrial revolution were. The entire episode was dogged by failing engines and unruly crowds. The well-known Member of Parliament for Liverpool, William Huskisson, got off the train and crossed the track to talk with the prime minister in his carriage only to be caught by an open door falling onto the path of the train pulled by the Rocket traveling the other way. He died of his injuries that evening, https://en.wikipedia.org/wiki/Opening_of_the_Liverpool_and_Manchester_Railway.

23. A nice illustration of this linking of wires to information is by Pamela Fox in her "Transporting bits over wires," at https://www.khanacademy.org/computing/ap -computer-science-principles/the-internet/wires-wifi-physical-network-connections /a/transporting-bits-over-wires/.

24. Standage, *The Victorian Internet.*

25. Charles Wheatstone began his research in acoustics but went on to invent several devices such as early electric motors and the stereoscope, https://www.kcl.ac.uk /aboutkings/history/famouspeople/charleswheatstone.

26. See Clarke's third law, https://en.wikipedia.org/wiki/Clarke%27s_three_laws.

27. The period of railway building that began after the first passenger railways were opened was frenetic. At one point in 1846, the British Parliament had 272 bills before it to construct new railways. The bubble burst in that year, but this simply signaled a period of intense railway construction that only tailed off in the late 1870s. See Jesse Colombo, "The British 'Railway Mania' bubble," *The Bubble Bubble*, April 19, 2012, http://www.thebubblebubble.com/railway-mania/. On another continent, Chicago was a boomtown in the 1830s with equally dramatic growth due to its establishment as a rail hub for the Midwest. William Cronon in his book *Nature's Metropolis: Chicago and the Great West*, W. W. Norton, New York, 1992, paints a remarkable picture of this mania, even more frenetic than what one can see in parts of modern-day China.

28. Duncan Geere, "How the first cable was laid across the Atlantic," *Wired Magazine*, January 18, 2011, https://www.wired.co.uk/article/transatlantic-cables, and in Standage's *The Victorian Internet.*

29. Alexander Graham Bell is accredited with the invention of the telephone in Boston in 1875–1876. Like many of the inventions portrayed in this book, there were many close antecedents and many alternative conceptions of such a device. A good summary is in the Wikipedia entry available at https://en.wikipedia.org/wiki /Invention_of_the_telephone.

30. Sharon Basaraba, "A guide to longevity throughout history, from the prehistoric onward increases in lifespan from prehistory through the modern era," *Very Well Health*, July 14, 2019, https://www.verywellhealth.com/longevity-throughout -history-2224054.

31. The death of distance is attributed to Frances Cairncross from the title of her book *The Death of Distance: How the Communications Revolution Is Changing Our Lives*, Harvard Business School Press, Cambridge, MA, 1997. *The Economist* published a special issue on the subject in 1995 in which her own article was entitled "The death of distance," *The Economist*, *336*, No. 7934, 63–63, September 30, 1995. The term, however, had appeared in many guises already, particularly as the "annihilation of distance" that was used in the early years of the nineteenth century. We will return to this in a later chapter when we deal explicitly with the impact of the death of distance on the form and functioning of the contemporary city.

32. Nicholas Negroponte, *Being Digital*, Vintage Books, New York, 1995.

33. Anon, "How far, how fast?," *Medieval Worldbuilding Information*, October 13, 2009, https://writemedieval.livejournal.com/4706.html.

34. We have already introduced Klaus Schwab's characterizations of the four industrial revolutions that comprise the period from the invention of machines (and the science of mechanics), namely the internal combustion engine, to the present day, which is dominated by digital technologies (and the science of biology). As we demonstrated in chapter 1, there are several variants on the division of the last 250 years into periods, and the one that best accords to the arguments in this book is from mechanical to electrical to digital eras, with the latter being that which has been elaborated into a fourth era, consistent with Schwab's argument in his *The Fourth Industrial Revolution*, World Economic Forum, Geneva, Switzerland, 2016.

35. Cairncross, *The Death of Distance*.

36. H. G. Wells wrote a wonderful essay in 1902 entitled "The Probable Diffusion of Great Cities." It is chapter 2 in his book *Anticipations*, Chapman and Hall Ltd., London, which sketches a remarkably prescient future with respect to the spread of cities and the death of distance.

37. I elaborate on this proposition and argument in my book *Inventing Future Cities*, endnote 8, chapter 1.

38. Negroponte, *Being Digital*.

39. I wrote about the invisibility of information technologies in an editorial entitled "Invisible cities," *Environment and Planning B*, *17*, 127–130, 1990. A much more eloquent argument defines the work of Italo Calvino in his book *Invisible Cities*, Secker and Warburg, London, 1974, where the notion of what is visible and invisible in cities is widely exploited in conversations between Kublai Khan and Marco Polo, purportedly in Venice or a fictional rendition thereof.

40. The argument is wider than cities of course. Rita Gunther McGrath points the direction in her article "The world is more complex than it used to be," *Harvard Business Review*, August 31, 2011, https://hbr.org/2011/08/the-world-really-is-more-compl.

CHAPTER 3

Turing, "Computing machinery," endnote 15, chapter 1.

1. There are many good biographies of Turing that recount his early life and later contributions in detail, but that by Andrew Hodges, *Alan Turing: The Enigma*, Hutchinson Publishing Co., London, 1983, is one of the most comprehensive. We also noted Turing's birthplace in the nursing home in London's Maida Vale in endnote 20, chapter 2. In Britain, he mainly lived in the south of the country in the London region, apart from the last eight years of his life where he lived in Manchester working at the university. We should also note that King's College Cambridge where Turing studied is not King's College London where James Clerk Maxwell produced his famous wave equations. The latter is part of the University of London system, whereas the former is a college of the University of Cambridge, see https://www.quora.com/Is-King-s -College-London-related-to-King-s-College-at-Cambridge-University.

2. Of the twenty-three problems posed by David Hilbert in 1900, it was the second that Gödel tackled—the decidability or the incompleteness problem—and this is what attracted the young Turing. There is a good account of these problems in the relevant Wikipedia entry, https://en.wikipedia.org/wiki/Hilbert%27s_problems. In George Dyson's book *Turing's Cathedral*, endnote 24, chapter 1, there is an excellent description of the environment in which Turing lived during his time in Princeton.

3. There is a useful and clear account of how a Turing machine functions on the University of Cambridge's Computer Laboratory's website at https://www.cl.cam.ac .uk/projects/raspberrypi/tutorials/turing-machine/one.html.

4. Alan M. Turing, "On computable numbers, with an application to the Entscheidungsproblem, 1936," *Proceedings of the London Mathematical Society, S.2, 42* No. 1, 1937, 230–265.

5. An excellent visual survey of his life and especially his years in the US and then at Bletchley Park where he began to work on his return to England is provided by Andrew Hodges in his "The Alan Turing Internet Scrapbook," https://www.turing .org.uk/scrapbook/oracle.html. Back in England, just before he went to Bletchley Park, there is some evidence that he worked near London's West End in one of the government's spy services, possibly an offshoot of MI6, which was then located in Victoria, and in this capacity, he was in the physical neighborhood of our information mile. If we knew exactly where the location was, we might even take it in on our walk, but spy services have a habit of covering their tracks.

6. On August 15, 1939, more than eighty years ago from the time of writing, the Government Code and Cypher School (GC&CS) moved to Bletchley Park (https:// bletchleypark.org.uk). Those who made that journey were told in no uncertain terms that the location was top secret. Among many other mandates, they were instructed that "The address for official correspondence and private letters will be Room 47 Foreign Office and the official Telephone Number—Whitehall 7947." There is an

excellent description of the address in Eleanor Jainey's memoir at https:// www .eliotsofporteliot.com/evjainey/18.html.

7. Campbell-Smith, *Masters of the Post*, endnote 2, chapter 2.

8. Hodges, *Alan Turing*, and Sinclair McKay, *The Lost World of Bletchley Park: The Illustrated History of the Wartime Codebreaking Centre*, Aurum Press, London, 2013. You can currently (June 2022) see one of the Enigma machines in the reception area of the Alan Turing Institute, 1st Floor, The British Library, London.

9. Joel Greenberg, *Gordon Welchman: Bletchley Park's Architect of Ultra Intelligence*, Frontline Books, London, 2014.

10. In fact, up until about the last quarter of the twentieth century, if you were selected to study the harder sciences at a British university, you would have had to learn German as a prerequisite to following various physics and chemistry courses.

11. Jerry Roberts, *Lorenz: Breaking Hitler's Top Secret Code at Bletchley Park*, History Press, Stroud, UK, 2017.

12. Bill Tutte is one of the great unsung heroes of World War II. The Lorenz cipher was only declassified in 2004, and by then, Bill was sadly departed. I met him once in the Faculty Club at the University of Waterloo, Ontario, in 1975, where he was, by then, one of the world's most eminent professors in graph theory, but he was entirely unknown for his wartime work. Slowly but surely, his contributions have come to light, and he is now well remembered in his hometown of Newmarket in Suffolk, UK, where there is a wonderful sculpture dedicated to his memory, which dominates the town center. See https://en.wikipedia.org/wiki/W._T._Tutte.

13. Copeland, *Colossus*, endnote 21, chapter 2. The instruction to destroy the as yet undelivered Colossus machines, nine of them, once the war ended, is attributed to Winston Churchill.

14. There are at least two movies about Bletchley Park. The first, starring Kate Winslet, was made in 2001 and is called *Enigma* based on Robert Harris's book (Random House, London, 1995). See https://www.imdb.com/title/tt0157583/. The second and more recent one, starring Benedict Cumberbatch and Keira Knightley, was made in 2014 and is entitled *The Imitation Game* after Turing's famous paper of the same name. See https://www.imdb.com/title/tt2084970/ and Turing, "Computing machinery," endnote 15, chapter 1 (https://www.csee.umbc.edu/courses/471/papers/turing.pdf).

15. Norman Macrae, *John Von Neumann: The Scientific Genius Who Pioneered the Modern Computer, Game Theory, Nuclear Deterrence, and Much More*, Pantheon Books, New York, 1992.

16. The Project Gutenberg EBook of Bacon and Shakespeare, by Albert F. Calvert, July 24, 2015 [EBook #49516], see https://www.gutenberg.org/files/49516/49516 -h/49516-h.htm

17. Tilly Blyth, *Information Age: Six Networks that Changed the World*, Science Museum, London, 2014.

18. A useful summary of Babbage's difference engine can be found at https://en .wikipedia.org/wiki/Charles_Babbage, but in some respects it was his collaborator, Ada Lovelace, who broke the mold in writing software for his mechanical engine. A nice account of her involvement can be found in James Essinger's, *Ada's Algorithm: How Lord Byron's Daughter Ada Lovelace Launched the Digital Age through the Poetry of Numbers*, Gibson Square, London, 2013.

19. Dyson, *Turing's Cathedral*, endnote 24, chapter 1.

20. Several excellent resources exist about the history of computing machines, and much of the background here is taken from these: The History of Computing Project, https://www.thocp.net/; The Computer History Museum, https://computerhistory .org/; and The National Museum of Computing, https://www.tnmoc.org/, among several others.

21. Dyson, *Turing's Cathedral*, endnote 24, chapter 1.

22. Simon Lavington has written extensively about these times. See his *Early British Computers*, Manchester University Press, Manchester, UK, 1986; *Early Computing in Britain: Ferranti Ltd. and Government Funding, 1948–1958*, Springer, Berlin, Germany, 2019.

23. See Alan M. Turing, "The chemical basis of morphogenesis," *Philosophical Transactions of the Royal Society of London B, 237*, No. 641, 37–72, 1952, and John von Neumann, *Theory of Self-Reproducing Automata*, University of Illinois Press, Champaign-Urbana, IL, 1966.

24. Ian McEwan, *Machines Like Me, And People Like You*, Jonathan Cape, London, 2019.

25. Maurice V. Wilkes, "Early computer developments at Cambridge: The EDSAC," *Radio and Electronic Engineer, 45*, No. 7, 332–335, 1975.

26. In 1943, Thomas Watson, president of IBM, said this, while Ken Olsen, founder of the Digital Equipment Corporation, made his remark in 1977. See Robert Strohmeyer, "The 7 worst tech predictions of all time," *PC World*, December 31, 2008; https://www.pcworld.com/article/155984/worst_tech_predictions.html.   See   also endnote 3, chapter 1.

27. Jay Forrester tells us this in the preface to his *Urban Dynamics*, MIT Press, Cambridge, MA, 1969.

28. Paul E. Ceruzzi, William Aspray, and Thomas J. Misa, *History of Modern Computing*, MIT Press, Cambridge, MA, 1998.

29. An excellent review of the early development of computers from before the Digital Age is contained in Charles Eames and Ray Eames, *A Computer Perspective: Background to the Computer Age*, Harvard University Press, Cambridge, MA, 1973, New Edition, 1990.

30. Jon Gertner, *The Idea Factory: Bell Labs and the Great Age of American Innovation*, Penguin Press, New York, 2013.

31. Michael Swaine and Paul Freiberger, *Fire in the Valley: The Birth and Death of the Personal Computer*, 3rd Edition, Pragmatic Bookshelf, Raleigh, NC, 2014.

32. Batty, "The computable city," endnote 18, chapter 1.

33. Anthony M. Townsend, *Smart Cities: Big Data, Civic Hackers, and the Quest for a New Utopia*, W. W. Norton, New York, 2013.

34. There is a surprising dearth of good accounts of the continued miniaturization of computer hardware, but the Wikipedia entry provides a decent summary at https://en.wikipedia.org/wiki/Miniaturization.

35. Computers were first used to simulate cities long before they were used to control and manage them in real time. There is a long tradition of urban modeling that we will explore extensively in part III of this book, but to get an impression of how this domain is structured, Anthony Townsend's *Smart Cities* deals with this history, as well as contrasting it with the current preoccupation of embedding computers into the very fabric of what we are modeling using those same computers—a classic demonstration of their universality.

36. Batty, *Inventing Future Cities*, endnote 8, chapter 1.

37. Gordon E. Moore, "Cramming more components onto integrated circuits," *Electronics*, *38*, No. 8, 114–117, April 19, 1965, https://newsroom.intel.com/wp-content/uploads/sites/11/2018/05/moores-law-electronics.pdf.

38. John W. Tukey is credited with coining the word "software" when he was at Bell Labs in 1953. This was long before the PC revolution that brought software to the attention of the general computer user in the 1980s. See David Leonhardt, "John Tukey, 85, statistician; coined the word 'software,'" *The New York Times*, July 28, 2000, Section A, 19.

39. Jay Forrester directed the Whirlwind project, which was ultimately housed at the MIT's Lincoln Laboratory. On Ed Morrow's TV show *See It Now*, he demonstrated computer graphics for the first time on an oscilloscope. You can find a clip of this movie on the MIT video site, along with a complete transcript of the show: https://infinitehistory.mit.edu/video/edward-murrows-see-it-now—jay-forrester-and-whirlwind-computer-1951.

CHAPTER 4

Bill Gates in a speech entitled "Software breakthroughs: solving the toughest problems in computer science," at the University of Illinois Urbana-Champaign, February 24, 2004, reported in https://www.computerworld.com/article/2534366/the-quotable-bill-gates--in-his-own-words.html.

1. There are many characterizations of the cults that surrounded the early computer installations. Joseph Weizenbaum in his book *Computer Power and Human Reason: From Judgment to Calculation*, W. H. Freeman, New York, 1976, talks at length about the priesthood, and so too did Steven Levy in his book *Hackers: Heroes of the*

Computer Revolution, O'Reilly Media, Sebastopol, CA, 2010, where he provides a stark contrast to the priesthood in the form of nerds and hippies who particularly dominated university computing centers.

2. A good visual account of the media associated with the early input and output of data for mainframe computers is contained in "The history of computer data storage, in pictures," April 12, 2019, available at https://royal.pingdom.com/the-history -of-computer-data-storage-in-pictures/. A similar visual history of PC hardware can be found at https://royal.pingdom.com/the-history-of-pc-hardware-in-pictures/.

3. There is a good discussion of Fred Williams's career and the role of Tom Kilburn, among others, in the construction of the Manchester Baby. See Gerard O'Regan's *Giants of Computing: A Compendium of Select, Pivotal Pioneers*, Springer, London, 2013. David Anderson reports on Tom Kilburn in his article "Historical reflections: Tom Kilburn: A tale of five computers," *Communications of the ACM*, 57, No. 5, 35–38, 2014. Simon Lavington's books *Early British Computers* and *Early Computing in Britain*, cited above in endnote 22, chapter 3, fill in the picture.

4. In the mid-1960s, I learned a programming language called Atlas Autocode, where our data and programs were punched onto paper tape. This often involved massive reels of such tape being spread out and filling whole rooms as we attempted to search for errors in the program, which needed fixing by splicing new punched tape in to correct the program on the paper tape. I do not have any pictures of those times, but there is a good one at https://www.dailymail.co.uk/sciencetech/article-2220065 /Flossie-worlds-oldest-starred-alongside-James-Bond-Dr-Who-brought-life-shed -Kent.html.

5. The first computers did not have operating systems as such, or at least they were not generally accessible to computer users. The first were introduced in the mid-1950s. By the mid-1960s, IBM were introducing such systems in their 360 series of machines, but operating systems only came of age with the development of mini-computers, where users were forced to engage directly with the machine just prior to the take-off of PCs. See https://en.wikipedia.org/wiki/History_of_operating_systems.

6. The history of these computers and their characterization as supercomputers is described in their relevant Wikipedia entries available at https://en.wikipedia.org/wiki /Manchester_computers and https://en.wikipedia.org/wiki/Atlas_(computer).

7. An excellent review of the early development of computers from the pre-digital age is contained in Eames and Eames, *A Computer Perspective*, endnote 29, chapter 3.

8. There were many somewhat idiosyncratic applications of computing in the early days. The Lyons Tea Company with its network of more than 250 teashops around Britain—the McDonald's of its day—had an enormous supply chain that they automated and managed using expertise and computers spun off from Bletchley Park. Georgina Ferry's excellent little book *A Computer Called Leo: Lyons Teashops and the World's First Office Computer*, Fourth Estate, London, 2003, paints an enticing picture

of what it was like in the early 1950s to be involved in making computers work back then.

9. See Kemeny's wider impact in https://history.computer.org/pioneers/kemeny.html.

10. The BASIC language—Beginner's All-purpose Symbolic Instruction Code—is contained in John G. Kemeny and Thomas E. Kurtz, *Basic: A Manual for BASIC, the Elementary Algebraic Language Designed for Use with the Dartmouth Time Sharing System*, 1st Edition, Dartmouth College Computation Center, Hanover, NH, 1964. This can be downloaded from http://bitsavers.trailing-edge.com/pdf/dartmouth /BASIC_Oct64.pdf.

11. The DTSS, as it was called, was developed by Kemeny and Kurtz in 1963–1964 at Dartmouth so they could teach computing to an army of undergraduates from diverse disciplines. They were widely lauded for this as one of the most participatory initiatives in the early history of the field. See http://dtss.dartmouth.edu/history.php.

12. The Whirlwind project and its use in the SAGE missile systems are recounted in Wayne E. Carlson's online book *Computer Graphics and Computer Animation: A Retrospective Overview*, Ohio State University Press, Columbus, OH, 2017. See specifically https://ohiostate.pressbooks.pub/graphicshistory/chapter/2-1-whirlwind-and-sage/.

13. Tracy Kidder's book *The Soul of a New Machine*, Little, Brown and Co., New York, 1981, recounts those heady times in which computer engineers were involved in the design of the Data General Eclipse machine—a machine that never really took the world by storm but showed the frenetic pace of engagement in the quest to build ever faster and more powerful minicomputers.

14. The rise and fall of the Blackberry as the most successful spin-off from the University of Waterloo's engineering program is told here by Jacquie McNish and Sean Silcoff, *Losing the Signal: The Untold Story Behind the Extraordinary Rise and Spectacular Fall of BlackBerry*, Flatiron Books, New York, 2015.

15. All this is reported in Harold Alkema and Kenneth McLaughlin's *40 Years of Computer Science at the University of Waterloo*, September 29, 2007, available at https://cs.uwaterloo.ca/40th/index.html. The chronology of the last sixty years is available at https://cs.uwaterloo.ca/40th/about/timeline.html.

16. Joel N. Shurkin, *Engines of the Mind: The Evolution of the Computer from Mainframes to Microprocessors*, W. W. Norton, New York, 1996.

17. Swaine and Freiberger, *Fire in the Valley*, endnote 31, chapter 3.

18. Alan R. Earls, *Digital Equipment Corporation (MA) (Images of America)*, Portsmouth, NH, 2004.

19. Not only was there horror in the administrative office that the machines they were about to acquire for the new art of word processing had been anticipated in the same department by the academic staff using them to do science, but the rather rapid realization that these same machines could be used for games was even more

mind wrenching. In those early days, a professor could occasionally be seen switching on a microcomputer and loading up a game in the administrative offices to the outrage of the secretarial staff. Whatever next!

20. Earls, *Digital Equipment.*

21. Joseph Schumpeter's notion of innovation as creative destruction, on which he elaborated in his many books, particularly his magnum opus *Business Cycles 1: A Theoretical, Historical, and Statistical Analysis of the Capitalist Process*, McGraw Hill, New York, 1939, is currently in fashion under the guise of disruptive technologies of which we will say more later in this book. In particular, Joseph L. Bower and Clayton M. Christensen's "Disruptive technologies: Catching the wave," *Harvard Business Review*, 73, No. 1, 43–53, 1995, is one of the first and seminal papers of this genre.

22. Swaine and Freiberger, *Fire in the Valley*, endnote 31, chapter 3.

23. Randy Alfred, "April 4, 1975: Bill Gates, Paul Allen form a little partnership," *Wired Magazine*, April 4, 2008, https://www.wired.com/author/randy-alfred.

24. From "The History of Computers" available at https://history-computer.com/ModernComputer/Personal/Altair.html.

25. Walter Isaacson, *Steve Jobs: The Exclusive Biography*, Little, Brown Book Group, New York, 2015.

26. What was remarkable about the PC revolution was the speed at which very innovative software was produced. The spreadsheet for the PC was designed by a Harvard Business School student, Dan Bricklin, in 1978, who then got his friend, Bob Frankston, from the MIT to develop production-level code for the Apple II. It was developed almost overnight and then marketed as VisiCalc ("visible calculator") by their company Software Arts. I first saw it at the University of Waterloo in 1981 when it was being used to run transport models, something that none of us ever envisaged could be done with such a generic piece of software, but it was all-purpose. Graphics was also fast expanding in parallel, of course, against a background of everything from word processing to games. See https://history-computer.com/ModernComputer/Software/Visicalc.html.

27. James W. Cortada, *IBM: The Rise and Fall and Reinvention of a Global Icon*, MIT Press, Cambridge, MA, 2019.

28. Sean Braswell, "The agreement that catapulted Microsoft over IBM," *OZY News for the Disruptive*, https://www.ozy.com/flashback/the-agreement-that-catapulted-microsoft-over-ibm/94437/.

29. Britain's answer to the PC or rather the Apple II was the BBC Micro, sponsored by the British Broadcasting Corporation and focused on education. These machines had operating systems that were close to their inputs and outputs in that their screens enabled users not only to type code at the command line but also to run it and produce outputs on the same screen as the code. In short, there were no separate

windows for graphics. This was much like the basic interface where users can type and display on the command line. This then made interacting with the operating system easier than with the IBM PC, which was based on DOS, and much easier than the Apple Macintosh, which, back then, rarely involved any user in seeing the operating system. Because of the focus on graphics, the PC could be used as a screen buffer for minicomputer outputs, and in this sense, it did enable a whole generation of programmers to begin to experience graphics.

30. There are several good references to the Xerox PARC project. Michael A. Hiltzik's *Dealers of Lightning: Xerox PARC and the Dawn of the Computer Age*, Harper Collins, New York, 1999, is a very comprehensive review of all the technologies developed there, while Robert C. Alexander and Douglas K. Smith's *Fumbling the Future: How Xerox Invented, Then Ignored, the First Personal Computer*, iUniverse Books, Lincoln, NB, 1999, is a somewhat less generous but nevertheless interesting review.

31. There are many descriptions of Doug Engelbart's contributions, but a concise and focused one is by Valerie Landau, "How Douglas Engelbart invented the future," *Smithsonian Magazine*, January 2018, https://www.smithsonianmag.com/innovation /douglas-engelbart-invented-future-180967498/. Ted Nelson's PROJECT XANADU, The Original Hypertext Project, was started in 1960 but never completed, like Babbage and many others before him. See http://www.xanadu.net/. A good account of Nelson's life and work can be found in Douglas R. Dechow and Daniele C. Struppa (Editors), *Intertwingled: The Work and Influence of Ted Nelson*, Springer, New York, 2015.

32. Mark Hall and John Barry, *Sunburst: The Ascent of Sun Microsystems*, Contemporary Books, Bel Air, CA, 1990, and Gertner, *The Idea Factory*, endnote 30, chapter 3.

33. We noted this first demonstration of computer graphics to a lay audience on Ed Morrow's CBS show *See It Now* in endnote 39, chapter 3, but Maurice Wilkes, the founder of the Computer Laboratory at the University of Cambridge, stole the show with his animated dancer two years earlier in 1949. See Annabel Jankel and Rocky Morton, *Creative Computer Graphics*, Cambridge University Press, Cambridge, 1984.

34. William Gibson, *Neuromancer*, Victor Gollanz, London, 1984, and the original movie in 1999 to which it relates, *The Matrix*, https://en.wikipedia.org/wiki/The _Matrix.

35. Robert Quigley, "Kitty: One of the first-ever computer animations," March 22, 2010, https://www.themarysue.com/kitty-computer-animation-russia-1968-video/. The YouTube movie is available at https://www.youtube.com/watch?v=sebqkS_EKBE.

36. There is a short note on Allan Schmidt's movie that he made using SYMAP-like techniques on my blog, http://www.complexcity.info/media/movies/early-computer -movies-1967-86/, and the movie is on YouTube available at https://www.youtube.com /watch?v=aySwJKK6i2s.

37. Nick Chrisman, *Charting the Unknown: How Computer Mapping at Harvard Became GIS*, ESRI Press, Redlands, CA, 2006.

38. The display technology for CATS is covered by Andrew V. Plummer in his report "The Chicago Area Transportation Study: Creating the first plan (1955–1962): A narrative," 2006, available at http://www.surveyarchive.org/Chicago/cats_1954-62.pdf.

39. This SOM archive video offers a look back at the early days of 3D visualization. See Vimeo https://vimeo.com/93315120 and the article at https://www.archdaily .com/778376/this-som-archive-video-offers-a-look-back-at-the-early-days-of-3d -visualization.

40. There is a useful history of early commercial computer graphics on the Auto-CAD Wikipedia page: https://en.wikipedia.org/wiki/AutoCAD.

CHAPTER 5

Eric Schmidt's speech at the Internet World Trade Show, New York, November 18, 1999. See https://www.oxfordreference.com/view/10.1093/acref/9780191826719.001 .0001/q-oro-ed4-00017947.

1. Hodges, *Alan Turing*, endnote 2, chapter 3, recounts this in more detail, and the oration at this place on the hundred-year anniversary of his birth: https://www.turing .org.uk/publications/oration.html.

2. The Royal Institution: Science Lives, available at https://www.rigb.org/our-history /people/f/michael-faraday.

3. Simon Winchester, *The Map That Changed the World: A Tale of Rocks, Ruin and Redemption*, Viking Penguin, London, 2002.

4. Steven Johnson, *The Ghost Map: The Story of London's Most Terrifying Epidemic— And How It Changed Science, Cities, and the Modern World*, Riverhead Books, New York, 2007.

5. The Old Shady Guide to London, available at http://www.shadyoldlady.com /location.php?loc=1583.

6. In the discovery of DNA, Francis Crick was an alumnus of UCL, but the main action in London took place at King's, where Maurice Wilkins and Rosalind Franklin produced the spectroscopy that confirmed the structure proposed by Crick and Watson. See https://www.kcl.ac.uk/aboutkings/history/famouspeople/wilkinsfrank lin and James Watson, *The Double Helix*, Touchstone Books, New York, 1998.

7. The discovery and invention of television. See https://en.wikipedia.org/wiki/John _Logie_Baird.

8. The first internet pioneers. See http://www0.cs.ucl.ac.uk/csnews/internet_pioneers .html.

9. Janet Abbate, *Inventing the Internet*, MIT Press, Cambridge, MA, 1999.

10. TCP/IP. See https://history-computer.com/Internet/Maturing/TCPIP.html.

11. Katie Hafner and Matthew Lyon, *Where Wizards Stay Up Late: The Origins of The Internet*, Touchstone, Simon & Schuster, New York, 1996. Although the first email was sent in 1971 in the US, probably the first in the UK was sent soon after. In 1973, Leonard Kleinrock, one of the internet pioneers, sent an email from his office at UCLA to his colleague Larry Roberts, who was still at the same conference in Brighton, UK, where he had just been. It asked him to ask him to bring back his electric razor, which he had forgotten, and he did this by managing to log onto Roberts's machine. See https://uk.pcmag.com/features/118038/meet-the-professor-who-was-there-when-the-internet-was-turned-on. However, it was Queen Elizabeth II who ranks as one of the first to send an email in the UK when she visited the Royal Signals and Radar Establishment (RSRE) in Malvern in 1976, nine years before my colleagues and I at the University of Wales in Cardiff began to use email, and in an era when there was little or no knowledge about the potential of this medium, see Cade Metz, "How the Queen of England beat everyone to the internet," *Wired Magazine*, December 2012, https://www.wired.com/2012/12/queen-and-the-internet/.

12. The development of packet switching more or less simultaneously at the UK's National Physical Laboratory by Davies and at the US RAND (Research And National Development) Corporation by Baran in the early to mid-1960s is yet another example in the history of technology where two or more inventions are made at much the same time but entirely independently of one another. In chapter 4, we made the same point about the development of the PC.

13. IPTO: the Information Processing Techniques Office.

14. The idea of many layers of computation that build on one another is absolutely generic to science and the modern world. In computing, there are almost too many layers to even categorize, but in terms of communications, the seven-layer network model is as good an illustration of this hierarchy as any. See Keith Shaw, "The OSI model explained: How to understand (and remember) the 7 layer network model," *Network World*, October 22 2018, https://www.networkworld.com/article/3239677/the-osi-model-explained-how-to-understand-and-remember-the-7-layer-network-model.html.

15. Adam Smith, *The Wealth of Nations*, Vols. 1 and 2, Strahan, Penguin Classics 1982 Printing, London, 1776. The idea that any organized whole built from the bottom up has an internal consistency that enables it to function effectively is intrinsic to many systems. Widespread computation in contemporary society is no exception. There is still a puzzle as to why societies do not fall apart without any strong central coordination, but it appears to be in the human condition that we all imperceptibly adapt to one another to keep the ship on course, so to speak.

16. The Network Encyclopedia. See https://networkencyclopedia.com/networking-history-1980/.

17. Geeks for Geeks: Types of Network Topology. See https://www.geeksforgeeks.org/types-of-network-topology/.

18. Hafner and Lyon, *Where Wizards Stay*.

19. Abbate, *Inventing the Internet*.

20. Notwithstanding the invention of the World Wide Web by Tim Berners-Lee at CERN in Geneva, the European and certainly the British effort in computing eventually went the same way as the British car industry. Arguably, even in the early 1960s, advances in Britain were still going hand in hand with those in the US, but soon the US began to draw away, as much because the scale of the industry was so different. The British had a penchant for much more amateur invention and development, and it was only a matter of time before the industry became intensively competitive, highly managed, and deeply professionalized in a way that did not seem to suit the British psyche.

21. Negroponte, *Being Digital*, endnote 33, chapter 2.

22. Tim Berners-Lee, *Weaving the Web: The Original Design and Ultimate Destiny of the World Wide Web*, Harper One, New York, 1999.

23. Hypertext, of course, emerged much earlier in the 1960s in Ted Nelson's XANADU project and in Douglas Engelbart's inventions of multimedia from mouse to net. These early developments are described in endnote 32, chapter 4, in our discussion of Xerox PARC, but a very readable summary is by Howard Rheingold, *Tools for Thought: The History and Future of Mind-Expanding Technology*, Simon & Schuster, Prentice Hall, Englewood Cliffs, NJ, 1985.

24. Bush, "As we may think," endnote 12, chapter 1

25. Leonard Kleinrock, who was at the heart of the internet from the word go, developed many novel ideas about communications networks and their physical and topological structure when he was a doctoral student at the MIT. His book, *Communication Nets: Stochastic Message Flow and Delay*, McGraw-Hill, New York, 1964, established the "foundational technology of the Internet."

26. This is sometimes thought of as a forerunner to the World Wide Web, but the program that Berners-Lee devised in 1980 did not take off, and in the mid-1980s, the disk on which the program was written was reused by some of his colleagues, who deleted it. So much for history! So much for the Internet! https://www.internet -guide.co.uk/ENQUIRE.html.

27. James Gillies and Robert Cailliau, *How the Web Was Born: The Story of the World Wide Web*, Oxford University Press, Oxford, 2000.

28. I spent two months at the National Center for Supercomputing Applications at the University of Illinois in Champaign-Urbana in 1986 working on visualizing urban models. Larry Smarr, who was then the director, had received a donation of IBM PCs with good graphics cards and accelerated processors, and he was anxious to develop scientific visualizations, which was the main focus of the center. My programs were also ported to Sun workstations, which were being used to visualize star systems, which was Larry's basic scientific interest and expertise. There was little hint

of networking in those days, but it was only six years later when Mosaic was developed and only three years to the time when the World Wide Web became a feasible research project at CERN. These quickly took the world by storm in the 1990s.

29. Jay Hoffmann, "The history of the browser wars: When Netscape met Microsoft," *The History of the Web*, October 1, 1997, https://thehistoryoftheweb.com/brow ser-wars/.

30. Anna Crowley Redding, *Google It: A History of Google*, Feiwel & Friends, New York, 2018. In fact, Google was still building its base when the dot-com bust occurred. Yet, it withstood the ravages of time, somewhat protected behind its wall of initial venture capital.

31. David G. Messerschmitt, "The future of computer-telecommunications integration," Department of Electrical Engineering and Computer Sciences, University of California, Berkeley, CA, https://pdfs.semanticscholar.org/c6b0/dd0532cfa194f 4581ca3149a279b0bd7e88b.pdf.

32. An up-to-date statement is in Robert M. Metcalfe's "It's all in your head," *Forbes*, April 20, 2007, https://www.forbes.com/forbes/2007/0507/052.html#504565 b047d3. Metcalfe's law is stated in many places by many people, but it was interpreted in the form used here by George Gilder in 1993 in "Telecosm: Metcalfe's law and legacy," *Forbes*, September 13, 1993, and in his book *Telecosm: The World After Bandwidth Abundance*, Simon & Schuster, New York, 2000. A somewhat dissenting comment is by Bob Briscoe, Andrew Odlyzko, and Benjamin Tilly, "Metcalfe's law is wrong," *IEEE Spectrum*, July 1, 2006, https://spectrum.ieee.org/computing/networks /metcalfes-law-is-wrong.

33. There are many different estimates, all based on different assumptions as to what a website is and whether it is active. For one estimate, see https://www.millforbusi ness.com/how-many-websites-are-there/. The number of web pages is much greater than the two billion or so websites, with some estimating that there are twenty-five times as many. See https://techjury.net/blog/how-many-websites-are-there/.

34. To define Wikipedia, look at its own Wikipedia entry, https://en.wikipedia.org /wiki/Wikipedia.

35. "Web 1.0 vs Web 2.0 vs Web 3.0 vs Web 4.0 vs Web 5.0—A bird's eye on the evolution and definition," . . . on how web (technology) is evolving; https://flatworld business.wordpress.com/flat-education/previously/web-1-0-vs-web-2-0-vs-web-3-0 -a-bird-eye-on-the-definition/ updated to Digital Evolution Past, Present, and future outlook of digital technology at https://flatworldbusiness.wordpress.com/digital -evolution/.

36. Margaret Rouse (Editor), "Definition—What does app mean?," https://www .techopedia.com/definition/28104/app.

37. Once the browser wars ended and Microsoft Explorer came out on top, the new upstart, which was started in 1998, was Google. Essentially, the two Google

founders—Sergy Brin and Larry Page—identified a basic method for examining the structure of a graph based on its roots, which they called "PageRank," and they used it to search the net conceived of as a network of links. The method, in fact, was widely known in graph/network theory, but they improved the algorithm many times. The comparative advantage that the company quickly gained through its focus on really massive databases of relevant information and the very neat algorithms they developed simply blew every other search engine out of the water. No wonder we now refer to search as "Googling."

38. This is one of the best of many websites cataloguing the size of the internet in various ways: https://www.internetlivestats.com.

39. According to Statista, only 2 percent of data are archived. See https://www.statista.com/statistics/871513/worldwide-data-created/.

40. James Martin was one of the great disseminators of the information revolution through his many books on computing and his philanthropic ventures. His legacy still funds the James Martin School at the University of Oxford, his alma mater, where his endowment focuses on science, society, and information technology. His book, *The Wired Society: A Challenge for Tomorrow*, Prentice-Hall, Englewood Cliffs, NJ was published in 1978, and it remains a focus for the mission he began before the PC was born, which is continued at Oxford. A useful summary of his contributions is in the interview with him by John Carroll in *Business Strategy Review*, *15*, 44–47, December 2004, https://papers.ssrn.com/sol3/papers.cfm?abstract_id=621871.

41. Batty, *Inventing Future Cities*, endnote 8, chapter 1.

42. There are many laws of computing that have been defined somewhat serendipitously. A good exposition is by David Delony, "The laws of computing," *Technopedia*, November 15, 2013, https://www.techopedia.com/2/28205/trends/the-laws-of-computing, where he identifies: Moore's law, Metcalfe's law, Reed's law, Beckstrom's law, Brooks' law, and Hofstadter's law. To those, we might add: Gilder's law, Sarnoff's law, and Zuckerberg's law. For even more, see Jim Gray's "Sixteen laws of computing" (under "Talks"), https://jimgray.azurewebsites.net.

43. Kai-Fu Lee, *AI Superpowers: China, Silicon Valley, and the New World Order*, Houghton Mifflin Harcourt, Boston, MA, 2018.

44. Walter Isaacson, *The Innovators: How a Group of Hackers, Geniuses and Geeks Created the Digital Revolution*, Simon & Schuster, London, 2014.

45. Ray Kurzweil has written many books about exponential and super-exponential growth, particularly focusing on the singularity. See *The Singularity*, endnote 10, chapter 1. See also his websites, especially https://www.kurzweilai.net/the-law-of-accelerating-returns.

46. Gilder, *Telecosm*. See also his law available at: http://ramrcblog.blogspot.com/2014/08/gilders-law.html.

47. Fred Brooks, *The Mythical Man-Month: Essays on Software Engineering*, Addison Wesley, Reading, MA, 1975.

48. Saul Hansell, "Zuckerberg's law of information sharing," *The New York Times, Bits*, November 6, 2008, at https://bits.blogs.nytimes.com/2008/11/06/zuckerbergs-law-of-information-sharing/.

CHAPTER 6

From Paul A. Samuelson, "Thünen at two hundred," *Journal of Economic Literature, 21*, No. 4, 1468–1488, 1983.

1. There are many commentaries on the evils of the nineteenth-century city—"the industrial city" as it came to be called. Lewis Mumford's *The Culture of Cities*, Martin Secker and Warburg, London, 1938, provides an excellent introduction, and he projects this onto a wider canvas in his *The City in History*, Harcourt, Brace and World, New York, 1961. Peter Hall's *Cities in Civilization: Culture, Innovation and Urban Order*, Weidenfeld and Nicolson, London, 1998, is a more recent contribution. These three books all deal with the sweep of urban history, particularly the emergence of the modern city since the beginning of the industrial revolution.

2. Conventional wisdom about the formation of cities until comparatively recently was based on the notion that first came the agricultural revolution, with man developing settled agriculture as early as 10,000 BCE, and then came urban living in cities from around 4000 BCE, but these dates are hedged in uncertainty. These assumptions are writ large in much of our thinking until about the middle of the last century. See, for example, Gordon Childe's *What Happened in History*, Penguin Books, London, 1942. However, in her book *The Economy of Cities*, Jane Jacobs turned this thinking on its head to some extent, arguing that pockets of urban life emerged much earlier in nomadic times.

3. Hannah Ritchie and Max Roser, "Urbanization," *Our World in Data*, September 2018. Here, there are various data sets that you can download that give a detailed picture of world urbanization since about 1500, available at https://ourworldindata.org/urbanization. See also https://www.worldbank.org/en/topic/urbandevelopment/overview.

4. I speculate extensively on this in my paper "Commentary: When all the world's a city," *Environment and Planning A, 43*, No. 4, 765–772, 2011 and in my book *Inventing Future Cities*, endnote 8, chapter 1.

5. William Morris, "Town and country," at a meeting sponsored by the Hammersmith Socialist Society, at Kelmscott House, Hammersmith, London, May 29, 1892, and portions thereof printed in *The Journal of Decorative Art*, April 13, 1893. See https://www.marxists.org/archive/morris/works/1892/town.htm.

6. See Jean-Paul Rodrigue, *The Geography of Transport Systems*, Routledge, New York, 2017, and the timeline at https://transportgeography.org/?page_id=1599.

7. Yves Charbit, "The Platonic city: History and utopia," *Population*, *57*, 207–235, 2002, https://www.persee.fr/doc/pop_1634-2941_2002_num_57_2_18391. Plato lived from 427 BCE to 347 BCE, and the original source of his speculations about ideal cities is his book, *The Republic*, Penguin Classics, London, 1976 (translation from the original c. 390 BCE). A useful perspective on the ideal city has been developed by Ruth Easton in her *Ideal Cities: Utopianism and the (Un)Built Environment*, Thames and Hudson, London, 2002.

8. Richard Wycherley, *How the Greeks Built Cities*, Macmillan, New York, 1949.

9. Isaacson, *Leonardo da Vinci*, endnote 16, chapter 1.

10. There are many commentaries on Leonardo's idealization of urban form. See, for example, Hilary Clarke, "Leonardo the city planner: Da Vinci's New Milan," *Engineering and Technology*, May 22, 2019, available at https://eandt.theiet.org/content/articles/2019/05/leonardo-the-city-planner-da-vinci-s-new-milan/. See also https://www.fastcompany.com/90163788/the-plague-inspired-da-vinci-to-design-a-city-we-should-steal-his-ideas.

11. Barone, *Leonardo da Vinci*, endnote 16, chapter 1.

12. Gordon Wood, *The American Revolution: A History*, Weidenfeld and Nicolson, London, 2002.

13. Kathy Barber, "Watts steam engine, 1776," *Stories of Change*, March 23, 2017, https://storiesofchange.ac.uk/node/40. See also Watts's biography at https://www.britannica.com/biography/James-Watt/.

14. Smith, *The Wealth of Nations*, endnote 16, chapter 5. An abridged version is available at https://www.adamsmith.org/the-wealth-of-nations.

15. His book was published in 1826 and only fully translated into English 140 years later. See Johann Heinrich von Thünen, *Der Isolierte Staat in Beziehung auf Landwirtschaft und Nationalökonomie*, Wirtschaft & Finan, Hamburg, Germany, 1826, published as *Von Thünen's Isolated State*, Pergamon Press, Oxford, 1966, translated by Carla M. Wartenberg; edited with an introduction by Peter Hall. Hall's summary along with commentary on the wider context of location theory is described in Basak Demires Ozkul's article "Von Thünen revisited," *Built Environment*, *41*, No. 1, 99–111. In fact, the original book from 1826 was only its first volume. A more developed argument, which relates to wider concepts in economics, was published in Volume 2, 1850, and Volume 3, 1863.

16. There is a good account of von Thünen's early life in Bernard Dempsey's *The Frontier Wage*, Loyola University Press, Chicago, IL, 1960, where there is also a translation of Volume 2 of *The Isolated State*.

17. Walter Isard, *Location and Space-economy: A General Theory Relating to Industrial Location, Market Areas, Land Use, Trade, and Urban Structure*, MIT Press, Cambridge, MA, 1956.

18. Samuelson, "Thünen at two hundred."

19. This quotation is attributed to Alfred Marshall by David Reismann in his *Alfred Marshall's Mission*, Palgrave Macmillan, London, 1990, where he is also quoted as saying: "I [referring to Marshall] loved von Thünen above all my other masters." The source is A. C, Pigou *Memorials of Alfred Marshall*, Macmillan, London, p 360.

20. Fernand Braudel, *The Wheels of Commerce: Civilization and Capitalism: 15th–18th Century*, Harper & Row, New York, 1982.

21. There are number of good websites on von Thünen. In 1997, I was introduced by John Nysteun of the University of Michigan, one of the doyens of the quantitative revolution in geography, to the Thünen Society, North American Division. See http://www-personal.umich.edu/~copyrght/thunen/thunen/. This has many interesting reflections on the man and his model. A more recent portal to his work is at Das Portal für die Verbreitung des wissenschaftlichen und humanistischen Thünen-erbes [The portal for the dissemination of the scientific and humanistic Thünen heritage] available at https://www.thuenen.info/index.php?id=40&L=1.

22. Some of these elaborations are given in Samuelson, "Thünen at two hundred."

23. There are several demonstrations of the model on the web. Look at one of my own web pages, http://www.complexcity.info/media/software/von-thunens-model/, where you can download Philip Steadman's version of the model and run it on a PC. Note that this is nearly forty years old and may not run on your PC now but Manishika Jain's YouTube movie, https://www.youtube.com/watch?v=lumvhK7CzDE, is a lucid exposition of the model, and there are other related movies referenced on this page.

24. There is a summary of the social-ecological approach pioneered in Chicago in Roderick McKenzie's "The ecological approach to the study of the human community," *American Journal of Sociology*, *30*, No. 3, 287–301, 1924. The definitive statement and its application to Chicago is in Robert Park and Ernest Burgess, *The City*, University of Chicago Press, Chicago, IL, 1925. The 1967 edition of the book includes McKenzie's article.

25. Martin Beckmann was in the vanguard of urban economics, first bringing his skills in astrophysics to economics and then these ideas to cities. His paper "On the distribution of rent and residential density in cities" was first presented at the *Interdepartmental Seminar in Mathematical Applications in the Social Sciences*, *19*, 1, Yale University, New Haven, CT, 1957. There is an excellent summary of the origins of urban economics by John McDonald in his paper "William Alonso, Richard Muth, resources for the future, and the founding of urban economics," *Journal of the History of Economic Thought*, *29*, 67–84, 2007. Beckmann's paper or a variant thereof was published twelve years after his first seminar presentation, as soon as urban economics came of age in 1969, and it appeared in the *Journal of Economic Theory*, *1*, 60–67, 1969.

26. William Alonso's PhD thesis was ultimately published in what became a seminal text: *Location and Land Use: Toward a General Theory of Land Rent*, Harvard University Press, Cambridge, MA, 1964.

27. Robert Solow's paper with William Vickrey symbolizes the way mainstream economists specializing in growth theory have contributed to urban economics. See the paper "Land use in a long narrow city," *Journal of Economic Theory*, 3, No. 4, 430–447, 1971.

28. Joel Garreau, *Edge Cities: Life on the New Frontier*, Anchor Books, Doubleday, New York, 1991.

29. There are many casual and empirical definitions of polycentricity. For an applied perspective, see Peter Hall and Kathy Pain (Editors), *The Polycentric Metropolis: Learning from Mega-city Regions in Europe*, Earthscan, London, 2006.

30. This is a general problem in physics "pertaining to the properties of microscopic systems made of many interacting particles." See https://en.wikipedia.org/wiki/Many -body_problem. Theory is able to work with two interacting particles but stumbles when more than two are involved.

31. Michael Batty, *The New Science of Cities*, MIT Press, Cambridge, MA, 2013, and Michael Batty, "Cities as systems of networks and flows," in T. Haas and H. Westlund (Editors), *In the Post-Urban World: Emergent Transformation of Cities and Regions in the Innovative Global Economy*, 56–69, Routledge, London, 2017.

32. Yaneer Bar-Yam, "Complexity rising: From human beings to human civilization, a complexity profile," *Encyclopedia of Life Support Systems* (EOLSS), developed under the Auspices of the UNESCO, Oxford, 2002, at https://necsi.edu/complexity-rising -from-human-beings-to-human-civilization-a-complexity-profile.

33. Catherine Bauer Wurster, "The form and structure of the future urban complex," in Lowdon Wingo, Jr.'s edited collection *Cities and Space: The Future Use of Urban Land*, Johns Hopkins University Press, Baltimore, MD, 1963.

34. One of the best statements of these forces is in the collection of articles edited by Melvin Webber in *Explorations into Urban Structure*, University of Pennsylvania Press, Philadelphia, PA, 1964.

35. Garreau, *Edge Cities*.

36. One of the earliest coherent commentaries on urban sprawl and the decline of the central city in the US was in William Whyte, Francis Bello, Seymour Freedgood, Daniel Seligman, and Jane Jacobs, *The Exploding Metropolis*, Doubleday, Garden City, New York, 1958.

37. Andres Duany, Elizabeth Plater-Zyberk, and Jeff Speck, *Suburban Nation: The Rise of Sprawl and the Decline of the American Dream*, North Point Press, New York, 2000.

38. Luis Bettencourt, *Introduction to Urban Science: Evidence and Theory of Cities as Complex Systems*, MIT Press, Cambridge, MA, 2021; Marc Barthelemy, *The Structure and Dynamics of Cities*, Cambridge University Press, Cambridge, 2016.

39. Michael Batty, "Defining urban science," in J. Shi, M. Goodchild, M. P. Kwan, and A. Zhang (Editors), *Urban Informatics*, Springer, Berlin and New York, 2021, pp. 15–28.

40. Negroponte, *Being Digital*, endnote 33, chapter 2.

41. The transition to the Digital Age is often marked by the move from print to screen, from physical to digital, and in some respects, this defines the central discontinuity in social life that this book is all about. In fact, it has only really happened since the millennium, for it took seventy years for all of us to acquire devices that would enable us to interact with information using screens rather than papers and books. The transition is still not complete, and in the time-honored fashion, like all the disruptive change we are cataloguing here, we will probably still retain a modicum of print media due to our physiological needs.

42. Standage, *The Victorian Internet*, endnote 22, chapter 2.

43. Claude S. Fischer, *America Calling: A Social History of the Telephone to 1940*, University of California Press, Berkeley, 1992.

CHAPTER 7

From an interview "Internet Time: Will the 'beat' go on?" with Nicholas Negroponte, Director of the MIT's Media Lab, February 26, 1999, available at http://edition.cnn .com/TECH/computing/9902/26/t_t/internet.time/.

1. Francis Wheen, *Karl Marx*, Fourth Estate, London, 1999.

2. There is a good deal of revisionism about the fact that Marx worked in the Round Reading Room of the British Museum for much of his life. He certainly worked in the museum in the years when he first came to London, but he left his lodgings at 28 Dean Street in London's Soho about the time the new Reading Room opened, and in his later years, he visited only infrequently as much due to ill-health as to his living in Hampstead, which was some miles away. See Colin Higgins, "Seeing 'sights' that don't exist: Karl Marx in the British Museum Round Reading Room," *Library & Information History*, *33*, No. 2, 81–96, 2017.

3. Ralph Waldo Emerson, "The young American," A Lecture read before the Mercantile Library Association, Boston, February 7, 1844, in his *The Complete Works*, Vol. I. Nature, Addresses and Lectures, 1904; https://www.bartleby.com/90/0110.html.

4. Theodore Levitt, "The globalization of markets," *Harvard Business Review*, May 1983, https://hbr.org/1983/05/the-globalization-of-markets. In fact, he was really the popularizer of the term "globalization," which was in existence well before his article. It had gained common currency by the 1990s. See, in particular, Manuel Castells, *The Rise of the Network Society: Economy, Society and Culture. Volume 1: The Information Age: Economy, Society and Culture*, Blackwell, Oxford, 1996.

5. Karl Marx and Frederick Engels, *Manifesto of the Communist Party*, February 1848, in Karl Marx, *Selected Works*, Vol. 1, Progress Publishers, Moscow, 1969, https:// www.marxists.org/archive/marx/works/download/pdf/Manifesto.pdf.

6. Karl Marx, *Grundrisse, Foundations of the Critique of Political Economy, 1861*, Penguin Books in association with New Left Review, London, 1973, see chapter 10, https://www.marxists.org/archive/marx/works/1857/grundrisse/ch10.htm.

7. Mumford, *City in History*, endnote 2, chapter 6. In fact, it is quite hard to date the first cities. The National Geographic Resource Library suggests they appeared some 7,500 years ago in 5500 BCE; https://education.nationalgeographic.org/resource/history-cities. The Wikipedia entry contains a good discussion and suggests that the elements we recognize as being an integral part of cities go back even further to the agricultural revolution itself, to Neolithic times, to about 8000 BCE. The timing is uncertain. See https://en.wikipedia.org/wiki/History_of_the_city.

8. Jean-Paul Rodrigue, "The emergence of mechanized transportation systems," in his book *Geography of Transport Systems*, endnote 7, chapter 6, https://transportgeography.org/.

9. Cairncross, *Death of Distance*, endnote 32, chapter 2.

10. Schwab, *Fourth Industrial Revolution*, endnote 35, chapter 2; and Klaus Schwab and Nicholas Davis, *Shaping the Future of the Fourth Industrial Revolution: A Guide to Building a Better World*, Penguin Books, London, 2018.

11. Nikolai Kondratieff, *The Major Economic Cycles*, 1925, republished as *The Long Wave Cycle*, translation by G. Daniels, E. P. Dutton, New York, 1984. Kondratieff himself called these cycles "long waves"; Kondratieff, 1926, 1935, translated by W. F. Stopper, "The long waves in economic life," *The Review of Economics and Statistics*, *17*, 105–115. However, it was Josef Schumpeter who coined the term "Kondratieff waves." See Schumpeter, *Business Cycles*, endnote 22, chapter 4.

12. Peter Hall, "The geography of the Fifth Kondratieff Cycle," *New Society*, *26*, March, 535–537, 1981.

13. Batty, *Inventing Future Cities*, endnote 8, chapter 1.

14. Kurzweil, *Singularity Is Near*, endnote 10, chapter 1.

15. David Harvey, "Between space and time: Reflections on the geographical imagination," *Annals of the Association of American Geographers*, *80*, No. 3, 418–434, 1990.

16. Jesse Ausubel and Cesare Marchetti, "The evolution of transport," *The Industrial Physicist*, American Institute of Physics, 20–24, April/May 2001, https://phe.rockefeller.edu/TIP_transport/transport.pdf.

17. Sara Jered, "The colossal amount of time spent consuming media may finally be flatlining, according to Nielsen Report," *Adweek40*, March 19, 2019, https://www.adweek.com/tv-video/the-colossal-amount-of-time-spent-consuming-media-may-finally-be-flatlining-according-to-nielsen-report/.

18. Marx and Engels, *Manifesto of the Communist Party*, p. 16.

19. Ray Kurzweil, *The Age of Spiritual Machines: When Computers Exceed Human Intelligence*, Penguin Press, New York, 2000.

20. Cairncross, *Death of Distance*, endnote 32, chapter 2, and her articles in the special issue of *The Economist*, *336*, No. 7934, published September 30, 1995.

21. The philosopher Martin Heidegger spoke about the abolition of distance, and although, as we have been at pains to point out, the term "the annihilation of distance" was used extensively in the media in the early to mid-nineteenth century, the term resonates with globalization, as we noted earlier. See https://plato.stanford.edu/entries/globalization/ and Google Answers http://answers.google.com/answers/threadview?id=553376.

22. Apparently, Twain did not actually say this, but he did say that an American journalist wrote incorrectly about his death while he was in London, and he had to refute the suggestion. See https://www.mentalfloss.com/article/562400/reports-mark-twains-quote-about-mark-twains-death-are-greatly-exaggerated.

23. There are many estimates of the speed and extent of travel in historical times that relate directly to the technology in play. We noted the work of Jesse Ausubel and Cesare Marchetti in their paper "Evolution of transport," but also see David Metz, "Peak car and beyond: The fourth era of travel," *Transport Reviews*, *33*, 255–270, 2013, and Yacob Zahavi, *The Unified Mechanism of Travel (UMOT) Model*, The World Bank, Washington, DC, 1976, http://www.surveyarchive.org/Zahavi/TheUMOTModel.pdf.

24. Cesare Marchetti and Jesse Ausubel, "Quantitative dynamics of human empires," *International Journal of Anthropology*, *27*, No. 1–2, 1–62, 2012, and http://phe.rockefeller.edu/docs/empires_booklet/.

25. Many eminent commentators on cities have pointed to the fact that the bigger the city, the more diverse, innovative, and creative it appears to be. Jane Jacobs in *Death and Life*, endnote 8, chapter 2, made much of this in her defense of the naturally evolving North American city, and more recently, Ed Glaeser in his *Triumph of the City*, endnote 8, chapter 2, has made similar points. These are also echoed in Luis Bettencourt and Geoff West's "A unified theory of urban living," *Nature*, *467*, 912–913, 2010.

26. We will come across the four Big Tech companies Google, Amazon, Facebook and Apple (GAFA) many times in this book, and, in fact, one might also add Microsoft (GAFAM). See https://en.wikipedia.org/wiki/Big_Four_tech_companies.

27. Batty, *Inventing Future Cities*, endnote 8, chapter 1.

28. Bell's telephone took a long time to take off. In fact, although invented in 1876, it took fifty years for it to reach 40 percent of the US population, and it was still less than 80 percent in 1960. It was comparatively expensive, and only with the revolution in broadband that we noted in earlier chapters on computer networking did its costs drop to enable almost complete penetration in the population of developed countries. See Claude Fischer and Glenn Carroll, "Telephone and automobile diffusion in the United States, 1902–1937," *American Journal of Sociology*, *93*, No. 5, 1153–1178, 1988.

29. See Kurzweil's Accelerating Intelligence weblog at https://www.kurzweilai.net /the-law-of-accelerating-returns.

30. Marx and Engels, *Communist Party*, chapter I, Bourgeois and Proletarians.

31. We outlined the logic of the way the city in history came to be ordered in concentric rings but punctuated by radial routes, taking people away to greener space toward the periphery of the city, using von Thünen's model. See von Thünen, *Der Isolierte Staat*, endnote 16, chapter 6, and Park and Burgess's model of Chicago's spatial ecology in their book *The City*, endnote 25, chapter 6. This model is still the dominant explanation and characterization of intra-urban structure, notwithstanding its generalization to polycentric urban landscapes.

32. London's population follows the classic example of the rise of the industrial city and then its comparative decline in the last fifty years, but in this case, immigration through the open labor market in the European Union and relaxing external immigration has seen its population rise back to levels not seen since the middle of the last century. See https://en.wikipedia.org/wiki/Demography_of_London.

33. Batty, *Inventing Future Cities*, endnote 8, chapter 1; and Michael Batty, "Defining cities and growth," in Edward Glaeser, Peter Nijkamp, and Karima Kourtit (Editors), *Urban Empires: Cities as Global Rulers in the New Urban World*, Routledge, London, 2021.

34. Paul Krugman, *The Self-Organizing Economy*, Blackwell, Oxford, 1996, and in Batty, *Inventing Future Cities*, endnote 8, chapter 1, p. 82.

35. Duncan Smith's weblog CityGeographics: Urban Form, Dynamics and Sustainability has some excellent imagery plotted from the Global Human Settlements Layer of the EU's Joint Research Council's population coverage. See his interactive map at http://luminocity3d.org/WorldPopDen/. The Greater Bay Area for example has upward of sixty million people as defined from this database.

36. Cairncross, *Death of Distance*, endnote 32, chapter 2, p. 16.

37. "Form ever follows function" was the phrase first defined by Louis Sullivan, the great modernist Chicago architect in his article "The Tall Office Building Artistically Considered," *Lippincott's Magazine*, March 23, 403–409, 1896. It became the mantra of the modern movement, and it is writ large in much thinking about how problematic the future form of cities will be. See Batty, *Inventing Future Cities*, endnote 8, chapter 1.

38. Kevin Kelly's broad sweep across the digital economy in *The Inevitable: Understanding the 12 Technological Forces That Will Shape Our Future*, Penguin Viking Books, New York, 2016, provides a comprehensive view of the new economy that is emerging, while Geoffrey Parker, Marshall Van Alstyne, and Sangeet Choudary's *Platform Revolution: How Networked Markets Are Transforming the Economy and How to Make Them Work for You*, W. W. Norton, New York, 2016, provides a useful summary of the way digital markets are changing the economy.

39. There are many commentaries in the popular press at the time of writing (2023) about the demise of the high street, even before the pandemic struck. See, for example, Cherry Reynard, "Death of the high street," *Forbes*, April 26, 2018, https://www.forbes .com/sites/cherryreynard/2018/04/26/death-of-the-high-street/#7793247b45cf, and Alex Marsh, "Crumbling Britain: The slow death of the high street," *New Statesman*, July 11, 2018, https://www.newstatesman.com/politics/uk/2018/07/crumbling-britain -slow-death-high-street. These trends have greatly accelerated due to the pandemic, and the future of the high street now is completely uncertain.

40. Jack Nilles, Roy Carlson, Paul Gray, and Gerhard Hanneman, *The Telecommuni cations-Transportation Tradeoff: Options for Tomorrow*, Booksurge, New York, 1976.

41. Edward Morgan Forster, *The Machine Stops*, Penguin Classics, London, 1909, reprinted in 2017; and Toffler, *Future Shock*, endnote 6, chapter 1.

42. This is quoted by Rich Karlgaard and Michael Malone in "City vs. country: Tom Peters and George Gilder debate the impact of technology on location," *Forbes ASAP Technology Issue*, *155*, No. 5, 56–61, February 27, 1995.

43. Ryan Browne, "70% of people globally work remotely at least once a week, study says," *MakeIt*, CNBC, May 30, 2018, https://www.cnbc.com/2018/05/30/70-percent -of-people-globally-work-remotely-at-least-once-a-week-iwg-study.html.

44. The book by Arthur C. Clarke, *2001: A Space Odyssey*, Penguin Putnam, New York, 1968, and the movie of the same name. See https://en.wikipedia.org/wiki /2001:_A_Space_Odyssey_(film).

45. Negroponte, *Being Digital*, endnote 33, chapter 2.

46. Martin Ford, *The Rise of the Robots: Technology and the Threat of Mass Unemployment*, Basic Books, New York, 2015.

47. David Kerrigan, *Life as a Passenger: How Driverless Cars Will Change the World*, CreateSpace, New York, 2017, and http://david-kerrigan.com.

48. The six levels of car autonomy range from "no automation" to "full automation," as defined by the International Society of Automotive Engineers. There are many websites describing these levels such as https://www.nhtsa.gov/technology -innovation/automated-vehicles-safety.

49. Taleb, *Black Swan*, endnote 27, chapter 1.

50. KPMG, "Connected and autonomous vehicles—The UK economic opportunity," March 2015, https://www.smmt.co.uk/wp-content/uploads/sites/2/Connected-and -Autonomous-Vehicles---The-UK-Economic-Opportu. . . . pdf

51. Karl Marx, "The fragment on machines," from *Grundrisse*, 690–712, and http:// thenewobjectivity.com/pdf/marx.pdf.

CHAPTER 8

The term "cyberspace" has multiple origins, although it was popularized by William Gibson in his short story "Burning Chrome," which is reproduced in his novel *Neuromancer*. See also https://en.wikipedia.org/wiki/Cyberspace.

1. The first underground—the Metropolitan Railway—was opened in 1863. It ran east from Paddington station along the Marylebone and Euston roads, taking in the mainline stations of Euston, St. Pancras, and King's Cross through to Farringdon. Its tunnel was frequently punctuated at ground level to let out steam from the engines; see https://www.historytoday.com/archive/months-past/first-day-london-tube. Marylebone station was never linked to the line, but in 1907, the new line from Elephant and Castle to Baker Street was extended to the mainline station, but it remained somewhat disconnected from the predominant east–west traffic along the Marylebone–Euston–King's Cross axis.

2. The telephone came to prominence through various inventions. The most significant was Alexander Graham Bell's 1876 invention discussed in the previous chapter; see endnote 29, chapter 7. There is a useful Wikipedia page available at https://en.wikipedia.org/wiki/Invention_of_the_telephone.

3. There are a number of somewhat humorous accounts of Karl Marx's drinking parties, particularly of their loutish behavior once they reached the end of Tottenham Court Road when the pubs ran out, but these relate to the late 1850s—in fact, about the time when the British Museum Reading Room opened. One can only speculate. Look at Matt Brown's "Do the Karl Marx pub crawl," *The Londonist*, https://londonist.com/london/drink/do-the-karl-marx-pub-crawl. For more background, see Wheen, *Karl Marx*, endnote 2, chapter 7.

4. Standage, *Victorian Internet*, endnote 22, chapter 2. Cooke and Wheatstone tested their line in 1837—see https://blog.sciencemuseum.org.uk/revealing-the-real-cooke-and-wheatstone-telegraph-dial/—just north of Euston station, not so far from the site of the Richard Trevithick's circular railroad, which he demonstrated in 1808 in Euston Square, laying the plans for that part of town to be dominated by railways throughout the nineteenth century and even today.

5. Berners-Lee, *Weaving the Web*, endnote 23, chapter 5.

6. The first internet connection at University College was established in 1973 by Peter Kirstein, as we noted; "ARPANET," endnote 19, chapter 2, and https://en.wikipedia.org/wiki/Internet_in_the_United_Kingdom.

7. The Post Office expanded in the early twentieth century and eventually acquired buildings in Fitzrovia adjacent to University College London. In fact, a large presence not so different from that at St. Martin Le Grand, somewhat ironically, given our introduction, was slowly established from the 1920s onward with the Post Office Tower being built in 1964 for various but usually secure telecoms traffic. Like

so many of the places associated with the Post Office, for security reasons, it was referred to as "Location 23." See Campbell-Smith, *Masters of the Post*, endnote 2, chapter 2, and https://en.wikipedia.org/wiki/BT_Tower.

8. Eva Pascoe established her café, arguably the first internet café, in 1994. The post on her blog from Eric Pfanner's article "A decade of internet cafes 'The world's first', Cafe Cyberia in London, takes a bow" is from the September 2, 2004, *International Herald Tribune*, which provides a concise history of the phenomena. See https://www.evapascoe .com/a-decade-of-internet-cafes-the-worlds-first-cafe-cyberia-in-london-takes-a-bow/.

9. Pascoe, "Internet cafes," https://www.evapascoe.com/, and personal communication, March 4, 2019.

10. Nick Fleming, "Ten years of Cyberia," *The Telegraph*, September 1, 2004, https:// www.telegraph.co.uk/news/uknews/1470674/Ten-years-of-Cyberia.html, and Susie Forbes, "All about Eva," *Wired Magazine*, No. 2.04, April 1996.

11. Around the fringes on the east side of the City of London, various high-tech start-ups began spontaneously in the 1990s, gradually gathering momentum and eventually being supported in gung-ho fashion by Prime Minister David Cameron's coalition government, which came to power in 2010. Eva Pascoe and her colleagues from Cyberia were one of the earliest of these. This is the nearest Britain has ever got to the nerd culture that dominates places such as Silicon Valley (San Francisco) and Silicon Alley (New York), notwithstanding the development of Silicon Fen (the Cambridge Phenomenon). Latterly, the Tech City has lost its glow due to property speculation and the difficulties of attracting venture capital in Britain. See Joshi Herrmann, "The failure of London's Tech City," *Spectator Life*, September 26, 2015, https://life.spectator.co.uk/articles/the-failure-of-londons-tech-city/.

12. Steve Coast, who first worked on mapping at our Centre for Advanced Spatial Analysis (CASA) at University College London, started OpenStreetMap in 2004 at a time when collecting data by crowdsourcing through the web was really taking off. See https://wiki.openstreetmap.org/wiki/History_of_OpenStreetMap. There are many articles on its evolution, and it now represents one of the most durable and evolving free services in the development of the web. There are more than five million registered users, with many millions more using the product simply to navigate in a casual manner. See Muki Haklay and Patrick Weber, "OpenStreetMap: User-generated street maps," *Pervasive Computing*, October–December 2008, 12–17, https://discovery.ucl.ac.uk/id/eprint/13849/1/13849.pdf.

13. Martin Dodge and Rob Kitchin's first book *Mapping Cyberspace*, Routledge, London, 2000, was quickly followed by their *Atlas of Cyberspace*, Prentice-Hall, London, 2001, where they portray the emergent cyberspace through many visualizations. This all began in UCL in 1996 in the shadow of Cyberia and the Post Office Tower, which was featured in Rob Kitchin's earlier book, *Cyberspace: The World in the Wires*, Wiley-Blackwell, London, 1997.

14. Gibson, *Neuromancer*, endnote 35, chapter 4, quoted in "'Cyberspace' Popularized," http://www.historyofinformation.com/detail.php?id=983.

15. Norbert Wiener, *Cybernetics: Or Control and Communication in the Animal and the Machine*, MIT Press, Cambridge, MA.

16. These are systems whose physical control is using "embedded software" extending to entire platforms and involving a convergence of human behaviors, machine intelligence, and big data. See https://en.wikipedia.org/wiki/Cyber-physical_system.

17. Packet switching is a method of transmitting a block of data by breaking it up into packages that are sent separately and reassembled when they reach their destination. The packages are sent independently of one another, over whatever route is optimum for each packet, and reassembled at the destination; https://en.wikipedia.org/wiki/Packet_switching. Latency is the delay occasioned during this process.

18. Telegeography is an early internet mapping company specializing in producing maps of telecoms traffic. See https://www2.telegeography.com/about-us.

19. Originally developed at Lucent (Bell Labs), the Internet Mapping Project spun off into the Lumeta Corporation. The project now appears to have ended. See http://www.cheswick.com/ches/map/.

20. CAIDA, the Center for Applied Internet Data Analysis, is part of the San Diego Super Computer Center at University of California, San Diego CA; https://www.caida.org/home/.

21. James P. Ronda, "How railroads forever changed the frontier," *American Heritage*, *58*, No. 4, https://www.americanheritage.com/how-railroads-forever-changed-frontier. For a detailed look at the influence of railroads in the Midwest during the mid-nineteenth century, see Cronon, *Nature's Metropolis*, endnote 28, chapter 2.

22. Duncan Geere, "How the first cable was laid across the Atlantic," endnote 28, chapter 2.

23. With perhaps the exception of Fischer, *America Calling*, endnote 42, chapter 6.

24. Dodge and Kitchin, *Atlas of Cyberspace*.

25. This historical evolution of the internet in maps is pictured in Timothy B. Lee, "Forty maps that explain the internet," *Vox*, June 2, 2014, at https://www.vox.com/a/internet-maps, while an early sketch of ARPANET's first four nodes is in *Scientific American*, https://www.scientificamerican.com/gallery/early-sketch-of-arpanets-first-four-nodes/.

26. Christopher Lee, "Where have all the Gophers gone? Why the web beat Gopher in the battle for protocol mind share," https://ils.unc.edu/callee/gopherpaper.htm.

27. Rheingold, *Tools for Thought*, endnote 24, chapter 5.

28. Alex Wright, "Mapping the Internet of Things," *Communications of the ACM*, *60*, No. 1, January 2017, 16–18; see also the Wikipedia entry https://en.wikipedia.org/wiki/Shodan_(website).

29. Shaw, "The OSI model explained."

30. Telegeography traditionally specialized in producing world telecoms maps and data, but increasingly they are into developing individual telecoms maps for clients while still producing global maps. See https://shop.telegeography.com/collections /telecom-maps.

31. Andrew Blum's excellent survey of the physicality of the internet in his book *Tubes: Behind the Scenes at the Internet*, Viking, New York, 2012, paints a picture of the new landscape where everyone is connected via routers and servers located in big anonymous boxes that are hard to recognize and even harder to get inside.

32. GAFA (Google, Amazon, Facebook, and Apple), but arguably Microsoft should be added to the quartet, thus GAFAM, endnote 27, chapter 7; see also Phil Simon, *The Age of the Platform: How Amazon, Apple, Facebook, and Google Have Redefined Business*, Motion Publishing, Henderson, NV, 2011, and https://en.wikipedia.org/wiki /Big_Tech.

33. Chris Churchey, "Stop virtual server sprawl," *IBM Systems Magazine*, December 2016, http://archive.ibmsystemsmag.com/power/systems-management/virtuali zation/stop-server-sprawl/.

34. Patrick Thibodeau, "Envisioning a 65-story data center," *Computer World*, https://www.computerworld.com/article/3054603/a-65-story-data-center-design -that-soars-with-ideas.html.

35. Google's first data center owned by the company is located where there is substantial hydroelectric power at The Dalles, OR. This center and those like it are akin to what was once called the "military-industrial complex," which dominated the age of atoms before bits. But their design is more Ballardian, and the lack of visible workers at such plants continually reminds us that the digital world will always be a world of automation. See https://www.google.com/about/datacenters/locations/the-dalles/. The number of data centers that Google has has now reached thirty. See https://dgtlinfra .com/google-cloud-data-center-locations/.

36. Analog devices such as loop counters are discussed in the Wikipedia page on traffic counts. See https://en.wikipedia.org/wiki/Traffic_count#Traffic_counter_devices.

37. "The history of CCTV—From 1942 to the present day," https://www.pcr-online .biz/2014/09/02/the-history-of-cctv-from-1942-to-present/. In fact, London is widely recognized as having the greatest density of CCTV cameras worldwide, with its citizens appearing to be relatively unconcerned about the invasion of their privacy in so obvious a way. See Jonathan Ratcliffe, "How many CCTV cameras are there in London 2019?," May 29, 2019, at https://www.cctv.co.uk/how-many-cctv-cameras-are -there-in-london/.

38. The new landscape of cell towers, some designed to look like trees, is yet another example of life imitating art, the anti-mimesis popularized by Oscar Wilde. Blum's book *Tubes* paints such a picture while the genre of science fiction popularized by

William Gibson in his book *Neuromancer*, endnote 35, chapter 4, reinforces this imageability.

39. Good surveys of urban dashboards are contained in Rob Kitchin, Tracey Lauriault, and Gavin McArdle's "Knowing and governing cities through urban indicators, city benchmarking and real-time dashboards," *Regional Studies, Regional Science*, 2, 1–28, 2015, and Steven Gray, Oliver O'Brien, and Stephan Hügel, "Collecting and visualizing real-time urban data through City Dashboards," *Built Environment*, 42, No. 3, 498–509.

40. Like Blum, Ingrid Burrington's *Networks of New York: An illustrated Field Guide to Urban Internet Infrastructure*, Melville House, Brooklyn, New York, 2016, is an excellent portrayal of the internet from the ground up, where she demonstrates how to explore the invisibility of the net from its small signs, which range from types of manhole covers to mobile-phone masts.

41. Uber and other platforms that allow their users to purchase services online inevitably change the behavior of their users, with indirect repercussions on the behavior of nonusers. This, in turn, not only changes form but also function, with these twin foci beginning to diverge. See also Ben Rossi, "The Uber-fication of everything: How Uber changed the world," *Information Age*, August 19, 2015, and https://www.information-age.com/uber-fication-everything-how-uber-changed-world-123460024/.

42. John Bradburn, David Williams, Rob Piechocki, and Kat Hermans, "Connected and autonomous vehicles: Introducing the future of mobility," https://www.snclavalin.com/~/media/Files/S/SNC-Lavalin/download-centre/en/report/connected-auto-vehicles.pdf.

43. ANPR is a technology that has advanced by stealth, as has CCTV in the UK, particularly in London. See https://en.wikipedia.org/wiki/Automatic_number_plate_recognition_in_the_United_Kingdom.

44. Even mobile phone "slow lanes" for pedestrians have been introduced, as reportedly up to 75 percent of walkers in denser parts of Central London are looking at their mobile phones as they walk; see https://www.ao-world.com/2019/08/27/uks-first-mobile-phone-slow-lane-created-for-distracted-walkers/.

45. The invisibility of information technologies is a key theme in this book, although the focus is mainly on information flows, human decision making, and digital interactions with information. But everywhere in this landscape, there are physical signs of the existence of ICT. Blum's *Tubes* and Burrington's *Networks of New York* admirably illustrate how this invisibility occasionally becomes visible.

CHAPTER 9

Asimov is quoted in Anon, "The computer society: The age of miracle chips, New microtechnology will transform society," *Time Magazine*, February 1978, 44–45. https://content.time.com/time/subscriber/article/0,33009,948017-3,00.html.

1. I quoted this immortal phrase in my book *Cities and Complexity*, MIT Press, Cambridge, MA, 2005, where it is taken from Patrick Geddes own collection of ideas about the dynamics of how cities evolve in his *Cities in Evolution*, Williams and Norgate, London, 1915, 139. When it comes to cities, the history of the last hundred years has been one where, slowly but surely, the notion of a city being in equilibrium has been eroded. Now, we think of cities as being in disequilibrium all the time, non-equilibrium, or even far-from-equilibrium, the mantra of complexity theory. See Jeff Johnson, Andre Nowak, Paul Ormerod, Bridget Rosewell, and Y.-C. Zhang (Editors), *Non-Equilibrium Social Science and Policy: Introduction and Essays on New and Changing Paradigms in Socio-Economic Thinking*, Springer, New York, 2017.

2. Batty, *Inventing Future Cities*, endnote 8, chapter 1.

3. This, of course, is what we called the "standard model" that preoccupied us in chapter 6. This is the monocentric radial structure that has dominated the spatial image of all cities hitherto and is the basis for the descriptive geography of the city, following Park and Burgess, *The City*, endnote 25, chapter 6, and the economic logic of how the cities determine the location of its functions as defined first by von Thünen, *Der Isolierte Staat*, endnote 16, chapter 6, all the way to Alonso, *Location and Land Use*, endnote 27, chapter 6.

4. The concept of the information or smart city as an embedding of automation through contemporary information technologies has been interpreted somewhat obliquely in the idea of the "electric city." See Ricky Burdett and Philipp Rode, "The electric city," *Electric City Conference*, 6–7, London School of Economics, 2–3.

5. Batty, "Computable city," endnote 18, chapter 1.

6. Matt Weinberger, "The worst things Bill Gates ever said," *Business Insider*, April 18, 2016, https://www.independent.co.uk/news/people/the-worst-things-bill-gates-ever-said-a6990046.html.

7. Readers may object to the fact that in a book on computers, smart cities, and big data, there is so little given over to questions of privacy, confidentiality, surveillance, and related issues. This is notwithstanding the importance of these ideas in the wider context of how the Digital Age appears to be dividing and polarizing populations in ways that can be clearly seen in the physical form and structure of cities. Even though we will not deal with these questions, they are implicit in much of what we say in this book, and we will raise them more explicitly in part IV. I hope interested readers are able to realize that any book such as this cannot be all-embracing.

8. The twenty-four-hour city goes back many years, but what we know and how we theorize about it is rather meagre largely due, one suspects, to the fact that comprehensive observational data in the past has been so poor. A relatively recent study in London identifies these issues. See Marion Roberts and Chris Turner, "Conflicts of liveability in the 24-hour city: Learning from 48 hours in the life of London's Soho," *Journal of Urban Design*, *10*, No. 2, 171–193, 2005.

9. An exception is F. Stuart Chapin and Philip Stewart's "Population densities around the clock," in H. H. Mayer and C. F. Kohn (Editors), *Readings in Urban Geography*, University of Chicago Press, Chicago, IL, 180–182, 1953, reprinted 1959, and reflected in Chapin's later book *Human Activity Patterns in the City: What People Do in Time and Space*, John Wiley, New York, 1974.

10. An excellent perspective on old and new data sources, on big data and digital data, is given by Rob Kitchin in his book *The Data Revolution: Big Data, Open Data, Data Infrastructures, and Their Consequences*, Sage, London, 2014. See also Rob Kitchin, Tracey Lauriault, and Gavin McArdle (Editors), *Data and the City*, Routledge, London, 2017.

11. We have made this distinction between low and high frequency several times in this book already, and several sources have elaborated on it. See, for example, the special issue of *Built Environment*, more particularly my introductory article, "Big data and the city," *Built Environment*, *42*, No. 3, 321–337, 2016.

12. One of the first and most comprehensive books on this subject of big data is Viktor Mayer-Schönberger and Kenneth Cukier, *Big Data: A Revolution That Will Transform How We Live, Work, and Think*, John Murray, London, 2013.

13. https://en.wikipedia.org/wiki/Smart_city.

14. Michael Batty, Kay Axhausen, Fosca Giannotti, Alex Pozdnoukhov, Armando Bazzani, Monica Wachowicz, Georgios Ouzounis, and Juval Portugali, "Smart cities of the future," *European Physical Journal Special Topics*, *214*, 481–518, 2012.

15. David Gibson, George Kozmetsky, and Rob Smilor, *The Technopolis Phenomenon: Smart Cities, Fast Systems, Global Networks*, Rowman and Littlefield, Lanham, MD, 1992.

16. William Dutton, John Blumler, and Ken Kraemer (Editors), *Wired Cities: Shaping the Future of Communications*, G. K. Hall, New York, 1987.

17. Martin, *Wired Society*, endnote 41, chapter 5.

18. We described the origins of email in chapter 5, where we noted that Ray Tomlinson, working on ARPANET with Bolt, Beranek, and Newman, is credited with sending the first email using the @ symbol to identify the owner of the message. See endnotes 9–12, chapter 5, and the discussion therein.

19. Julien Mailland and Kevin Driscoll, *Minitel: Welcome to the Internet*, MIT Press, Cambridge, MA, 2017. See also https://en.wikipedia.org/wiki/Minitel.

20. Tom Lean, "Prestel: The British internet that never was," August 23, 2016, https://www.historytoday.com/history-matters/prestel-british-internet-never-was.

21. There are many interchangeable terms for the smart city. We prefer "computable city," as this reflects our emphasis here, but the terms "virtual city," "intelligent city," "information city," "electric city," "wired city," "ubiquitous city," and so on are all used; the list is almost endless, and usage depends on context.

22. The biological analogy between the city and our body has recurred throughout history from the time of Leonardo da Vinci, if not before, back to the classical era.

Victor Gruen's book *The Heart of Our Cities: The Urban Crisis. Diagnosis and Cure*, Thames and Hudson, London, 1964, reflects this tradition in urban planning, while Geoff West's magnum opus *Scale: The Universal Laws of Life and Death in Organisms, Cities and Companies*, Weidenfeld and Nicolson, London, 2017, describes in more detail how the analogy can be pursued through networks rather than organs of the body.

23. Bahar Gholipour, "What is a normal heart rate?," *Live Science*, January 12, 2018, https://www.livescience.com/42081-normal-heart-rate.html.

24. The discussion of temporal cycles in cities is sparse, again due to a lack of time series data, but a focus on understanding declining or shrinking cities is developing. See Timothy Taylor, "The cycles of cities," *Conversable Economist*, July 20, 2017, https://conversableeconomist.blogspot.com/2017/07/the-cycles-of-cities.html. Cycles of a longer duration than those of the twenty-four-hour city are discussed in my book *Inventing Future Cities*, endnote 8, chapter 1, particularly those dealing with long waves, and we note these here in endnote 12, chapter 7.

25. The massive growth of personal monitoring using digital watches and trackers is part of this explosion of technologies for sensing the environment as well as ourselves. This links to various "quantified self" projects. See, for example, the Quantified Self Institute headquartered in Groningen, The Netherlands, available at https://qsinstitute.com/about/what-isquantified-self/.

26. There are various tap-in, tap-out payment systems on different fixed transit systems such as trains and buses, with a typical one described by Jon Reades, Chen Zhong, Ed Manley, Richard Milton, and Michael Batty in their paper "Finding pearls in London's oysters," *Built Environment*, *42*, No. 3, 365–381, 2016.

27. Jonathan Hall, Craig Palsson, and Joseph Price, "Is Uber a substitute or complement for public transit?," *Journal of Urban Economics*, *108*, 36–50, 2018.

28. Michael Batty, "Visualizing aggregate movement in cities," *Philosophical Transactions of the Royal Society B*, *373*, July 2, 2018.

29. The peak hour is often called the "rush hour," and of course, it lasts longer than an hour, with congestion on most traffic systems being greatest at the peaks in the early morning and late afternoon. The weekend days normally do not show distinct morning and early evening peaks, for few people go to and then return from work but rather use transport systems mainly for leisure purposes, entertainment, and shopping for example. See https://en.wikipedia.org/wiki/Rush_hour.

30. Reades et al., "London's oysters." See also see the video "UCL engineering—oyster gives up pearls" available at https://www.youtube.com/watch?v=9sAugcb2Qj4.

31. Data exhaust is the by-product from the data used to control whatever functions in the city are being automated in digital terms using sensors, which deliver data in real time. See https://en.wikipedia.org/wiki/Data_exhaust.

32. Oliver O'Brien, James Cheshire, and Michael Batty, "Mining bicycle sharing data for generating insights in sustainable transport systems," *Journal of Transport Geography*, *34*, 262–273, 2014.

33. These features of urban travel are explored in more detail in my book *Inventing Future Cities*, endnote 8, chapter 1.

34. eMarketer Report, "US time spent with media: eMarketer's updated estimates and forecast for 2014–2019," April 27, 2017, https://www.emarketer.com/Report/US -Time-Spent-with-Media-eMarketers-Updated-Estimates-Forecast-20142019/2002021.

35. In the first iPhones, you could unlock the data that was recorded anywhere the phone was located in an application called iPhoneTracker that produced a map "for you" of the sequence of locations visited in quite fine detail. This degree of open sur- veillance was soon removed by Apple but not before some used it to plot their own movements anywhere, in my own case across the globe. See "The trips of a wander- ing academic tracked from November 2010," https://vimeo.com/25362904.

36. Francesco Calabrese, Giusy Di Lorenzo, Liang Liu, and Carlo Ratti, "Estimating origin-destination flows using mobile phone location data," *IEEE Pervasive Comput- ing*, *10*, No. 4, 36–44, 2011.

37. One of the first projects utilizing mobile-phone data to figure out transportation patterns was the seminal paper by Marta Gonzalez, Cesar Hidalgo, and Albert-Laszlo Barabasi, "Understanding individual human mobility patterns," *Nature, 453*, 779–782, 2008. Many subsequent papers have taken these ideas further, notwithstanding the great difficulties in inferring trips from phone calls. See Serdar Çolak, Lauren Alexan- der, Bernardo Guatimosim Alvim, Shomik Mehndiretta, Tatiana Peralta Quiros and Marta Gonzalez, "Analyzing cell phone location data for urban travel: Current meth- ods, limitations and opportunities," *Conference of European Statisticians: Workshop on Statistical Data Collection: Riding the Data Deluge*, 29 April–1 May, Washington, DC, 2015, and Vincent Blondel, Adeline Decuyper, and Gautier Krings, "A survey of results on mobile phone datasets analysis," *European Physics Journal, Data Science, 4*, No. 10, 2015.

38. Anahid Nabavi Larijani, Ana-Maria Olteanu-Raimond, Julien Perret, Mathieu Brédif, and Cezary Ziemlicki, "Investigating the mobile phone data to estimate the origin destination flow and analysis; case study: Paris region," *Transportation Research Procedia, 6*, 64–78, 2015.

39. There are many sites that record usage and statistics, and these often differ, but for typical data, see Riley Panko, "Mobile app usage statistics 2018," https://themanifest .com/app-development/mobile-app-usage-statistics-2018 and https://www.businessof apps.com/data/app-statistics/.

40. User-generated content means any data that you, the user, generate using a digi- tal platform in some way, which largely means the use of Web 2.0 technologies. See https://en.wikipedia.org/wiki/User-generated_content.

41. As with all the figures quoted here, these are highly variable and only provide an order of magnitude estimate. For example, see the number of monthly active Twitter users worldwide from the first quarter of 2017 to the fourth quarter of 2021, *Statista*,

https://www.statista.com/statistics/970920/monetizable-daily-active-twitter-users
-worldwide/. Note that the other statistics of Twitter use in the main text are for 2019
when Twitter stopped publishing full metrics. Of course, at the time of writing, the
Twitterverse has been thrown into chaos by the purchase of the platform by Elon
Musk, who is changing the conditions of membership and the limits on its content
in ways that are controversial. The platform has now been relabelled X.

42. Fabian Neuhaus, "Urban rhythms: Habitus and emergent spatio-temporal dimensions of the city," unpublished PhD thesis, Centre for Advanced Spatial Analysis,
University College London, 2012.

43. There are many estimates of the number of geotagged tweets, but these appear
to be a rather small proportion of all tweets (between 1 and 5 percent). Other ways
of tagging geographical locations are discussed in Bo Han, Paul Cook, and Timothy
Baldwin, "Text-based Twitter user geolocation prediction," *Journal of Artificial Intelligence Research*, *49*, 451–500, 2014, https://people.eng.unimelb.edu.au/paulcook/live
-4200-7781-jair.pdf.

44. Batty, *Post-Urban World*, endnote 32, chapter 6.

45. Ajay Agrawal, Joshua Gans, and Avi Goldfarb, *Prediction Machines: The Simple
Economics of Artificial Intelligence*, Harvard Business Review Press, Cambridge, MA,
2018; and Eric Beinhocker, *The Origin of Wealth: Evolution, Complexity, and the Radical Remaking of Economics*, Harvard Business Review Press, Cambridge, MA, 2006.

46. One of the first to articulate crowdsourcing was clearly Clay Shirky in his book
Here Comes Everybody: The Power of Organizing Without Organizations, Penguin Press,
New York, 2008. Crowdsourced geographic data, of which the example of Open-
StreetMap is the best (see endnote 13, chapter 8), is a good example of what Michael
Goodchild has called "volunteered geographic information" (VGA). See his paper
"Citizens as sensors: The world of volunteered geography," *GeoJournal*, *69*, No. 4,
211–221, 2007.

47. Charlie Catlett, Urban Center for Computation and Data (UCCD), http://www
.urbanccd.org/about#about-urbanccd.

48. Argonne National Laboratory has always been at the forefront of applications
of computing to large-scale simulations of which cities are one of the most recent,
spinning off from grid and exascale computing. See https://www.anl.gov/argonne
-national-laboratory and https://en.wikipedia.org/wiki/Exascale_computing.

49. There is a good YouTube movie on the Array of Things, and I quote, the AoT
"is designed as a 'fitness tracker' for the city, collecting new streams of data on Chicago's environment, infrastructure, and activity. This hyper-local, open data can help
researchers, city officials, and software developers study and address critical city challenges, such as preventing urban flooding, improving traffic safety and air quality,
and assessing the nature and impact of climate change." See https://www.youtube
.com/watch?time_continue=16&v=BHrsllHJHeo&feature=emb_logo.

50. The website http://arrayofthings.github.io/ defines the project as "a networked urban sensor project that's changing our understanding of cities and urban life." It is a veritable treasure trove of material from which the quotes in this section are taken.

51. There has been substantial media coverage of the project. See http://arrayof things.github.io/media.html. See also Aamer Madhani, "Chicago begins building 'fitness tracker' to check its vitals," *USA Today*, https://www.usatoday.com/story /news/2016/08/29/chicago-begins-building-fitness-tracker-check-its-vitals/89434 620/ and Whet Moser, "The Array of Things is coming to Chicago (and the world)," https://www.chicagomag.com/city-life/September-2016/The-Array-of-Things-Is-Co ming-to-Chicago-and-the-World/.

52. It is worth noting that there are several other urban laboratory projects that are intent on providing demonstrations of how the city might be wired. In the UK, the project at the University of Newcastle is one of the first. See "UK's first Urban Observatory provides unique understanding of how our cities work," https://phys .org/news/2017-09-uk-urban-observatory-unique-cities.html. A very different one in east London at the Queen Elizabeth Park is more focused on very detailed wiring of a small area developing 3D visualizations, which are linked to sensor data delivered in real time. See "UCL's Tales of the Park to bring talking digital creatures to Queen Elizabeth Olympic Park," https://www.queenelizabetholympicpark.co.uk/news/news -articles/2017/09/ucls-tales-of-the-park-to-bring-talking-digital-creatures-to-queen -elizabeth-olympic-park.

CHAPTER 10

George E. P. Box, "Robustness in the strategy of scientific model building," in R. L. Launer and G. N. Wilkinson (Editors), *Robustness in Statistics*, Academic Press, New York, 201–236, 1979.

1. Sajal Lahiri, "Professor Wassily W. Leontieff 1905–1999," *The Economic Journal*, *110*, F695–F707, 2000, and UBS Nobel Perspectives, "How is the global economy interconnected?," https://www.ubs.com/microsites/nobel-perspectives/en/laureates /wassily-leontief.html.

2. Wassily Leontief biography, available at https://www.thefamouspeople.com/pro files/wassily-leontief-299.php.

3. Leontieff won the Nobel Prize in Economic Sciences (officially the Sveriges Riks-bank Prize in Economic Sciences) in 1973, while his notion of the circular economy has seen something of a resurrection in the last decade, particularly with respect to the idea that waste might be converted back to productive use, thus establishing feedback effects that enable the economy to be self-sustaining. See, for example, https://www .ellenmacarthurfoundation.org/circular-economy/what-is-the-circular-economy.

4. Samuelson, "Thünen at two hundred," endnote 1, chapter 6.

5. The notion of equilibrium plays a key role in economic science, essentially as a basis for examining the way the economy responds to marginal change. See Robert E. Kuenne, "Walras, Leontief and the interdependence of economic activities," *Quarterly Journal of Economics*, *68*, 323–354, 1954.

6. The Wilbur Machine, *The MIT 150 Exhibition*, MIT Museum, Cambridge, MA, January 8, 2011.

7. Quoted in Charles and Ray Eames, *Computer Perspective*, endnote 29, chapter 3.

8. UBS Nobel Perspectives, 1998.

9. We recounted the early history of Aitkin's efforts in developing the digital computer at Harvard in an earlier chapter, but see Dyson's book *Turing's Cathedral*, endnote 24, chapter 1 et seq., for a deeper history.

10. After he was awarded the Nobel Prize, Leontieff left Harvard in 1973 for New York University, but he continued his work on developing input–output analysis in many guises. I had the privilege of organizing a meeting in New York City in 1993, which brought together three Nobel prize winners—Leontieff, Phil Anderson, and Lawrence Klein—to discuss the development of economic modeling and geographic information systems, which are discussed in chapter 11. The interconnections between computing, economics, spatial analysis, and urban planning ran deep during those early years.

11. There are many good histories of Britain's railways. See, for example, Julian Holland, *The Times History of Britain's Railways*, Times Books, London, 2015.

12. George P. Landow, "Railway mania," in *The Victorian Web*, March 2014, http://www.victorianweb.org/technology/railways/fraud.html. Dan Bogart, Leigh Shaw-Taylor, and Xuesheng You, "The development of the railway network in Britain 1825–1911," The Cambridge Group for the History of Population and Social Structure, https://www.campop.geog.cam.ac.uk/research/projects/transport/onlineatlas/railways.pdf.

13. Arthur H. Robinson, "The 1837 maps of Henry Drury Harness," *The Geographical Journal*, *121*, 440–450, 1955.

14. We discussed this extensively in chapter 6 when we introduced the standard model. Although von Thünen did not specifically invoke the idea of gravity in his discussion of how a simple economic system is organized, he implicitly introduced the notion that distance and transportation cost were the key determinants of what activities were located in different places. See von Thünen, *Isolated State*, endnote 16, chapter 6.

15. Tobler's law, sometimes called the "first law of geography," is named after Waldo Tobler, who first introduced it in a paper in 1970, "A computer movie simulating urban growth in the Detroit region," *Economic Geography*, *46* (Supplement), 234–240, although the law had been implicit in much of location theory from the

eighteenth century onward. See also the Wikipedia entry https://en.wikipedia.org
/wiki/Tobler%27s_first_law_of_geography.

16. There are several speculations that Newton's second law of motion—the gravity law—came to be applied to many social phenomena quite soon after Newton
had articulated it in the late seventeenth century. See http://www.eoht.info/page
/Social+gravitation. In terms of the railways and their impact on location, Andrew
Odlyzko's 2015 paper "The forgotten discovery of gravity models and the inefficiency
of early railway networks" provides an excellent history that suggests that almost as
soon as new forms of transportation such as the railways were developed, ideas about
gravitational attraction in social systems were developed. See http://www.dtc.umn
.edu/~odlyzko/doc/mania09.pdf.

17. Herapath's work on gravity quickly moved to a long-term interest in the
morphology of the railway system, founding what ultimately became *The Railway
Gazette*. See https://en.wikipedia.org/wiki/John_Herapath.

18. Odlyzko, "Gravity models."

19. Henry Charles Carey, *Principles of Social Science*, 3 Volumes, J. B. Lippincott &
Co., Philadelphia, PA, 1858–1860. See also http://www.eoht.info/page/Henry+Carey.

20. Ernst G. Ravenstein, "The laws of migration," *Journal of the Statistical Society*, *48*,
167–227, 1885, and "The laws of migration," *Journal of the Statistical Society*, *52*,
214–301, 1889.

21. William J. Reilly, *The Law of Retail Gravitation*, Pilsbury, New York, 1931, republished in 1953. For a discussion relating to retail models and central place theory, see
my article "Reilly's challenge: New laws of retail gravitation which define systems of
central places," *Environment and Planning A*, *10*, 185–219, 1978.

22. Anon, "US states by vehicles per capita," https://www.worldatlas.com/articles
/us-states-by-vehicles-per-capita.html. See also Anon, "Was the rise of car ownership
responsible for the mid-century homeownership boom in the US?," *Old Urbanist*,
February 2013, https://oldurbanist.blogspot.com/2013/02/was-rise-of-car-ownership
-responsible.html.

23. Congress approved the Federal Highway Act (in which Al Gore, Sr., played a major
part) on June 26, 1956. See https://www.history.com/this-day-in-history/congress
-approves-federal-highway-act. For the contribution of his son to the internet, see
https://www.theregister.co.uk/2000/10/02/net_builders_kahn_cerf_recognise/.

24. A detailed history of transport modeling is presented in David Boyce and Huw
William's book *Forecasting Urban Travel: Past, Present and Future*, Edward Elgar, Cheltenham, UK, 2015, where the Detroit (DMATS) study is identified as the first large-
scale transportation modeling project.

25. Douglass Carroll formulated several early ideas in modeling transportation
built around the gravitational hypothesis. See his "Spatial interaction and the

urban-metropolitan regional description," *Papers and Proceedings of the Regional Science Association, 1,* No. 1, 59–73, 1955.

26. There are several in-depth discussions of CATS, including Boyce and Williams, *Forecasting Urban Travel.* In particular Alan Black's "The Chicago Area Transportation Study: A case study of rational planning," *Journal of Planning Education and Research, 10,* No. 1, 27–37, 1990, presents a brief overview. A longer report is Andrew V. Plummer, *Chicago Area Transportation,* endnote 39, chapter 4.

27. Plummer, *Chicago Area Transportation,* endnote 39, chapter 4.

28. I had the privilege of seeing the model being run in batch mode. Once the computer center at Bedfordshire County Council closed for the day, the model builders—Cripps and Foot—took over the machine to run their programs, having written the code, collected the data, punched the cards, loaded the mag tapes, and switched the machine to run the batch programs. The model that they built and applied is reported in their paper Eric Cripps and David Foot, "Strategy for the South East—5: Evaluating alternative strategies," *Official Architecture and Planning, 31,* No. 7, 928–941, 1968.

29. Alan M. Voorhees, "A general theory of traffic movement," *Proceedings,* Institute of Traffic Engineers, New Haven, CT, 46–56, reprinted in *Transportation, 40,* 1105–1116, 2013.

30. Walter G. Hansen, "Accessibility and residential growth," MCP degree thesis, MIT, Cambridge, MA, June 1959, and "How accessibility shapes land use," *Journal of the American Institute of Planning, 25,* No. 2, 73–76, 1959.

31. The 1959 special issue of the *Journal of the American Institute of Planners* was devoted to transportation modeling, while six years later, in 1965, Britton Harris edited another special issue of the same journal on urban development models. See Britton Harris, "New tools for planning," *Journal of the American Institute of Planners, 31,* 90–95, 1965.

32. We outlined what Forrester did in the first publicly broadcast demonstration of a computer graphic on Ed Morrow's show *See It Now* at Christmas in 1951. See endnote 39, chapter 3.

33. J. Douglas Carroll and Garred P. Jones, "Interpretation of desire line charts made on a Cartographatron," *Highway Research Board Bulletin, 253,* Highway Research Board, Washington, DC, 86–108, 1960, at https://trid.trb.org/view/120687.

34. Leo P. Kadanoff, Jerrold R. Voss, and Wendell J. Bouknight, "Computer display and analysis of urban information through time and space," *Technological Forecasting and Social Change, 2,* No. 1, 77–103, 1970.

35. Forrester, *Urban Dynamics,* endnote 27, chapter 3.

36. https://en.wikipedia.org/wiki/We_choose_to_go_to_the_Moon.

37. The most comprehensive of all the land-use models introduced in the 1960s was Ira S. Lowry's model of Pittsburgh, which he reported in his monograph *A Model of*

Metropolis, Memorandum RM-4035-RC, The RAND Corporation, Santa Monica, CA, 1964, and in his paper "Location parameters in the Pittsburgh model," *Papers and Proceedings of the Regional Science Association, 11*, 145–165. 1963. The spatial interaction models that were developed by Lowry and others were put on a solid theoretical statistical framework by Wilson in his paper "A statistical theory of spatial distribution models," *Transportation Research, 1,3*, 253–269, 1967, also see his book A. G. Wilson *Entropy in Urban and Regional Modelling*, Pion Press, London, 1970.

38. The special issue of the *Journal of the American Institute of Planners, 31*, 1965, edited by Britton Harris, "Quantitative models," contained summaries of most of these early models, while there were other review papers of the US experience at that time. Harris himself wrote several such papers. See, for example, "Quantitative models of urban development: Their role in metropolitan policy making," in Harvey S. Perloff and Lowdon Wingo, Jr. (Editors), *Issues in Urban Economics*, Resources for the Future, Johns Hopkins University Press, Baltimore, MD, 363–410, 1968.

39. A summary of the efforts to make comparative static models dynamic particularly the Pittsburgh model is contained in my book *Urban Modelling: Algorithms, Calibrations, Predictions*, Cambridge University Press, Cambridge, 1976.

40. F. Stuart Chapin and Shirley F. Weiss, "A probabilistic model for residential growth," *Transportation Research, 2*, No. 4, 375–390, 1968.

41. Kenneth Schlager, "A land use plan design model," *Journal of the American Institute of Planners, 31*, 103–111, 1965.

42. Gregory K. Ingram, John Kain, and J. Royce Ginn, *The Detroit Prototype of the NBER Urban Simulation Model*, National Bureau of Economic Research, Columbia University Press, New York, 1972.

43. The Penn Jersey Transportation Study (PJTS) spawned various model types, in particular the housing market models accredited to John Herbert and Ben Stevens from the University of Pennsylvania. See their paper "A model for the distribution of residential activity in urban areas," *Journal of Regional Science, 2*, No. 2, 21–36, 1960. Britton Harris, as research director of the PJTS, wrote many papers supporting these models. See, for example, his "Linear programming and the projection of land uses," PJ Paper 20, The PJTS Commission, Philadelphia, NJ, 1963.

44. We have not detailed here the British experience with land-use transportation models that followed closely on the heels of these developments in the US. I have summarized this experience in several papers, two of which record the early days and later developments. See Alan Wilson, 1970, Michael Batty, "Recent developments in land use modelling: A review of British research," *Urban Studies, 9*, 151–177, 1972, and "Fifty years of urban modelling: Macro statics to micro dynamics," in S. Albeverio, D. Andrey, P. Giordano, and A. Vancheri (Editors), *The Dynamics of Complex Urban Systems: An Interdisciplinary Approach*, Physica, Heidelberg, Germany, 1–20, 2008.

45. The Lincoln Laboratory and all the various developments in computing that clustered around it must have been a remarkable place when it was set up in 1951 at the beginning of the Cold War where systems thinking was writ large. In fact, in our context here, the laboratory spawned some amazing software solutions to various highly applied decision problems, ranging from systems dynamics and game theory to missile trajectories and cybersecurity. See https://en.wikipedia.org/wiki/MIT _Lincoln_Laboratory.

46. Jay W. Forrester, *Industrial Dynamics*, MIT Press, Cambridge, MA, 1961. See also the genesis of the field at https://www.systemdynamics.org/origin-of-system-dynamics.

47. SYMAP, defined as the SYnographic MApping System (or more colloquially as SYmbol MAPping), was used extensively at the Harvard Laboratory for Computer Graphics and Spatial Analysis from the late 1960s. See Nick Chrisman, *Charting the Unknown*, endnote 38, chapter 4. But the idea of overprinting text on a line printer, which is what SYMAP does, goes back to the beginning of digital computing itself. I saw the first instances in Cambridge in 1969 in the form of the "Marilyn Monroe" picture, but they must go back to the early 1950s as soon as line printer technology emerged for digital computers, but even before World War II for tabulation in accounting. See https://en.wikipedia.org/wiki/Line_printer.

48. Forrester, *Urban Dynamics*, endnote 27, chapter 3.

49. After the relative success of *Urban Dynamics*, Forrester quickly developed his systems dynamics for the world system. See his book *World Dynamics*, The Wright-Allen Press, Cambridge, MA, 1971, which preceded the "Limits to growth" report.

50. Donella H. Meadows, Jorgen Randers, Dennis L. Meadows, and William W. Behrens, "The limits to growth: A report for the Club of Rome's project on the predicament of mankind," Earth Island, London, 1972.

51. There were several attempts to ground Forrester's model in the spatial nexus of the city systems. Leo Kadanoff's was one of the first. See his paper "From simulation model to public policy: An examination of Forrester's 'urban dynamics,'" *Simulation*, 16, 261–268, 1971. A collection of such attempts was edited by Kan Chen as *Urban Dynamics: Extensions and Reflections*, San Francisco Press, San Francisco, CA, 1972.

52. Forrester, *Urban Dynamics*, endnote 27, chapter 3

53. Quoted in the article by Kristian Hvidtfelt Nielsen, "From monstrosity to laptop: The story of the personal computer," which is as good an origin story as one can find with respect to the invention of personal computing. See https://sciencenordic.com /computers-denmark-forskerzonen/from-monstrosity-to-laptop-the-story-of-the-perso nal-computer/1452258.

54. Quoted by Kevin Fogarty, "Tech predictions gone wrong, A romp through some of the worst tech predictions from people who should have known better," https:// www.computerworld.com/article/2492617/tech-predictions-gone-wrong.html.

55. These were heady times for a youngster such as myself to get involved in the cutting edge of systems science. In those days, you could meet everyone who had been involved in inventing the field of computer models of cities—yes, everyone— and you could read everything that had ever been written on the subject. I had the privilege of meeting many of those who were instrumental in forging the field in May 1970 when Eric Cripps and I spent two weeks in the US visiting Penn, the MIT, Harvard, the RAND Corporation, Berkeley, and North Carolina, where all the action was. But this was also an auspicious time for on our visit. The National Guard were firing tear gas on the campus at Berkeley, it was the height of the Vietnam War, and the tragic events at Kent State University took place in the two weeks we were there. The "Requiem" which the next section of our chapter and the next endnote recounts was fast dispelling the optimism of those times.

56. Douglass B. Lee, "Requiem for large-scale models," *Journal of the American Institute of Planners*, *39*, 163–178, 1973. The various quotes in this section from Lee are taken from this paper.

57. There are many good summaries of this experience. See, for example, Garry Brewer's *Politicians, Bureaucrats, and the Consultant: A Critique of Urban Problem Solving*, Basic Books, New York, 1973.

58. "Wicked problems" seem to have first been articulated by C. West Churchman in an editorial in *Management Science*, *14*, No. 4, B141–B142, 1967, but he credits Horst Rittel with inventing the term or at least inspiring him to invent it. See Andrejs Skaburskis, "The origin of 'wicked problems,'" *Planning Theory and Practice*, *9*, No. 2, 277–280, 2008. However, the best and most accessible discussion is Horst Rittel and Mel Webber's "Dilemmas in a general theory of planning," *Policy Sciences*, *4*, No. 2, 155–169, 1973.

59. Dick Klosterman organized a special session at the American Planning Association Annual Meeting in Chicago in May 1993, and the papers were published in the special issue in the winter of 1994. Klosterman's introduction argues that the field was beginning to emerge from its long sleep. See his "Large-scale urban models: Retrospect and prospect," *Journal of the American Planning Association*, *60*, No. 1, 3–6, 1994. But Lee was unrepentant, even though the PC revolution had happened, GIS was being widely applied, and new digital data sources were making it easier to represent the city system. See his "Retrospective on large-scale urban models," *Journal of the American Planning Association*, *60*, No. 1, 35–40, 1994, in the same issue.

60. These experiences are summarized in my commentary "Can it happen again? Planning support, Lee's Requiem and the rise of the smart cities movement," *Environment and Planning B*, *41*, 388–391, 2014, which complements Marco te Brömmelstroet, Peter Pelzer, and Stan Geertman's "Forty years after Lee's Requiem: Are we beyond the seven sins?," *Environment and Planning B*, *41*, 381–391, 2014. As yet, new generations of simulation models reflecting to an extent the new urban science have not made an impact on urban planning, largely due to the fact that there is little consensus about their general practical relevance.

CHAPTER 11

Edsger Dijkstra, the inventor of the formal algorithm that first solved the shortest-route problem in the theory of graphs, is accredited with this quotation from a paper in 1977, but others appear to have made similar observations. See https://amturing.acm.org/award_winners/dijkstra_1053701.cfm.

1. Martin Gayford, "David Hockney's iPad art," *The Telegraph*, October 20, 2010, https://www.telegraph.co.uk/culture/art/art-features/8066839/David-Hockneys-iPad-art.html.

2. Klint Finley, "Tech time warp of the week: Watch Andy Warhol paint Debbie Harry on an Amiga 1000 Computer," *Wired Magazine*, March 10, 2014, https://www.wired.com/2014/10/warhol-blondie-amiga/.

3. The TV advertisement, one of the greatest of all time, began with the caption: "On January 24th, Apple will introduce the Macintosh and you will see why 1984 won't be like '1984.'" It then proceeded to show "Androgynous, drone-like humans march in unison as a Big Brother-like figure tells them it's time to celebrate the first glorious anniversary of the Information Purification Directives." It was a parody on the IBM PC, and, of course, it painted the world of IBM as akin to the Orwellian dystopia of 1984, with Apple being cast as the upstart who would change the computer industry from the bottom up, as indeed it is still doing—just—from our current vantage point of 2022. See the video on YouTube at https://www.youtube.com/watch?v=axSnW-ygU5g and the wider context at https://www.youtube.com/watch?v=PsjMmAqmblQ. Of course, if we dig a little deeper into the locations and events that most influenced some of the ideas in this book, we find George Orwell camped out not far from where Karl Marx lived at the bottom of Tottenham Court Road. Marx, Darwin, and Orwell walked the same streets in Soho and Bloomsbury, as did Turing and some of the early internet pioneers.

4. I referred to Wilkes's animation in my book *Microcomputer Graphics: Art, Design, and Creative Modelling*, Chapman and Hall, London, 1986, which, in turn, I took from the quotation in Annabel Jankel and Rocky Morton's book *Creative Computer Graphics* that we noted in endnote 34, chapter 4. A brief reference to these kinds of animations is in the early history of the Computer Laboratory at the University of Cambridge. See https://www.cl.cam.ac.uk/relics/chron.html. We have also quoted Jay Forrester's demo in previous chapters. See endnote 39, chapter 3.

5. This is a famous quotation noted in many places https://en.wikiquote.org/wiki/Computer_science, although it is relevant to any branch of knowledge that seeks to elevate theory above the tools that are used to advance that theory, computer science or rather computation being no exception. The same might be said about the idea of the computable versus the smart city.

6. Ben Laposky, "Oscillons: Electronic abstractions," in Ruth Leavitt, *Artist and Computer*, Creative Computing Press, Morristown, NJ, 21–22, 1976.

7. Short histories of computer graphics are given in my book *Microcomputer Graphics*, and "Computer art: Definition, history, characteristics of new media art" at http://www.visual-arts-cork.com/computer-art.htm.

8. "Andy Warhol and Marshall McLuhan: The artist and the visionary," https://mcluhangalaxy.wordpress.com/2011/12/08/andy-wharhol-marshall-mcluhan-the-artist-the-visionary/, and Guy Hepner, "An art history of Andy Warhol," https://guyhepner.com/an-art-history-of-andy-warhol/.

9. "Marilyn Monroe," https://knowyourmeme.com/photos/592796-ascii-art, and Paul Tupaczewski, "Printer pictures go hi-res," https://www.atarimagazines.com/v6n3/asciiart.html. There is more ASCII art as "Mickey Mouse." See Batty, *Microcomputer Graphics*.

10. Chrisman, *Charting the Unknown*, endnote 38, chapter 4.

11. Fil Salustri, "A brief history of computer graphics," *DesignWIKI MEC222*, Engineering Graphical Communications, https://deseng.ryerson.ca/dokuwiki/mec222:brief_history_of_computer_graphics.

12. There are many histories of graphics and computer displays, and these are often mixed together, but currently, most histories do not go beyond about 2012. See https://www.computerhistory.org/timeline/graphics-games/.

13. Jo Marchant, "A journey to the oldest cave paintings in the world," *Smithsonian Magazine*, December 2016, https://www.smithsonianmag.com/history/journey-oldest-cave-paintings-world-180957685/, and Jennifer Nguyen, "Archaeologists figured out that some of the world's oldest cave drawings don't just depict animals—they're constellations of stars," *Business Insider*, December 20, 2018, https://www.businessinsider.com/ancient-cave-drawings-are-constellations-of-stars-2018-12?r=US&IR=T.

14. There are many definitions of "vectors" versus "rasters" and the associated graphics that comes from defining maps in terms of vectors based on points, lines, and polygons and rasters based on pixelated grid-like regular tessellations of locations, with each raster usually defined by points that locate the grid in question. For a textbook introduction, see Paul A. Longley, Michael F. Goodchild, David J. Maguire, and David W. Rhind, *Geographic Information Science and Systems*, 4th Edition, John Wiley, New York, 2015. A computer science definition is given at https://www.geeksforgeeks.org/vector-vs-raster-graphics/.

15. We have already noted SYMAP. See Nick Chrisman, *Charting the Unknown*, endnote 38, chapter 4. But the origins of this kind of line-printer map go back further to the origins of information systems and computer cartography at the University of Washington, where Edgar Horwood and Duane Marble worked on such maps in parallel to Alan Schmidt in Michigan, who produced roughly similar map graphics. At the System Development Corporation (SDC) in 1970, similar line-printer maps were produced to show the geography of optimal school locations. See endnote 55, chapter 10, and Barry Wellar (Editor), *Foundations of Urban and Regional Information*

Systems and Geographic Information Systems and Science, Special Publication in Celebration of URISA's 50th Anniversary Conference 1963–2012, https://www.urisa.org /clientuploads/directory/Documents/Books%20and%20Quick%20Study/Founda tions_FINAL.pdf.

16. Ivan Sutherland's seminal contribution to computer graphics in general, interactive graphics in particular, and computer-aided design was noted in chapter 3, and there are many good summaries of his contributions such as "i-Programmer: Ivan Sutherland— Father of computer graphics," February 21, 2019, https://www.i-programmer.info/his tory/people/329-ivan-sutherland.html.

17. MIT School of Science, "A natural fit: Science's intersection with computing," https://science.mit.edu/natural-fit/ and the influence of Bell Labs in all this is recounted in Gertner, *Idea Factory*, endnote 30, chapter 3.

18. Paul Gabrielsen, "Contributions to computer graphics," https://attheu.utah .edu/facultystaff/contributions-to-computer-graphics/, and Carlson, *Computer Graphics*, endnote 13, chapter 4.

19. William Warntz, *The Geography of Price: A Study in GeoEconometrics*, University of Pennsylvania Press, Philadelphia, PA, 1959, and his "A new map of the surface of population potentials for the United States," *Geographical Review*, 54, No. 2, 170–184, 1964. This dates back to the astronomer John Quincy Stewart's pioneering work "Empirical mathematical rules concerning the distribution and equilibrium of population." *Geographical Review*, 37, 461–485, 1947.

20. Nick Chrisman, *Charting the Unknown*, endnote 38, chapter 4.

21. The group that were graduate students around Carl Steinitz at the Harvard Graduate School of Design (GSD), namely Jack Dangermond, who founded ESRI, Lawrie Jordon, who founded ERDAS, and many others who were involved in the Lab, celebrated its impact on the field of GIS in May 2015—the 50th anniversary. See https://cga-download.hmdc.harvard.edu/publish_web/CGA_Conferences/2015_ Lab_Legacy/2015_CGA_Conference_Program.pdf. See also Joe Francica, "Fifty years of commercial GIS—Part 1: 1969–1994," December 31, 2018, https://www.directions mag.com/article/8410.

22. The group that came together at the University of Washington, Seattle, in the late 1950s was truly remarkable. It preceded the latter group at Harvard in GIS by more than a decade, and it contrasted with the group of land-use transportation modelers that we discussed with respect to the developments of urban models in chapter 10. A good summary of the Washington group's influence and work is in Brian J. L. Berry, "Geography's quantitative revolution: Initial conditions, 1954–1960. A personal memoir," *Urban Geography*, 14, No. 5, 434–441, 1993.

23. David Rhind, "Computer-aided cartography," *Transactions of the Institute of British Geographers, New Series*, 2, No. 1, 71–97, 1977.

24. John T. Coppock, and David W. Rhind, "The history of GIS," in David J. Maguire, Michael F. Goodchild, and David W. Rhind (Editors), *Geographical Information Systems*, Longman, London, 21–43, 1991.

25. Roger Tomlinson returned to London as a part-time graduate student in the early 1970s to pursue a PhD in GIS. Peter Wood, an urban geographer who worked on early computer mapping in the West Midlands, acted as his advisor. As Roger was frequently referred to as the "father of GIS," Peter has the rather dubious title of "grandfather," which is amusing in that he is younger. The thesis is available at https://discovery.ucl .ac.uk/id/eprint/1563584/. In fact, the department that Roger graduated from is the same one that spawned Brian Berry, the doyen of quantitative geography referred to earlier in this chapter in endnote 23. These guys also walked in the steps of Marx, Darwin, Orwell and, of course, John Snow (see endnote 30 below) without their ever knowing it.

26. Roger Tomlinson and Michael A. G. Toomey, "GIS and LIS in Canada," in Gerald McGrath and Louis Sebert (Editors), *Mapping a Northern Land: The Survey of Canada 1947–1994*, McGill-Queen's University Press, Montreal, 1999. See also Roger Tomlinson, "Origins of the Canada geographic information system," *ArcUser Fall 2012*, ESRI, Redlands, CA, https://www.esri.com/news/arcnews/fall12articles/origins-of-the-can ada-geographic-information-system.html.

27. Roger Tomlinson, *Thinking About GIS: Geographic Information Systems Planning for Managers*, ESRI Press, Redlands, CA, 2003.

28. Edward Tufte in his *The Visual Display of Quantitative Information*, Graphics Press USA, New Haven, CT, 1983, popularized Napoleon's March on Moscow map by Charles Joseph Minard, whose more complete graphics can be seen at Jared Green, "Charles-Joseph Minard: A legacy of beautiful data-based maps," in *The Dirt: Uniting the Built and Natural Environments*, October 3, 2018, https://dirt.asla.org/2018/10/03 /charles-joseph-minard-a-legacy-of-beautiful-data-based-maps/.

29. Snow's insights and contributions are presented by Steve Johnson in his book *Ghost Map*, endnote 5, chapter 5, while the link to GIS and spatial analysis is contained in many texts on GIS. See Longley, Goodchild, Maguire, and Rhind, *Geographic Information*. Also see the original work by John Snow, *On the Mode of Communication of Cholera*, Cosimo Classics, London, 1848 (2020).

30. Francis Wheen, *Karl Marx*, endnote 2, chapter 7.

31. The idea of map layers and their combination using various forms of operation that enable their synthesis, through addition, sieving common locations, and similar additive and exclusivity operators, has for a long time been at the core of identifying best locations for particular activities. It was adopted widely for physical planning problems, particularly in landscape. See Carl Steinitz, Paul Parker, and Lawrie Jordan, "Hand-drawn overlays: Their history and prospective uses," *Landscape Architecture*, *66*, 444–55, 1976. In the 1960s when GIS first began to develop, both

Christopher Alexander and Marvin Manheim at Harvard-MIT and Ian McHarg at Penn developed variants of this method. See Chris Alexander and Marvin Manheim, "The use of diagrams in highway route location: An experiment," R62–3, Civil Engineering Systems Laboratory, MIT, Cambridge, MA, 1962, and Ian McHarg, *Design with Nature*, Doubleday, Garden City, NY, 1967.

32. Carl Steinitz popularized the method of overlay analysis that became central to the way GIS developed as combinations of map layers. Dana Tomlin, also at the Graduate School of Design, developed the idea more formally as map algebra in his Map Analysis Package (MAP), which he further developed at Yale University before moving to Penn. The legacy in these developments is clearly more design than geography and spatial analysis, although in the last decade, the idea of geography by design—geodesign—has extended the overlay model into group planning and planning support systems. See Carl Steinitz, *A Framework for Geodesign: Changing Geography by Design*, ESRI Press, Redlands, CA, 2013.

33. Open-source GIS has roots going back to the 1970s, although the most-high profile system came from the US Army Corp of Engineers in the form of the GRASS system (Geographic Resource Analysis Support Software). As more and more functions have been added to GIS, GIS has broken up on the desktop, and now there are literally thousands of functions that can be assembled using various open-source languages to let users form their own methods of geoprocessing. The Wiki entry summarizes some of these: https://wiki.osgeo.org/wiki/Open_Source_GIS_History.

34. There is a considerable amount of very focused research on the problems of reading maps and interpreting them in various ways. It is not our purpose here to review this, but it is important to say that a very large proportion of the population find using a map difficult. This, however, has begun to change, we speculate, with the provision of online maps, GPS receivers in cars and phones, and the general notion of locating oneself using geocoordinates, which are now much more widely used and understood than at any time hitherto. For a short discussion, see Graham Southorn, "Why are some people better at map reading than others?," https://360 .here.com/2014/10/06/people-better-map-reading-others/.

35. Barbara P. Buttenfield and Robert B. McMaster (Editors), *Map Generalization*, John Wiley, New York, 1992.

36. Alfred Korzybski, "A non-Aristotelian system and its necessity for rigor in mathematics and physics," *American Association for the Advancement of Science*, New Orleans, LA, 1931 reprinted in *Science and Sanity*, International Non-Aristotelian Library, Oxford, 747–761, 1933.

37. Lewis Carroll, *The Complete Illustrated Works*, Gramercy Books, New York, 727, 1893, latest edition 1982.

38. Jorges L. Borges, "On exactitude in science," *Los Anales de Buenos Aires*, *1*, No. 3, 53, 1946.

39. Michael Batty, "Digital twins," *Environment and Planning B: Urban Analytics and City Science*, *45*, No. 5, 817–820, 2018.

40. The Xerox PARC Map Server was introduced in 1993 as soon as the web was open enough for such delivery of map graphics. It was a simple demonstration of how maps might be served across the web with conversion from raster to display as GIFS. See https://en.wikipedia.org/wiki/Xerox_PARC_Map_Viewer.

41. NASA WorldWind is an "open source virtual globe API (which) . . . allows developers to quickly and easily create interactive visualizations of 3D globe, map and geographical information," https://worldwind.arc.nasa.gov.

42. For web mapping, Wikipedia has a comprehensive list of map products and their dates of introduction. See https://en.wikipedia.org/wiki/Web_mapping. For a more detailed analysis, see Steve Feldman, "The history of web mapping," https://www.slideshare.net/stevenfeldman/history-of-web-mapping.

43. Michael F. Goodchild, "Citizens as sensors," endnote 46, chapter 9.

44. The history of OSM at https://wiki.openstreetmap.org/wiki/History_of_OpenStreetMap recorded five million users as of November 2018. Many developments focus on how the mapping system can be tailored to projects that require specialist expertise. See, for example, Haklay and Weber, *IEEE Pervasive Computing*, endnote 13, chapter 8.

45. Google Earth originated from Keyhole, a company that developed terrestrial streaming data for the globe. It was used by the US Military, and on the rising tide of interest in the US with respect to locational services, Google acquired the company in 2004 and then launched Google Earth the following year. Other Google products are now integrated with the system, and it is easy to populate the globe with one's own data in 2D and 3D form. See https://en.wikipedia.org/wiki/Google_Earth.

46. Anon, "What is remote sensing? The definitive guide," https://gisgeography.com/remote-sensing-earth-observation-guide/.

47. Mark Sullivan, "A brief history of GPS," *PC World*, August 9, 2012, https://www.pcworld.com/article/2000276/a-brief-history-of-gps.html. For a fuller account, see Scott Pace, Gerald Frost, Irving Lachow, David Frelinger, Donna Fossum, Donald K. Wassem, and Monica Pinto, *The Global Positioning System Assessing National Policies*, Critical Technologies Institute, The Rand Corporation, Santa Monica, CA, 1995, https://www.rand.org/content/dam/rand/pubs/monograph_reports/MR614/MR614.pdf.

48. Google Maps and Bing Maps are the main services from the major platform software companies, Google and Microsoft, but there is a useful summary at the following Wiki page https://en.wikipedia.org/wiki/Comparison_of_web_map_services.

49. Andrew Hudson-Smith, Michael Batty, Andrew Crooks, and Richard Milton, "Mapping for the masses: Accessing Web 2.0 through crowdsourcing," *Social Science Computer Review*, *27*, No. 4, 524–538, 2009.

50. MapTube is a free resource for viewing, sharing, mixing, and mashing maps online. Created by Richard Milton of UCL's Centre for Advanced Spatial Analysis, users can select any number of maps to overlay and view. See http://www.maptube .org/.

51. Colouring London is a knowledge exchange platform designed by Polly Hudson that is directed at collecting information on every building in the city to help make the city more sustainable. Like OpenStreetMap, it is essentially a crowdsourcing site that is focused on building data infrastructure from the bottom up. It is being developed in the same stable from which OSM originated, CASA, UCL, and is currently part funded by the Alan Turing Institute. See http://colouringlondon.org/.

CHAPTER 12

Andrew McAfee quoted at https://www.informationweek.com/strategic-cio/mit-s-mca fee-smart-machines-pick-up-the-pace.

1. The remarkable history of this company, so key to urban life in Britain in the 1950s, is told by Georgina Ferry in her book *Computer Called LEO*, endnote 9, chapter 4. The first Lyons teashop was established in Piccadilly in 1909, a little west from the end of our walk along the information mile where we began our history of information technologies in chapter 2.

2. John Graham-Cumming's keynote speech at the Strata Conference in London in 2012 was entitled "The great railway caper: Big data in 1955." You can watch his wonderful exposition of how the team of engineers at the Lyons Tea Company wrestled not only with big data in the early 1950s but also with the arcane methods of operating a massive mainframe computer, running various programs, and solving the shortest routes problem well before Dijkstra published his proof in 1959. I wonder whether anyone had invented it before the Lyons engineers. See https://www .youtube.com/watch?v=pcBJfkE5UwU.

3. Tim Greening-Jackson's paper "LEO I and the BR job" can be downloaded from John Graham-Cumming's blog at https://blog.jgc.org/2012/10/the-great-railway -caper-big-data-in-1955.html or directly from the Google drive https://drive.google .com/file/d/0BwUohGCPTAlANlFrQ3M3TnIyZjg/view/. But how long it will stay there, no one knows—for as in all the web links in these endnotes, the permanence of the record can never be assured.

4. Edsger Dijkstra's two-page paper "A note on two problems in connexion with graphs," *Numerische Mathematik*, *1*, 269–271, 1959, must be one of the shortest papers of all time to influence anything to do with computing the properties of connectivity in networks. Google Scholar reveals it to have more than 31,000 citations since the early 1990s.

5. For a recent polemic on the British problem of capitalizing on our innovations, see Ryan Khurana, "Why is the UK so hostile to innovation?," *CAPX*, April 3, 2017,

https://capx.co/why-is-the-uk-so-hostile-to-innovation/. For a deeper economic analysis, see Michael Kitson and Jonathan Michie, "The deindustrial revolution: The rise and fall of UK manufacturing, 1870–2010," Centre for Business Research, University of Cambridge Working Paper No. 459, Cambridge, UK, https://www.cbr.cam .ac.uk/fileadmin/user_upload/centre-for-business-research/downloads/working-papers /wp459.pdf.

6. There are many definitions of the 5Vs—see https://www.bbva.com/en/five-vs -big-data/—and the field is continually attempting to define more. The basic five are velocity, volume, value, variety, and veracity, with two more—variability and visualization—sometimes being added to this list.

7. Kitchin's *Data Revolution*, endnote 11, chapter 10, covers traditional and new data, big data and little data, and all variants in between.

8. The details can be found in Greening-Jackson "LEO I," with the computational background also in Frank Land, "Remembering LEO," London School of Economics and Political Science, January 2012, https://www.researchgate.net/publication/2920 55297.

9. Batty, *New Science of Cities*, endnote 32, chapter 6.

10. Batty, *Post-Urban World*, endnote 32, chapter 6.

11. Christopher A. Kennedy et al., "Energy and material flows of megacities," *Proceedings of the National Academy of Sciences, 112*, No. 19, 5985–5990, 2015.

12. Chris Anderson, "The end of theory: Will the data deluge make the scientific method obsolete?," *Wired Magazine*, July 16, 2007, http://www.wired.com/science /discoveries/magazine/16-07/pb_theory.

13. We used this quotation from George Box as the preface to chapter 10. Box said this several times, but the easiest source to acquire is from his book with Norman Draper, *Empirical Model Building and Response Surfaces*, Wiley-Blackwell, New York, 1987. John Sterman, who is one of Jay Forrester's disciples in systems dynamics, addressed the same issues in his paper "All models are wrong: Reflections on becoming a systems scientist," *Systems Dynamics Review, 18*, No. 4, 501–531, 2002 (but oddly there is no attribution to George Box!).

14. Chris Anderson, '"The end of theory."

15. John W. Tukey, "The future of data analysis," *The Annals of Mathematical Statistics, 33*, No. 1, 1–67, 1962.

16. Geoff Boeing, Michael Batty, Shan Jiang, and Lisa Schweitzer, "Urban analytics: History, trajectory, and critique," in Rachel Franklin and Serge Rey (Editors), *Handbook of Spatial Analysis in the Social Sciences*, 503- 516, Edward Elgar, Cheltenham, UK, 2022.

17. Quoted in Alex Singleton, Seth Spielman, and David C. Folch, *Urban Analytics*, Sage, London, 2018.

18. A beginner's guide to the bizarre world of machine learning and neural nets with respect to extracting robust patterns is contained in Tariq Rashid, *Make Your Own Neural Network*. Amazon Media, London, n.d.

19. AI began as soon as computers were invented, spurred on by the optimistic conviction that we would soon be able to design and program machines that could replace human thinking. We discussed some of this background in chapter 1. The field promised much but yielded little, and there was a strong reaction against the kind of strong AI that was being pushed until the 1980s. Once computer memory became almost limitless, the notion of searching for patterns in data became possible, and the field began to abandon the idea that the methods used to extract data should mimic human thought. A weak AI emerged, which now dominates the field based on searching for deep patterns that may have little to do with the way we think but which nevertheless can be very useful—wrong but useful, as George Box so presciently said, op. cit. and the quote to chapter 10. In fact, the current hype associated with the field is based on the notion that the patterns embedded in big data can be used to condition routine behavior but do not usually need any strong theoretical rationale, as Chris Anderson implies in his "End of theory" *Wired* article.

20. Check out any of the online translation systems if you have not experienced such pattern recognition before, for example https://translate.google.com/.

21. Lee, "Large Scale Models," endnote 56, chapter 10.

22. Two recent reviews of these developments are. (1) for CA models, Yan Liu, Michael Batty, Siqin Wang, and Jonathan Corcoran, "Modelling urban change with cellular automata: Contemporary issues and future research directions," *Progress in Human Geography*, *45*, 3–24, 2021; and (2) for ABM models, Alison Heppenstall, Andrew Crooks, Nick Malleson, Ed Manley, Jiaqi Ge, and Michael Batty, "Agent-based models for geographical systems: A review," *Geographical Analysis*, *53*, 76–91, 2021.

23. Box, "All models are wrong," endnote 1, chapter 10.

24. His SYMAP time-lapse movie depicting the growth of Lansing, MI, from 1850 to 1965 was pieced together from line-printer maps and then videoed. See https://www.youtube.com/watch?v=aySwJKK6i2s. Jack Dangermond acknowledged Alan Schmidt for his movie and its insights at the ESRI Users Conference in 2004. See https://www.youtube.com/watch?v=CDeRKYo3lHU. See also endnote 37, chapter 4.

25. Chapin and Weiss, "Probabilistic model for residential growth," endnote 40, chapter 10.

26. Tobler, *Economic Geography*, endnote 16, chapter 10, with the movie Animated Cartography of Detroit available at https://www.youtube.com/watch?v=kRsF9S8JqBI.

27. Conway proposed his "Game of Life" as a physical board game that he enticed his colleagues at his Cambridge college (Gonville & Caius) to play in 1970. For the prize and challenge, see Martin Gardner, "Mathematical games: The fantastic combinations of John Conway's new solitaire game 'Life,'" *Scientific American*, *223*, 120–123, 1970.

28. Martin Gardner has a lucid explanation of the game in his book *Wheels, Life and Other Mathematical Amusements*, W. H. Freeman, San Francisco, CA, 1985.

29. Stanislav Ulam, *Adventures of a Mathematician*, Charles Scribner's Sons, New York, 1976.

30. Stephen Wolfram has done much to promote the idea of cellular automata through his various books, such as *Cellular Automata and Complexity: Collected Papers*, CRC Press, Taylor and Francis, Boca Raton, FL, 1994.

31. The application of CA to cities is presented in my book *Cities and Complexity*, endnote 2, chapter 9, while a review of the field is contained in Liu et al., "Modelling urban change with cellular automata."

32. George Chaudhuri and Keith C. Clarke, "The SLEUTH land use change model: A review," *The International Journal of Environmental Resources Research*, 1, 88–104, 2013. See also the SLEUTH Master Class by Keith Clarke available at https://slideplayer .com/slide/6990931/.

33. Roger White, Guy Engelen, and Inge Uljee, *Modeling Cities and Regions as Complex Systems: From Theory to Planning Applications*, MIT Press, Cambridge, MA, 2015.

34. Applications to cities are contained in my own book, Batty, *Cities and Complexity*, endnote 2, chapter 9. Applications of CA to cities, building on examples in this book, are available in the Urban Suite in the Netlogo Library available at http://ccl .northwestern.edu/netlogo/resources.shtml.

35. We have not detailed the development of complexity in this book, but it represents an important backdrop on which many computer models of cities—indeed, the very idea of the computable city—rests. A comprehensive and detailed tutorial by Melanie Mitchell at the Santa Fe Institute provides a good introduction at https:// www.complexityexplorer.org.

36. Itzhak Benenson and Paul M. Torrens, *Geosimulation: Automata-Based Modeling of Urban Phenomena*, John Wiley, Chichester, UK, 2004.

37. Andrew Crooks, Nicolas Malleson, Ed Manley, and Alison Heppenstall, *Agent-Based Modelling and Geographical Information Systems: A Practical Primer*, Sage, London, 2019.

38. Mohammad-Reza Namazi-Rad, Lin Padgham, Pascal Perez, Kai Nagel, and Ana Bazzan (Editors), "Agent based modelling of urban systems," First International Workshop, ABMUS 2016, Notes in Computer Science, Springer, Berlin, Germany, 2017.

39. Paul Waddell, "UrbanSim: Modeling urban development for land use, transportation, and environmental planning," *Journal of the American Planning Association*, 68, No. 3, 297–314, 2002, and Paul Waddell, "Integrated land use and transportation planning and modeling: Addressing challenges in research and practice," *Transport Reviews*, 31, No. 2, 209–229, 2011.

40. Chris Barrett et al., *TRANSIMS (TRansportation ANalysis SImulation System)*, 6 Volumes, Los Alamos National Laboratory, Los Alamos, NM, 1999.

41. Andrew Horni, Kai Nagel, and Kay Axhausen (Editors), *The Multi-Agent Transport Simulation MATSim*, Ubiquity Press, London.

42. Joshua M. Epstein and Robert L. Axtell, *Growing Artificial Societies: Social Science from the Bottom Up*, Bradford Books, Cambridge, MA, 1996.

43. The notion of deep structure, which is now being used as the essence of the search for pattern in big data, has a very long lineage, particularly in language, music, philosophy, and sociology. The notion that there are meaningful structures buried in big data is, to an extent, still an act of faith, awaiting theory to catch up. See Jean Piaget, *Structuralism*, Routledge, London, 1971.

44. Anderson, "End of theory."

45. Jo Best, "IBM Watson: The inside story of how the Jeopardy-winning supercomputer was born, and what it wants to do next," *TechRepublic*, September 9, 2013, https://www.techrepublic.com/article/ibm-watson-the-inside-story-of-how-the-jeopardy-winning-supercomputer-was-born-and-what-it-wants-to-do-next/.

46. The line between weak and strong AI is inevitably blurred, as we intimated in chapter 1. But this contrast is not the main one, for that relates to patterns that can be explained in hindsight from good theory and those that cannot be so explained. The former tends to be more acceptable than the latter, and it is an open question as to what the most profound challenges are in this debate. The development of machine learning in the company DeepMind raises such challenges. See https://deepmind.com/research.

CHAPTER 13

Quoted by Mark Zuckerberg when Facebook acquired Oculus Rift in 2014. See https://www.theverge.com/2014/3/25/5547456/facebook-buying-oculus-for-2-billion.

1. Patrick Geddes wrote what arguably is the first theoretical exposition of town planning in the Modern Age. His *Cities in Evolution*, endnote 2, chapter 9, mirrors his eclectic approach to cities, which was founded on ideas about social evolution drawn from Darwin's theory of natural selection and fashioned in ways in which the public at large might contribute to the planning and reconstruction of their own communities. His book is hardly a reasoned and considered philosophy. It is, to an extent, a ragbag of ideas but all demonstrating ideas about past, present, and future cities through his various exhibitions, particularly in his long-term project in Edinburgh, the Outlook Tower. One cannot be but struck by Geddes' anticipations of what we now call "crowdsourcing."

2. The concept of the civic survey—which became enshrined in the cliché "survey then plan" or "survey-analysis-plan"—was promoted relentlessly by Geddes

throughout his life. See his paper "A suggested plan for a civic museum and its associated studies," *Sociological Papers*, *3*, 197–240, 1906. For a contemporary rendition, see Mark Tewdwr-Jones, Dhruv Sookhoo, and Robert Freestone, "From Geddes' city museum to Farrell's urban room: Past, present, and future at the Newcastle City Futures exhibition," *Planning Perspectives*, 35, 277–297, 2019.

3. The implications of the Outlook Tower and his many exhibitions for the future city are detailed in Joshua F. Cerra, Brook Weld Muller, and Robert F. Young, "A transformative outlook on the twenty-first century city: Patrick Geddes' Outlook Tower revisited," *Landscape and Urban Planning*, *166*, 90–96, 2017, while the basic ideas of communicating such information are presented by Marco Amatia, Robert Freestone, and Sarah Robertson in "'Learning the city': Patrick Geddes, exhibitions, and communicating planning ideas," *Landscape and Urban Planning*, *166*, 97–105, 2017.

4. The Chinese have a fascination with scaled-down artifacts of the real thing. The fact that there were so many city models built of traditional physical materials prior to China's opening up, which began in 1979, provides a massive archive of past plans and models that have now been incorporated into the many city exhibition centers that exist in countless Chinese cities.

5. The biggest and most elaborate exhibition center to date appears to be that in Guangzhou, where a new building has been designed specifically for the many city models and plans that define the planning of the city and the region from early dynastic times. See Anon, "A brand new city landmark now opens to public," *Life of Guangzhou*, April 28, 2018, https://www.lifeofguangzhou.com/wap/knowGZ/content.do?contextId=7458&frontParentCatalogId=175.

6. Michael Batty, "At the crossroads of urban growth," *Environment and Planning B: Planning and Design*, *41*, 951–953, 2014.

7. Anon, "The Citizen's House of Wuhan opens its doors," *Arte Charpentier*, February 28, 2013, http://www.arte-charpentier.com/en/the-citizens-house-of-wuhan-opens-its-doors/.

8. Wuhan Urban Planning Exhibition Hall (WUPEH), *16th International Conference on Computers in Urban Planning and Urban Management*, July 8–12, 2019, School of Urban Design Wuhan University, China, http://cupum2019.aconf.org/news/2606.html.

9. There are several good summaries of the Abercrombie and Paton-Watson Plan for Plymouth. See the blog http://www.cyber-heritage.co.uk/history/city.htm and, in *National Geographic* magazine, Harvey Klemmers's study "A city that refused to die, a study of blitzed and fire bombed Plymouth in 1941" and "The making of a modern city, Plymouth City Museum and Art Gallery," April 27–June 29, 2013, https://plymhearts.org/. See also "A plan for Plymouth: Our first great welfare-state city," https://municipaldreams.wordpress.com/2013/01/15/a-plan-for-plymouth-our-first-great-welfare-state-city/.

10. Michael F. Goodchild, "Reimagining the history of GIS," *Annals of GIS*, *24*, No. 1, 1–8, 2018.

11. Bill Fetter at Boeing worked on vector graphics throughout the 1950s, and it is likely that the first 3D city models were tried out at this time. Fetter also popularized the term "computer graphics." See Dave Kasik and Chris Senesac, "Visualization: Past, present, and future at Boeing," http://on-demand.gputechconf.com/gtc/2013/presentations/S3440-Boeing-Visualization-Past-Present-Future.pdf. Paul Heckbert has a brief but thorough description in his "History of Computer Graphics (CG)," which goes up until 1999. See https://www.cs.cmu.edu/~ph/nyit/masson/history.htm.

12. Ivan Edward Sutherland, "Sketchpad: A man-machine graphical communication system," PhD Thesis, MIT, 1963, available as Technical Report, Number 574, UCAM-CL-TR-574, Computer Laboratory, University of Cambridge, UK, https://www.cl.cam.ac.uk/techreports/UCAM-CL-TR-574.pdf.

13. Negroponte, "Architecture machines," endnote 19, chapter 1.

14. A very early demonstration of the power of such graphics is from researchers at the Leeds School of Town Planning reproduced in my paper "Computers and design," Papers in Planning Research, 2, UWIST Department of Town Planning, Cardiff, 1980. See also G. R. Beacon and P. G. Boreham, "Computer-aided architectural design at Leeds Polytechnic," Computer-Aided Design, 10, 325–331, 1978.

15. Nicholas Negroponte, "The return of the Sunday Painter," in Michael L Dertouzos and Joel Moses (Editors), The Computer Age: A Twenty-Year View, MIT Press, Cambridge, MA, 21–37, 1979.

16. SOM, "A look back at the early days of 3D visualization," nine cities by SOM produced in 1984 and scanned from the original 16-mm film, https://www.som.com/news/this_som_archive_video_offers_a_look_back_at_the_early_days_of_3d_visualization, and on Vimeo at https://vimeo.com/93315120.

17. This is the main method used in early microcomputing on Apple, IBM, and other personal computers. I demonstrate it in my book Batty, Microcomputer Graphics, endnote 5, chapter 11.

18. The level of detail (LOD) can be controlled so that the finest detail from the viewpoint is progressively aggregated to maximize the speed of rendering or the detail is excluded entirely from the scene. See David Luebke, Martin Reddy, Jonathan D. Cohen, Amitabh Varshney, Benjamin Watson, and Robert Huebner, Level of Detail for 3D Graphics, Morgan Kaufmann, San Francisco, CA.

19. The early computer games movement was dominated by unconventional computing and various hacks. See Levy, Hackers, endnote 2, chapter 4, and Anon, "Video game history," History, September 1, 2017, https://www.history.com/topics/inventions/history-of-video-games.

20. Olly Richards, "Best idea wins: How Pixar grew up," The Telegraph, November 21, 2015, https://www.telegraph.co.uk/film/what-to-watch/pixar-history-good-dinosaur-toy-story/. One of the first demonstrations of the power of graphics in film was the use of fractal geometry by Loren Carpenter of Pixar to illustrate how a dead

planet comes alive in the movie *Star Trek II: The Wrath of Kahn*. See http://www
.historyofinformation.com/detail.php?id=3141.

21. Michael Batty, David Chapman, Steve Evans, Muki Haklay, Stefan Kueppers, Naru Shiode, Andy Smith, and Paul Torrens, "Visualizing the city: Communicating urban design to planners and decision-makers," in Richard Brail and Dick Klosterman (Editors), *Planning Support Systems: Integrating Geographic Information Systems, Models, and Visualization Tools*, ESRI Press, Redlands, CA, 405–443, 2001; also see the Wiki page https://en.wikipedia.org/wiki/3D_city_models.

22. Michael Batty and Andy Hudson-Smith, "Urban simulacra: From real to virtual cities, back and beyond," *Architectural Design*, *75*, No. 6, 42–47, 2005.

23. Oliver Dawkins, Adam Dennett, and Andrew Hudson-Smith, "Living with a digital twin: Operational management and engagement using IoT and mixed realities at UCL's Here East Campus on the Queen Elizabeth Olympic Park," Centre for Advanced Spatial Analysis (CASA), University College London, https://www.researchgate.net /publication/324702983_Living_with_a_Digital_Twin_Operational_management_ and_engagement_using_IoT_and_Mixed_Realities_at_UCL's_Here_East_Campus_on_ the_Queen_Elizabeth_Olympic_Park/link/5adddd11a6fdcc29358ba112/download. The Virtual London project (now called ViLo) is a great example of using a 3D model as a framework for the metaverse that contains much of this media in a form that imparts a richness of detail that is essential to good urban design.

24. There are many reviews of this field where the reader can get a quick feel for the media. See "Immersion, virtual reality, mixed reality, augmented reality, what are the differences?," http://www.immersion.fr/en/virtual-reality-mixed-reality-augmented -reality-what-are-the-differences/, and Dragana Nikolic, "Headsets, caves, AR and VR— What's the reality for construction?," *BIM+*, February 8, 2017, http://www.bimplus .co.uk/technology/hea8dsets-cav6es-ar-and-vr-w7hats-reality/#.

25. Michael Batty, Hui Lin, and Min Chen, "Virtual realities, analogies and technologies in geography," in Barney Warf (Editor), *Handbook on Geographies of Technology*, Research Handbooks in Geography Series, Edward Elgar, Cheltenham, UK, 96–112, 2017.

26. The Arizona State University Decision Theater: A real-time, multi-display environment forms the core of the theater, which also has a clone at the hub of US policy making in Washington, DC, https://dt.asu.edu/facilities.

27. Sean P. Egen, "The history of avatars," *iMedia*, June 22, 2005, https://www.ime diaconnection.com/articles/ported-articles/red-dot-articles/2005/jun/the-history -of-avatars/, and Brian Coleman, *Hello Avatar. Rise of the Networked Generation*, MIT Press, Cambridge, MA, 2011.

28. Andrew Hudson-Smith, Richard Milton, Joel Dearden, and Michael Batty, "Virtual cities: Digital mirrors into a recursive world," Working Paper 125, Centre for Advanced Spatial Analysis, University College London, December 2007.

29. Howard Rheingold, *Virtual Reality: The Revolutionary Technology of Computer-Generated Artificial Worlds—And How It Promises to Transform Society*, Simon & Schuster, New York, 1992.

30. Jaron Lanier, *Dawn of the New Everything: A Journey Through Virtual Reality*, Picador, London, 2017.

31. Dawkins et al., "Digital twin."

32. Virtual Reality Society, "History of virtual reality," https://www.vrs.org.uk/virtual-reality/history.html.

33. Unity is an all-purpose VR and 3D graphics web-based platform. See https://unity.com.

34. Virtual geographic environments are defined as "embracing three scientific requirements of Geographic Information Science (GIScience)—multi-dimensional visualization, dynamic phenomenon simulation, and public participation." See Hui Lin, Min Chen, Guonian Lu, Qing Zhu, Jiahua Gong, Xiong You, Yongning Wen, Bingli Xu, and Mingyuan Hu, "Virtual geographic environments (VGEs): A new generation of geographic analysis tools," *Earth-Science Reviews*, *126*, 74–84, 2013.

35. Autodesk 3D Studio Max: 3D modeling and rendering software for design visualization, games, and animation. See https://www.autodesk.co.uk/products/3ds-max/overview/.

36. Leah A. Wasse, "The basics of LiDAR." LiDAR stands for Light Detection and Ranging. It is a technique for determining height or distance based on the line of sight which uses a pulsed laser. See https://www.neonscience.org/lidar-basics.

37. Frank Taylor, "First review of new Google Earth," *Google Earth Blog*, April 18, 2017, https://www.gearthblog.com/blog/archives/2017/04/first-review-new-google-earth.html.

38. Nicholas Malleson, Alison Heppenstall, and Andrew Crooks, "Place-based simulation modelling: Agent-based modelling and virtual environments," *Oxford Research Encyclopaedia of Criminology and Criminal Justice*, Oxford University Press, Oxford, 2018.

39. Michael Batty, "The new urban geography of the third dimension," *Environment and Planning B: Planning and Design*, *27*, 483–484, 2000; Steven Evans, Ron Liddiard, and Philip Steadman, "3DStock: A new kind of three-dimensional model of the building stock of England and Wales, for use in energy analysis," *Environment and Planning B: Planning and Design*, *44*, No. 2, 227–255, 2017; and Steven Evans, Ron Liddiard, and Philip Steadman, "Modelling a whole building stock: Domestic, non-domestic and mixed use," *Building Research and Information*, *47*, No. 2, 156–172, 2019.

40. Benjamin Watson, Pascal Müller, Oleg Veryovka, Andy Fuller, Peter Wonka, and Chris Sexton, "Procedural urban modeling in practice," *IEEE Computer Graphics and Applications*, *28*, No. 3, 18–26, 2008.

41. Abercrombie and Paton-Watson's Plan for Plymouth.

42. Brian A. White, *Second Life: A Guide to Your Virtual World*, Que, Hoboken, NJ, 2007.

43. "Unity is an all-purpose VR defined as "Unity is a real-time 3D development platform for building 2D and 3D application, like games and simulations, using .NET and the C# programming language. Unity can target 25+ platforms across mobile, desktop, console, TV, VR, AR, and the web." https://dotnet.microsoft.com/en-us/apps/games /unity#:~:text=Unity%20is%20a%20real%2Dtime,%2C%20AR%2C%20and%20 the%20web.

44. Jonathan Taylor, "The emerging geographies of virtual worlds," *Geographical Review*, *87*, No. 2, 172–192, 1997.

45. Hui Lin and Michael Batty (Editors), *Virtual Geographic Environments*, Science Press, Beijing, China, 2009.

46. Batty, Lin, and Chen, "Virtual realities."

47. Sand tables and touch tables enable the user to interact with the phenomena on a map-like surface by incorporating physical objects with digital interaction. Simtable also incorporates agent-based modeling into its interface to the sand table, while the same has been done with touch tables at CASA, UCL. For touch tables, see Batty, Lin, and Chen, "Virtual realities," and for sand tables, see http://www.simtable.com and https://www.youtube.com/watch?v=W-Pn-aV5DtE.

48. Donna R. Berryman, "Augmented reality: A review," *Medical Reference Services Quarterly*, *31*, No. 2, 212–218, 2012.

CHAPTER 14

Berners-Lee, *Weaving the Web*, endnote 23, chapter 5. See https://quotepark.com/ quotes/1921417-tim-berners-lee-in-an-extreme-view-the-world-can-be-seen-as-only/.

1. Castells, *Network Society*, endnote 5, chapter 7.

2. The term "platform" in information technology is a basic framework on which a number of related and integrated hardware–software applications run. The platform originally pertained to the PC itself, but as computation has diffused to the far corners of the web, platforms are now defined as more generic systems of hardware and software that tend to be quite extensive but integrated, often global in their import and reach. See https://www.techopedia.com/definition/3411/platform, and for a traditional definition, see https://home.kpmg/xx/en/home/insights/2018/02 /rise-of-digital-platforms-fs.html for more business-orientated applications.

3. The IoT was originally defined by Kevin Ashton in 1999 who directs the Auto-ID Center at the MIT. It is essentially an ecology of computable devices that are networked together and, in its broadest conception, is the set of all devices that are able in principle, if not practice, to communicate with one another. The IoT was anticipated by many commentators and, to an extent, is similar to ubiquitous computing.

See Weiser's "The computer for the 21st century," endnote 22, chapter 1. See also the writings of Bruce Sterling in his book (with Lorraine Wild and Peter Lunenfeld) *Shaping Things*, MIT Press, Cambridge, MA, 2005, and https://en.wikipedia.org/wiki /Internet_of_things.

4. By the early 1990s, there were several email systems that populated the emerging internet. The Ethernet was being put in place globally, and systems such as BITNET (Because It's Time NETwork) were being widely used by university academics and administrators, mainly at first for email. No longer did professional computer users need to couple their terminals to the net using modems, although this practice continued into the late 1990s. Once the internet became widespread by the millennium, the kind of email system being used no longer mattered, and with the widespread use of Wi-Fi, which had come of age by the end of the first decade of the twenty-first century, the jigsaw that we assembled in chapter 5 was complete. See https://en .wikipedia.org/wiki/History_of_email and https://en.wikipedia.org/wiki/BITNET.

5. A clear exposition of network effects, which essentially are generated when a device is developed that can, in principle, link to any other of the same devices such as telephone, a PC, a smartphone, and so on, is contained in Carl Shapiro and Hal Varian's *Information Rules: A Strategic Guide to the Network Economy*, Harvard Business School Press, Cambridge, MA, 1998.

6. Alina Selyukh provides a good history of these early companies in her post "The big internet brands of the '90s—Where are they now?," https://www.npr.org/sec tions/alltechconsidered/2016/07/25/487097344/the-big-internet-brands-of-the-90s -where-are-they-now?t=1572597015091.

7. After a long and somewhat tortuous history, Yahoo was purchased by Verizon in 2017 and finally entered the graveyard of the earliest internet companies. Jeremy Ring provides a good history in *We Were Yahoo!: From Internet Pioneer to the Trillion Dollar Loss of Google and Facebook*, Post Hill Press, New York, 2018. The Yahoo name lives on, but it remains a shadow of its former self. Its Wikipedia page provides a recent history. See https://en.wikipedia.org/wiki/Yahoo!.

8. Berners-Lee, *Weaving the Web*, endnote 23, chapter 5.

9. See https://companiesmarketcap.com/tech/largest-tech-companies-by-market-cap/ and Tony DeGennaro, "10 most popular social media sites in China," *Dragon Social*, January 31, 2019, https://www.dragonsocial.net/blog/social-media-in-china/.

10. Berners-Lee, *Weaving the Web*, endnote 23, chapter 5.

11. The relevant Wikipedia page is an excellent summary of the history of the web browser, https://en.wikipedia.org/wiki/History_of_the_web_browser.

12. Eric Schmidt and Jonathan Rosenberg, *How Google Works*, Grand Central Publishing, New York, 2014.

13. The mathematics of PageRank are explained at https://en.wikipedia.org/wiki/Page Rank. In essence, the idea goes back many years to working out how close different

nodes are in a network graph and using this to rank the relative positions of the search terms that are associated with the nodes. At one level, such page ranking is simply a matter of solving a set of simultaneous equations, but the size of this set is enormous, almost too large to comprehend, and it is growing exponentially by the day.

14. This is a continually moving target, but numbers are given at https://skai.io/mon day-morning-metrics-daily-searches-on-google-and-other-google-facts/.

15. Redding, *Google It*, endnote 31, chapter 5.

16. Pay per click is nicely explained by Larry Kim in "What is PPC? Learn the basics of pay-per-click (PPC) marketing" at https://www.wordstream.com/ppc.

17. Dodge and Kitchin, *Atlas of Cyberspace*, endnote 14, chapter 8.

18. Joseph Menn, *All the Rave: The Rise and Fall of Shawn Fanning's Napster*, Crown Business, New York, 2003.

19. The history of file sharing from FTP (file transfer protocols) onward was very fragmented, with many systems such as BitTorrent skirting legal restrictions of various kinds. Many of these systems introduced in the early days of file sharing on the internet have declined in the face of professional and legal challenges. See https://en.wikipedia.org/wiki/BitTorrent_(company).

20. Kevin Kelly, *New Rules for the New Economy*, Viking Penguin, New York, 1998.

21. Natalie Berg and Miya Knights, *Amazon: How the World's Most Relentless Retailer Will Continue to Revolutionize Commerce*, Kogan Page, London, 2019.

22. The obvious victim of selling on the internet was the neighborhood and high-street bookshop. See https://www.telegraph.co.uk/culture/books/booknews/10654506/Dec line-of-the-independent-bookshop-as-UK-figures-fall-below-1000-for-first-time.html. After a quite catastrophic decline in such shops in many places around the world, the number has stabilized and is now rising slightly as bookshops reinvent themselves to meet a different level and type of demand. A wonderful example of this reinvention is Foyles, one of the oldest bookshops in Central London, which is located on Char- ing Cross Road at the center of what used to be the book quarter. The high street, however, is undergoing more radical change. See https://www.independent.co.uk/voi ces/high-street-arcadia-group-philip-green-amazon-delivery-a8961771.html.

23. Levy, *Hackers*, endnote 2, chapter 4.

24. Stewart Brand is one of the most celebrated members of Silicon Valley's counter- culture, being responsible for the *Whole Earth Catalog*, first published around 1968. His book went through many versions before it became a regular publication as the *Last Whole Earth Catalogue*, Random House, New York, 1971. The WELL (https://en .wikipedia.org/wiki/The_WELL) was first set up to complement the catalog. His books on the Media Lab, endnote 19, chapter 1, and *The Clock of the Long Now: Time and Responsibility*, Doubleday, New York, 1999, are essential reading as background to the ideas developed more generally in this book. His biography *Whole Earth: The Many*

Lives of Stewart Brand, Penguin Press, New York, 2022, written by John Markoff, provides a fascinating glimpse of the culture of Silicon Valley from the 1960s until today.

25. Howard Rheingold's book *The Virtual Community: Homesteading on the Electronic Frontier*, MIT Press, Cambridge, MA, 2000, is a good summary of these early community networks.

26. Duncan Watts, *Six Degrees: The Science of a Connected Age*, Random House, New York, 2003.

27. See, for example, Robert Dahl, *Who Governs?: Democracy and Power in the American City*, Yale University Press, New Haven, CT, 1961.

28. Albert-Laszlo Barabasi, *Linked: How Everything Is Connected to Everything Else and What It Means*, Perseus Books, New York, 2002.

29. Web 2.0 was first defined by Darcy DiNucci in 1999 and popularized by Tim O'Reilly. See https://en.wikipedia.org/wiki/Web_2.0. In essence, Web 1.0 is the passive web, referred to sometimes by Tim Berners-Lee as the "read-only web," while Web 2.0 is the interactive web, where users not only can read content but also can manipulate it. The line between the two is somewhat blurred, but the "semantic web," again defined by Berners-Lee, has not yet emerged, for this would be the truly intelligent web, which in itself is a nebulous kind of concept. Maybe the metaverse is Web 3.0, but I doubt it.

30. Information and data about Facebook (Meta) can be found at https://datarepor tal.com/essential-facebook-stats. There are some good books summarizing the heady rise and the conundrums and paradoxes that Facebook has revealed, particularly with respect to privacy. Two that are worth looking at are David Kirkpatrick's *The Facebook Effect: The Real Inside Story of Mark Zuckerberg and the World's Fastest Growing Company*, Simon & Schuster, New York, 2010, and, more recently, Stephen Levy, *Facebook: The Inside Story*, Penguin Press, New York, 2020.

31. Facebook has courted controversy from episodes such as Zuckerberg's refusal to respond appropriately to a call from a Select Committee of the British Parliament. This was to discuss the scandal over Cambridge Analytica, which used extensive personal data from Facebook, without the authority of those whose data was used, but which was reportedly condoned by Facebook. See https://www.theguardian.com/technol ogy/2019/mar/17/the-cambridge-analytica-scandal-changed-the-world-but-it-didnt -change-facebook.

32. Wikipedia explains the genesis and growth of TikTok, https://en.wikipedia.org /wiki/TikTok.

33. Nick Bilton, *Hatching Twitter: A True Story of Money, Power, Friendship, and Betrayal*, Portfolio Penguin Books, New York, 2013. Statistics are at https://www.internetlivestats .com/twitter-statistics/.

34. DeGennaro, *Dragon Social*.

CHAPTER 15

Marc Andreessen, "Why software is eating the world," *The Wall Street Journal*, August 11, 2011, http://www.wsj.com/articles/SB10001424053111903480904576512 250915629460.

1. A long line of futurists anticipated the information society, from E. M. Forster's portrait of a completely but remotely connected world in his short story *The Machine Stops*, 1909, https://librivox.org/the-machine-stops-by-e-m-forster/ and endnote 42, chapter 7, to the predictions made by Arthur C. Clarke in his *Profiles of the Future: An Inquiry into the Limits of the Possible*, Weidenfeld and Nicolson, London, 1962; Martin's *The Wired Society*, endnote 41, chapter 5; and Toffler's *The Third Wave*, endnote 5, chapter 1.

2. Andreessen, "Eating the world."

3. Counting the number of computers existing at any point in time is, of course, problematic, largely because the very definition of what is a computer is increasingly ambiguous. PCs, desktops, and laptops are easiest to count. Casual estimates suggest that in mid-2019, there were some two billion, including servers, desktops, and laptops, but the installed base of personal computers (PCs) in use at any one time worldwide from 2013 to 2019 is estimated by Statista to be about 1.5 billion. This excludes smartphones and iPads. See https://www.statista.com/statistics/610271/world wide-personal-computers-installed-base/.

4. Apps, short for "applications," in one sense, are entirely generic, although in common parlance, apps are small programs that are often mobile or portable from one device to another. Techopedia defines an app as "computer software, or a program, most commonly a small, specific one used for mobile devices. The term app originally referred to any mobile or desktop application" but "the term has evolved to refer to small programs that can be downloaded and installed all at once." See https://www.techopedia.com/definition/28104/app.

5. Even though layers of hardware, software, orgware, and dataware now dominate the evolution of technologies, piling one on top of another in ever-increasing profusion, the closest we have got to visualizing this picture is in the diffusion of technologies through time where waves of new technologies are increasingly close to the ones already invented. See my *Inventing Future Cities*, endnote 8, chapter 1, p. 196.

6. Chirag Rabari and Michael Storper, "The digital skin of cities: Urban theory and research in the age of the sensored and metered city, ubiquitous computing and big data," *Cambridge Journal of Regions, Economy and Society*, 8, No. 1, 27–42, 2014. See also Chirag Rabari, PhD thesis, UCLA, Los Angeles, CA, 2013.

7. Andreessen, "Eating the world."

8. Andrew McAfee and Erik Brynjolfsson, *Machine, Platform, Crowd: Harnessing Our Digital Future*, W. W. Norton, New York, 2017.

9. Transactions processing "is information processing in computer science that is divided into individual, indivisible operations called transactions." Traditionally, it was confined to accounting and business operations, but it is now widespread with respect to many computable functions involving big data. See https://en.wikipedia.org/wiki/Transaction_processing.

10. Thomas L. Friedman, *Thank You for Being Late: An Optimist's Guide to Thriving in the Age of Accelerations*, Penguin Press, New York, 2016. An earlier foray into this whole question of a world speeding up is in James Gleick's book *Faster: The Acceleration of Just about Everything*, Random House, New York, 2000.

11. Amelia Heathman, "iPhone through the ages: How Apple smartphones have evolved from 1st generation to the new iPhone 11," *Evening Standard*, September 10, 2019, https://www.standard.co.uk/tech/apple-iphone-2019-through-the-ages-iphone-to-iphone-xs-a4232841.html; Devanshi Adhvaryu, "Before the age of the iPhone: The story of how our cell phones came to be," September 24, 2017, *The Varsity*, https://thevarsity.ca/2017/09/24/before-the-age-of-the-iphone/. For a more detailed history, see Brian Merchant, *The One Device: The Secret History of the iPhone*, Bantam Press, New York, 2017.

12. Rob Kitchin and Martin Dodge, *Code/Space: Software and Everyday Life*, MIT Press, Cambridge, MA, 2014.

13. Geoffrey D. Austrian, *Herman Hollerith: Forgotten Giant of Information Processing*, Columbia University Press, New York, 1982. The volatility of the company IBM, which acquired Hollerith's technology, is presented in Cortada, *IBM*, endnote 28, chapter 4.

14. Alan R. Hevner and Donald J. Berndt, "Eras of business computing," *Advances in Computers*, *52*, 1–90, 2000.

15. The series of books by Kenneth L. Kraemer such as his *Computers and Local Government*, Praeger, New York, 1977, with John Leslie King (Editor), *Computers and Local Government: Volume 2*, Praeger, New York. 1978, and with James L. Perry, *Technological Innovation in American Local Governments: The Case of Computing*, Pergamon Press, New York, 1980, provide a comprehensive account of how US local government responded to the development of computer information systems from the 1960s onward.

16. These are "killer applications." One of the most high profile is defined in Wikipedia as follows: "One of the first recognized examples of a killer application is generally agreed to be the VisiCalc spreadsheet for the Apple II series computer," https://en.wikipedia.org/wiki/Killer_application. A more detailed list of such apps can be found Michael Lewis's "Killer apps explained—history, examples, impacts & future applications," *Money Crashers*, https://www.moneycrashers.com/killer-apps-explained-history-examples-impacts-future/.

17. Client-server architecture is defined in https://www.sciencedirect.com/topics/computer-science/client-server-architecture from the book by Jan L. Harrington,

Relational Database Design, Clearly Explained, 3rd Edition, Morgan Kaufmann, New York, 2009, and see also https://en.wikipedia.org/wiki/Client–server_model.

18. We defined a platform, and we need to elaborate on it further as a group of technologies that are used as a base upon which other applications, processes, or technologies are developed. See, for example, Parker, Alstyne, and Choudary, *Platform Revolution*, endnote 39, chapter 7. A platform can also be defined as essentially the computer system and its organization that enables different software applications to be run. The concept is both generic and specific, and there is considerable ambiguity about the term, as is the case with much of the definitions and semantics in the computer industry. See https://www.techopedia.com/definition/3411/platform. This accords more to our earlier comments on the meaning of the term platform.

19. Shapiro and Varian, *Information Rules*, endnote 6, chapter 14, and Agrawal, Gans, and Goldfarb, *Prediction Machines*, endnote 45, chapter 9.

20. McAfee and Brynjolfsson, *Machine, Platform, Crowd*.

21. Barns, *Platform Urbanism*, endnote 21, chapter 1.

22. Stephen Watts and Muhammad Raza, "SaaS vs PaaS vs IaaS: What's the difference and how to choose," *BmcBlogs*, June 15, 2019, https://www.bmc.com/blogs/saas-vs-paas-vs-iaas-whats-the-difference-and-how-to-choose/.

23. Bower and Christensen, "Disruptive technologies," endnote 22, chapter 4, and their follow-up article, where they warn that a lot of what has been defined as disruptive is not the case. See Clayton M. Christensen, Michael Raynor, and Rory McDonald, "What is disruptive innovation?," *Harvard Business Review*, *93*, No. 12, 44–53, 2015.

24. Brad Stone, *The Upstarts: How Uber, Airbnb, and the Killer Companies of the New Silicon Valley Are Changing the World*, Little, Brown and Company, New York, 2017.

25. Tom Goodwin, "The battle is for the customer interface," *TechCrunch*, March 3, 2015, https://techcrunch.com/2015/03/03/in-the-age-of-disintermediation-the-battle-is-all-for-the-customer-interface/.

26. Stephen McBride, "Uber's nightmare has just begun," *Forbes*, September 4, 2019, https://www.forbes.com/sites/stephenmcbride1/2019/09/04/ubers-nightmare-has-just-started/#42be598b7e03.

27. Zahratu Shabrina, "The impact of the platform economy in cities: The case of Airbnb," unpublished PhD Thesis, Centre for Advanced Spatial Analysis, University College London, 2019.

28. Felix Richter, "Smartphones cause photography boom," *Statista*, August 31, 2017, https://www.statista.com/chart/10913/number-of-photos-taken-worldwide/.

29. Apps short for "applications," endnote 5, this chapter.

30. See John Clement, "Number of apps available in leading Apps 2019," October 9, 2019, *Statista*, where he says "As of the third quarter of 2019, Android users were able to choose between 2.47 million Apps, making Google Play the app store with biggest

number of available Apps. Apple's App Store was the second-largest app store with 1.8 million available apps for iOS." See https://www.statista.com/statistics/276623 /number-of-apps-available-in-leading-app-stores/.

31. Mansoor Iqbal, "App download and usage statistics (2019)," *BusinessofApps*, November 19, 2019, https://www.businessofapps.com/data/app-statistics/.

32. John Clement, "Percentage of all global web pages served to mobile phones from 2009 to 2018," *Statista*, July 22, 2019, https://www.statista.com/statistics/241462/glo bal-mobile-phone-website-traffic-share/.

33. Mansoor Iqbal, "App download and usage statistics (2019)," *BusinessOfApps*, November 19, 2019, https://www.businessofapps.com/data/app-statistics/, and Jory MacKay, "Here's how much you use your phone during the workday," *Rescue Time Blog*, March 21, 2019, https://blog.rescuetime.com/screen-time-stats-2018/.

34. Joei Chan, "All you need to know about the Chinese social media landscape in 2019," *Linkfluence*, https://www.linkfluence.com/blog/chinese-social-media-landscape -2019, and Kate Chernavina, "Top 20 Chinese social media sites of 2019," *Hi-Com*, December 11, 2019, https://www.hicom-asia.com/chinese-kol-top-10-social-media-plat forms-they-use/.

35. Mark Sweney, "Britons hang up the landline as call volumes halve: Rise in mobile usage causes rapid decline of traditional telephones over last six years," *The Guardian*, January 5, 2019, https://www.theguardian.com/business/2019/jan/05/brit ons-hang-up-landline-call-volumes-halve.

36. The use of money in contemporary society is rapidly declining. See Mike Orcutt, "An elegy for cash: The technology we might never replace," *MIT Technology Review*, January 3, 2020, https://www.technologyreview.com/s/614998/an-elegy-for-cash-the -technology-we-might-never-replace/.

37. "Cryptocurrency" is the term used for virtual money—that which is not backed up by nationally agreed currencies of the nation-state. See Jake Frankenfield, "Bit-coin," *Investopedia*, October 26, 2019, https://www.investopedia.com/terms/b/bit coin.asp and also the Wiki definition https://en.wikipedia.org/wiki/Cryptocurrency.

38. Julia Kagan, "Financial technology—Fintech," *Investopedia*, June 25, 2019, https://www.investopedia.com/terms/f/fintech.asp.

39. Sukumar Ganapati, "Using mobile apps in government," *IBM Center for the Business of Government*, http://www.businessofgovernment.org/sites/default/files/Using %20Mobile%20Apps%20in%20Government.pdf.

40. Dashboards are often used to deliver access to these kinds of applications. See Kitchin, Lauriault, and McArdle, "Knowing and governing cities," endnote 40, chapter 8.

41. Janice Morphet and Robin Morphet, "New urban agenda: New urban analyt-ics," Centre for Advanced Spatial Analysis, The MacArthur Project Research Report, University College London, August 2019.

42. Shoshana Zubroff, *The Age of Surveillance Capitalism*, Profile Books, London, 2019.

43. Andreessen, "Eating the world."

CHAPTER 16

Lewis Mumford, *The Myth of the Machine*, Harcourt Brace Jovanovich, New York, 1970.

1. Yuval Harari, *Homo Deus: A Brief History of Tomorrow*, Vintage, London, 2017.

2. Harari's thesis about the way we have evolved in terms of our abilities to shape our planet is elaborated in his first book, *Sapiens: A Brief History of Humankind*, Vintage, London, 2015.

3. Helen Rosenau, *The Ideal City: Its Architectural Evolution*, Studio Vista, London, 1974, and Wycherley, *Greeks Built Cities*, endnote 9, chapter 6, https://www.questia .com/read/6476103/how-the-greeks-built-cities/.

4. The standard model reflects the central focus of the city, its radial transportation structure, and the succession of circular bands of different land uses, which we named after Johann Heinrich von Thünen, *Der Isolierte Staat*, endnote 16, chapter 6.

5. Kathleen Mary Kenyon, "Jericho Town, West Bank," *Encyclopaedia Britannica*, https://www.britannica.com/place/Jericho-West-Bank, and the Wiki page for Jericho, https://en.wikipedia.org/wiki/Jericho.

6. For an analysis of changes in the world's largest cities, see my paper, "Rank clocks," *Nature*, *444*, No. 30, 592–596, 2006; Tertius Chandler, *Four Thousand Years of Urban Growth*, Edwin Mellen Press, New York, 1989; and George Modelski, *World Cities: –3000 to 2000*, Faros 2000, Washington, DC, 2003. For a recent updating of these data sets, see the paper by M. Reba, E. Reitsma, and K. C. Seto, "Spatializing 6,000 years of global urbanization from 3700 BC to AD 2000," *Scientific Data*, *3*, 160034, 2016.

7. The EU JRC (European Union Joint Research Centre, Ispra, Italy) data set defines agglomerations rather than distinct cities. These agglomerations are portrayed in my book *Inventing Future Cities*, endnote 8, chapter 1.

8. Indeed, Mumford in his book *City In History*, endnote 2, chapter 6, notes that Plato said that the ideal size was "the number of citizens who might be addressed by a single voice," that is, a population of 5,040 citizens. This argument is detailed in Yves Charbit's paper "The Platonic city," endnote 8, chapter 6.

9. Leonardo da Vinci, *Notebooks of Leonardo da Vinci 1452–1519*, translated and arranged by Edward MacCurdy, Braziller, New York, 1955, available at https:// archive.org/details/noteboo00leon.

10. A thorough account of the efforts of the Victorian philanthropists is given by William Ashworth in his book *The Genesis of Modern British Town Planning*, Routledge and Kegan Paul, London, 1954, and a stronger link to urban studies is made by

Gordon E. Cherry in his paper "The town planning movement and the late Victorian city," *Transactions of the Institute of British Geographers*, *4*, No. 2, 306–319, 1979.

11. Ebenezer Howard's ideas for the garden city were first published in 1898 in his book *Tomorrow: A Peaceful Path to Reform*, Routledge, London, 1898, revised 2009. It was revised soon after in 1902 with a different title, *Garden Cities of Tomorrow*.

12. The development of small new towns of a similar modest size in North America developed in equivalent circumstances of philanthropy in industrial society is outlined by John Reps in his book *The Making of Urban America: A History of City Planning in the United States*, Princeton University Press, Princeton, NJ, 1965.

13. Many of these ideas and innovations are described in more detail in my book *Inventing Future Cities*, endnote 8 chapter 1.

14. Alfred Marshall, *Principles of Economics*, 8th Edition, Macmillan, London, 1890 and 1920, is accredited with the principle associated with economies of scale—agglomeration economies—which he fashioned through examples in his seminal book, saying: "When an industry has thus chosen a locality for itself, it is likely to stay there long: So great are the advantages which people following the same skilled trade get from near neighborhood to one another . . . if one man starts a new idea, it is taken up by others and combined with suggestions of theirs" (Book IV, chapter 10).

15. Ernst Friedrich Schumacher, *Small Is Beautiful: A Study of Economics as if People Mattered*, Vintage, London, 1973.

16. West, *Scale*, endnote 22, chapter 9; Bettencourt, *Urban Science*, endnote 39, chapter 6.

17. Plato, *The Republic*, endnote 8, chapter 6.

18. Singapore's intelligent island program is outlined in my article "Technology highs," *The Guardian*, June 22, p. 29, 1989, and at http://smartisland.com/singapore -the-smart-island-smart-nation/. The wider context of the development of IT in cities is in my editorial "Intelligent cities: Using information networks to gain competitive advantage," *Environment and Planning B*, *17*, 247–256, 1990.

19. The NCB was formed on September 1, 1981, to implement the computerization of the civil service, to coordinate computer education and training, and to develop and promote the computer services industry. See http://eresources.nlb.gov .sg/history/events/f499072d-330f-4763-a105-dd9940f1890b.

20. ISDN is "a set of communication standards for simultaneous digital transmission of voice, video, data, and other network services over the traditional circuits of the public switched telephone network," from Wikipedia https://en.wikipedia.org /wiki/Integrated_Services_Digital_Network.

21. Manuel Castells and Peter Hall, *Technopoles of the World: The Making of Twenty-First-Century Industrial Complexes*, Routledge, London, 1994; and Annalee Saxenian, *Regional Advantage—Culture and Competition in Silicon Valley and Route 128*, Harvard University Press, Cambridge, MA, 1996.

22. Edward A. Feigenbaum, *The Fifth Generation: Artificial Intelligence and Japan's Computer Challenge to the World*, Addison Wesley Longman, Reading, MA, 1983.

23. The multimedia super corridor (MSC) is "a Special Economic Zone and high-technology business district in central-southern Selangor, Malaysia," https://en .wikipedia.org/wiki/MSC_Malaysia. See also Bala Ramasamy, Anita Chakrabarty, and Madelaine Cheah, "Malaysia's leap into the future: An evaluation of the multimedia super corridor," *Technovation*, *24*, No. 11, 871–883, 2004.

24. A brief history of NSF and the internet can be found at https://www.nsf.gov/news /news_summ.jsp?cntn_id=103050.

25. Andy Hudson-Smith, Richard Milton, Joel Dearden, and Michael Batty, "The neogeography of virtual cities: Digital mirrors into a recursive world," in Marcus Foth (Editor), *Handbook of Research on Urban Informatics: The Practice and Promise of the Real-Time City*, Information Science Reference, IGI Global, Hershey, PA, 270–290, 2008. These first portals were cartoon like, much inspired by Japanese media representing very simple and obvious ways of placing graphics on these kinds of early web pages.

26. Martin Dodge, "Explorations in AlphaWorld: The geography of 3D virtual worlds on the Internet," in Peter Fisher and David Unwin (Editors), *Virtual Reality in Geography*, Taylor and Francis, London, 305–331, 2002.

27. Virtual Singapore "is a dynamic three-dimensional (3D) city model and collaborative data platform, including the 3D maps of Singapore" from National Research Foundation, Prime Minister's Office, Government of Singapore, https://www.nrf.gov .sg/programmes/virtual-singapore.

28. Batty, "Digital twins," endnote 40, chapter 11, and also my editorial, "A map is not the territory, or is it?," *Environment and Planning B*, *19*, No. 2, 121–124, 2019, and Matthew Wall, "Virtual cities: Designing the metropolises of the future," *BBC Technology of Business*, January 18, 2019, https://www.bbc.co.uk/news/business-46880468.

29. See Stephen Marshall, *Streets and Patterns: The Structure of Urban Geometry*, Routledge, London, 2004, and Michael Batty and Paul Longley, *Fractal Cities: A Geometry of Form and Function*, Academic Press, London, 1994, http://www.fractalcities.org/.

30. Jacobs, *Death and Life*, endnote 8, chapter 2.

31. There has been something of a revival in searching for measures of community that build on ideas about diversity, particularly due to the fact that many digital providers are now adding points of interest—activities and land uses—to their digital maps. See, for example, Patricia Sulis, Ed Manley, Chen Zhong, and Michael Batty, "Mobility data as proxy for measuring urban vitality," *Journal of Spatial Information Science*, *16*, 137–162, 2018.

32. There are many good histories of new towns before the digital era. In the UK, see, for example, Dennis Hardy, *From Garden Cities to New Towns: Campaigning for Town*

and Country Planning 1899–1946, Spon, London, 1991. In Asia-Pacific, see David Phillips and Anthony Gar-on Yeh (Editors), *New Towns in East and South-east Asia: Planning and Development*, Oxford University Press, Hong Kong, 1987.

33. Reps, *Urban America*.

34. Mumford, *City in History*, endnote 2, chapter 6, and Hall, *Cities in Civilization*, endnote 2, chapter 6.

35. Reps, *Urban America*.

36. Masdar is planned as an eco-city and a digital new town. See the International New Towns Institute, http://www.newtowninstitute.org/newtowndata/newtown .php?newtownId=1479. However, Masdar's zero-carbon dream could become the world's first green ghost town, https://www.theguardian.com/environment/2016/feb /16/masdars-zero-carbon-dream-could-become-worlds-first-green-ghost-town.

37. Lucy Williamson's article in 2013 "Tomorrow's cities: Just how smart is Songdo?," *BBC News Technology*, http://www.bbc.co.uk/news/technology-23757738; Jane Wakefield's article in 2013 "Tomorrow's cities: Do you want to live in a smart city?," *BBC News Technology*, http://www.bbc.co.uk/news/technology-22538561; and Linda Poo, "Sleepy in Songdo, Korea's smartest city," *CityLab*, https://www.city lab.com/life/2018/06/sleepy-in-songdo-koreas-smartest-city/561374/.

38. Le Corbusier, *The City of Tomorrow and Its Planning*, Dover Publications, New York, 1929.

39. Myungjun Jang and Soon-Tak Suh, "U-City: New trends of urban planning in Korea based on pervasive and ubiquitous geotechnology and geoinformation," in David Taniar, Osvaldo Gervasi, Beniamino Murgante, Eric Pardede, and Bernady Apduhan (Editors), *Computational Science and Its Applications—ICCSA 2010. ICCSA Lecture Notes in Computer Science*, 6016, 262–270, Springer, Heidelberg, Germany, 2010.

40. For the number of mobile phone and smartphone users, see https://www.bank mycell.com/blog/how-many-phones-are-in-the-world.

41. Simon Marvin and Andrés Luqu Ayala, "Urban operating systems: Diagramming the city," *International Journal of Urban and Regional Research*, 41, No. 1, 84–103, 2017.

42. Katia Moskvitch, "Smart cities get their own operating system," *BBC News Technology*, September 30, 2011, https://www.bbc.co.uk/news/technology-15109403.

43. Robert Eccles, Amy G. C. Edmondson, Susan Thyne, and Tiona Zuzul, "Living PlanIT," Harvard Business School Case 410–081, February 2010, Revised November 2013, https://www.hbs.edu/faculty/Pages/item.aspx?num=38370.

44. Norbert Wiener's book *Cybernetics*, endnote 16, chapter 8, sketched the idea that any system could be managed using appropriate control mechanisms. Indeed the subtitles to his book is "The scientific study of control and communication in the animal and the machine." Brian McLoughlin in his book *Control and Urban Planning*, Faber and Faber, London, 1973, speculated that this sort of control could be done

for the urban planning system. But the argument remains as to whether these ideas are to be considered and implemented literally or whether they are simply conceptual issues for discussion and reflection.

45. Herman van den Bosch, "PlanIT Valley: The smartest city never been built," *Smart City Hub* https://smartcityhub.com/governance-economy/planit-valley-the -smartest-city-never-been-built/.

46. Maanu Saadia recounts "How 'Blade Runner' and sci-fi made everything dystopian. Science fiction, especially Blade Runner, has spawned so many dystopias that dystopia itself has become banal. We need a new utopianism that embraces the city," *CityLab*, November 1, 2019, https://www.citylab.com/perspective/2019/11/dys topian-cities-science-fiction-blade-runner-books-movies/598624/.

47. Sidewalk Labs set up the Toronto project for "Reimagining cities to improve quality of life," https://www.sidewalklabs.com/, and the company is "experiment-ing" with part of the waterfront in Toronto https://www.sidewalktoronto.ca/.

48. Lucien Begault and Jessika Khazrik, "Smart cities: Dreams capable of becoming nightmares," *Amnesty Tech*, June 28, 2019, https://www.amnesty.org/en/latest/res earch/2019/06/smart-cities-dreams-capable-of-becoming-nightmares/.

49. Shoshana Zuboff, in *#BlockSidewalk*, January 2020, https://www.blocksidewalk .ca/supporters.

CHAPTER 17

Brand, *The Media Lab*, 9.

1. Edward Glaeser and David Cutler, *Survival of the City: Living and Thriving in an Age of Isolation*, Basic Books, London, 2021.

2. Batty, *Inventing Future Cities*, endnote 8, chapter 1.

3. Skype was first developed in the Baltic states in early 2003 and was eventually acquired by Microsoft in 2011. Microsoft also developed Teams for conferencing in 2011, while Zoom emerged around the same time. Google Meet grew out of Google Hangouts in 2017. There has also been massive growth in other proprietary, less generic systems tailored to specific contexts and events. For more detail and for a comparison of three of these, see https://gcloud.devoteam.com/blog/comparing-zoom -microsoft-teams-and-google-meet/.

4. Cairncross, *Death of Distance*, endnote 32, chapter 2; but the notion of such remote working goes back to E. M. Foster's essay *Machine Stops*, endnote 42, chapter 7, to futurists such as Toffler, *Third Wave*, endnote 5, chapter 1, and more focused discussions in transportation such as that by Nilles, Carlson, Gray, and Hanneman, *Telecommunications-Transportation*, endnote 41, chapter 7; and Vicky Gan and CityLab, "What telecommuting looked like in 1973: A vision of remote work from before the personal computer," *The Atlantic*, December 3, 2015, at https://www.theatlantic.com /technology/archive/2015/12/what-telecommuting-looked-like-in-1973/418473/.

5. The ultimate list of remote work statistics for 2022 is available at https://findstack .com/remote-work-statistics/.

6. There are several good sources for this kind of data. In particular, the Google and Apple mobility reports take mobile-phone traces in different locations to gauge the sense of changing volumes with respect to different patterns of trip making. See Google Mobility Reports (2022), https://www.gstatic.com/covid19/mobility /Region_Mobility_Report_CSVs.zip. Apple discontinued their reports on April 22, 2022, and Google on October 15, 2022. For analysis of these changed transport patterns, see Michael Batty, Roberto Murcio, Iacopo Iacopini, Maarten Vanhoof, and Richard Milton, "London in lockdown: Mobility in the pandemic city," in Abbas Rajabifard, Greg Foliente, and Daniel Paez (Editors), *COVID-19 Pandemic, Geospatial Information, and Community Resilience: Global Applications and Lessons*, CRC Press, Boca Raton, FL, 228–244, 2021; and Roberto Murcio, Richard Milton, and Michael Batty, "The impact of lockdowns on mobility in city systems," *Scienze Regionali*, 3, 373–390, 2021.

7. Uber and Lyft are getting less unprofitable, but COVID-19 is still a drag on their business, https://www.theverge.com/2021/2/11/22277043/uber-lyft-earnings-q4-2020 -profit-loss-covid.

8. 5G is the fifth-generation standard for broadband cellular networks that are expected to take over from the fourth generation by 2025. See https://en.wikipedia .org/wiki/5G. These networks work at much higher frequency in terms transmission, and it is expected that a 5G network for a typical domestic subscriber could be up to a hundred times as fast as 4G. See https://www.thalesgroup.com/en/world wide-digital-identity-and-security/mobile/magazine/5g-vs-4g-whats-difference.

9. Moore, *Electronics*, endnote 37, chapter 3.

10. Taleb, *Black Swan*, endnote 27, chapter 1.

11. Karl Popper develops this thesis about prediction unerringly in his many books. See, for example, *The Logic of Scientific Discovery*, Routledge and Kegan Paul, London, 1937, translation 1959; *The Poverty of Historicism*, Routledge and Kegan Paul, London, 1957; *Conjectures and Refutations: The Growth of Scientific Knowledge*, Routledge and Kegan Paul, London, 1963; and *The Myth of the Framework: In Defence of Science and Rationality*, edited by M.A. Notturno, Routledge, London, 1995, from which the quote at the frontispiece of this book is taken. A useful reflection on the relationship of his ideas to modeling is by Henk Tennekes, "Karl Popper and the accountability of scientific models," in Johan Grasman and Gerrit van Straten (Editors), *Predictability and Nonlinear Modelling in Natural Sciences and Economics*, Springer Nature, Switzerland, AG, 6–11, 2019.

12. In their book *Superforecasting*, Philip Tetlock and Dan Gardner take a wider view of prediction, suggesting that there are some events that are highly personal and meet Popper's criterion of closure that can indeed be predicted with some accuracy, although they do not dispute the wider thesis that the future is unpredictable. See endnote 30, chapter 1.

13. Michael Batty, "Creative destruction, long waves and the age of the smart city," in Robert D. Knowles and Celine Rozenblat (Editors), *Sir Peter Hall: Pioneer in Regional Planning, Transport and Urban Geography*, Springer Briefs on Pioneers in Science and Practice No. 52, Springer, Mosbach, Germany, 81–97; Batty, *Inventing Future Cities*, endnote 8, chapter 1.

14. Kurzweil, *Singularity*, endnote 10, chapter 1.

15. The standard model was extensively discussed on chapter 7, where we introduced Johann Heinrich von Thünen's *Der Isolierte Staat*, which reflects the logic of a central market for the exchange of products that are produced at different distances from that market and pay a rent that reflects the travel cost of the produce that is sold at the center. See endnotes 16, 17, chapter 6.

16. Edge cities were first defined by Joel Garreau in his book *Edge City*. See endnote 29, chapter 6.

17. Decentralization began due to diseconomies of scale, which were generated through a change in transport technology in the second industrial revolution when we moved from mass transit to private transport. This led to the phenomena of out-of-town (literally out-of-center) locations set within a growing sea of urban peripheral development loosely termed "sprawl." See Whyte, *Exploding Metropolis*, endnote 37, chapter 6.

18. Bauer Wurster, "Form and Structure," endnote 34, chapter 6.

19. See Hall and Pain, *Polycentric Metropolis*, endnote 30, chapter 6, and for a more theoretical urban economic focus, see Alex Anas, Richard Arnott, and Kenneth A. Small, "Urban spatial structure," *Journal of Economic Literature*, 36, 1426–1464, 1998.

20. Most of chapter 8 deals with Frances Cairncross's idea of the death of distance, but see also endnote 32, chapter 2.

21. The idea of the electronic cottage goes back to the rural idyll, which was consistent, of course, with the Arts and Crafts movement in the late nineteenth century. The notion of the self-contained workplace, remote or detached from the industrial city, is implicit in E. M. Forster's short story *The Machine Stops* that we quoted in endnote 42, chapter 7. In the Digital Age, some of the first stirrings of the notion that computers would change the nature or work and residence are contained in Toffler's *Future Shock*, endnote 6, chapter 1, and in James Martin and Adrian Norman's *The Computerized Society*, Prentice-Hall, Englewood Cliffs, NJ, 1970, but are also questioned by Tom Forester in his article "The myth of the electronic cottage," *Futures*, 3, 227–240, June 20, 1988.

22. Glaeser and Cutler, *Survival of the City*.

23. Michael Batty, "The post-pandemic city: Speculation through simulation," *Cities*, 124, 103594, May 2022.

24. Judith Evans and Nic Fildes, "BT agrees £210m sale of historic London site: Private equity buys Newgate St headquarters that was home to GPO's telegraph office,"

The Financial Times, July 17, 2019, https://www.ft.com/content/5dc97e0a-a880 -11e9-b6ee-3cdf3174eb89. In fact, in the old Post Office buildings in Carter Lane, south of the cathedral, there is still a BT presence, as there is in the Faraday Building on Queen Victoria Street.

25. 1890—GPO Headquarters—St. Martin's Le Grand, London, UK—Buildings on Waymarking.com, https://www.waymarking.com/waymarks/WMK9KA_1890_GPO_ Headquarters_St_Martins_le_Grand_London_UK. For the scale of the building, see https://blog.quintinlake.com/2010/10/23/nomura-house-formerly-north-range-of -the-general-post-office-headquarters-by-sir-edward-tanner/.

26. There are several eloquent expositions of the ways in which the constraints on industrial society are becoming unstuck due to new technologies and increasing flexibility in where, when, and how to locate in space and time. This increasing liquidity in society is captured in various expositions of contemporary arts from Zygmunt Bauman' book *Liquid Life*, Polity Press, Cambridge, UK, 2005, to that by Rem Koolhaas, Stefano Boeri, Sanford Kwinter, Nadia Tazi, and Hans-Ulrich Obrist, *Mutations*, ActarD, New York, 2001.

27. The concept of orgware originated sometime in the 1970s on the tail of software being distinguished from hardware. More recently, notions about dataware have emerged, but all these elements of computer and network technology define what we call "information infrastructure." See Gennady M. Dobrov, "The strategy for organized technology in the light of hard-, soft-, and org-ware interaction," *Long Range Planning*, *12*, No. 4, 79–90, 1979.

REFERENCES

Abbate, J. (1999) *Inventing the Internet*, MIT Press, Cambridge, MA.

Adhvaryu, D. (2017) "Before the age of the iPhone: The story of how our cell phones came to be," September 24, *The Varsity*, https://thevarsity.ca/2017/09/24/before -the-age-of-the-iphone/.

Agrawal, A., Gans, J., and Goldfarb, A. (2018) *Prediction Machines: The Simple Economics of Artificial Intelligence*, Harvard Business Review Press, Cambridge, MA.

Alexander, C., and Manheim, M. (1962) "The use of diagrams in highway route location: An experiment," R62-3, Civil Engineering Systems Laboratory, MIT, Cambridge, MA.

Alexander, R. C., and Smith, D. K. (1999) *Fumbling the Future: How Xerox Invented, Then Ignored, the First Personal Computer*, iUniverse Books, Lincoln, NB.

Alfred, R. (2008) "April 4, 1975: Bill Gates, Paul Allen form a little partnership," *Wired Magazine*, https://www.wired.com/author/randy-alfred.

Alkema, H., and McLaughlin, K. (2007) *40 Years of Computer Science at the University of Waterloo*, September 29, https://cs.uwaterloo.ca/40th/index.html.

Alonso, W. (1964) *Location and Land Use: Toward a General Theory of Land Rent*, Harvard University Press, Cambridge, MA.

Amatia, M., Freestone, R., and Robertson, S. (2017) "'Learning the city': Patrick Geddes, exhibitions, and communicating planning ideas," *Landscape and Urban Planning*, *166*, 97–105.

Anas, A., Arnott, R., and Small, K. A. (1998) "Urban spatial structure," *Journal of Economic Literature*, *36*, 1426–1464.

Anderson, C. (2007) "The end of theory: Will the data deluge make the scientific method obsolete?," *Wired Magazine*, July 16, http://www.wired.com/science/discov eries/magazine/16-07/pb_theory.

Anderson, D. (2014) "Historical reflections: Tom Kilburn: A tale of five computers," *Communications of the ACM*, *57*, No. 5, 35–38.

Andreessen, M. (2011) "Why software is eating the world," *The Wall Street Journal*, August 11, http://www.wsj.com/articles/SB10001424053111903480904576512250915629460.

Anon (1978) "The computer society: The age of miracle chips, new microtechnology will transform society," *Time Magazine*, February, 44–45, https://content.time.com/time/subscriber/article/0,33009,948017-3,00.html.

Anon (2009) "How far, how fast?," *Medieval Worldbuilding Information*, October 13, https://writemedieval.livejournal.com/4706.html.

Anon (2013) "The Citizen's House of Wuhan opens its doors," *Arte Charpentier*, February 28, http://www.arte-charpentier.com/en/the-citizens-house-of-wuhan-opens-its-doors/.

Anon (2015) "What is remote sensing? The definitive guide," November 2, https://gisgeography.com/remote-sensing-earth-observation-guide/.

Anon (2017a) "The birth of pioneering electrical engineer Guglielmo Marconi," *The Daily Telegraph*, April 28, https://www.telegraph.co.uk/technology/connecting-britain/guglielmo-marconi-birth/.

Anon (2017b) "Video game history," *History*, September 1, https://www.history.com/topics/inventions/history-of-video-games.

Anon (2018) "A brand new city landmark now opens to public," *Life of Guangzhou*, April 28, https://www.lifeofguangzhou.com/wap/knowGZ/content.do?contextId=7458&frontParentCatalogId=175.

Anon (2019) "The history of computer data storage, in pictures," April 12, https://royal.pingdom.com/the-history-of-computer-data-storage-in-pictures/.

Anon (2022a) "Was the rise of car ownership responsible for the mid-century home-ownership boom in the US?," *Old Urbanist*, February 2013, accessed January 11, 2023 https://oldurbanist.blogspot.com/2013/02/was-rise-of-car-ownership-responsible.html.

Anon (2022b) "Masdar," International New Towns Institute, http://www.newtowninstitute.org/newtowndata/newtown.php?newtownId=1479.

Anon (2022c) "US states by vehicles per capita," https://www.worldatlas.com/articles/us-states-by-vehicles-per-capita.html.

ASCII Art (2022) "Marilyn Monroe," https://knowyourmeme.com/photos/592796-ascii-art.

Ashworth, W. (1954) *The Genesis of Modern British Town Planning*, Routledge and Kegan Paul, London.

ASU (2022) Arizona State University Decision Theater, https://dt.asu.edu/facilities.

Austrian, G. D. (1982) *Herman Hollerith: Forgotten Giant of Information Processing*, Columbia University Press, New York.

Ausubel, J., and Marchetti, C. (2001) "The evolution of transport," *The Industrial Physicist*, 20–24, American Institute of Physics, April/May, https://phe.rockefeller.edu/TIP_transport/transport.pdf.

Autodesk (2022) 3D Studio Max, https://www.autodesk.co.uk/products/3ds-max /overview/.

Bar-Yam, Y. (2002) "Complexity rising: From human beings to human civilization, a complexity profile," *Encyclopedia of Life Support Systems* (EOLSS), UNESCO, Oxford, https://necsi.edu/complexity-rising-from-human-beings-to-human-civilization-a-co mplexity-profile.

Barabasi, A.-L. (2002) *Linked: How Everything Is Connected to Everything Else and What It Means*, Perseus Books, New York.

Barber, K. (2017) "Watts steam engine, 1776," *Stories of Change*, March 23, https:// storiesofchange.ac.uk/node/40.

Barns, S. (2019) *Platform Urbanism: Negotiating Platform Ecosystems in Connected Cities*, Palgrave Macmillan, London.

Barone, J. (Editor) (2019) *Leonardo da Vinci: A Mind in Motion*, British Library Press, London.

Barrett, C. et al. (Editors) (1999) *TRANSIMS (TRansportation ANalysis SIMulation System)*, 6 Volumes, Los Alamos National Laboratory, Los Alamos, NM.

Barthelemy, M. (2016) *The Structure and Dynamics of Cities*, Cambridge University Press, Cambridge.

Basaraba, S. (2019) "A guide to longevity throughout history, from the prehistoric onward increases in lifespan from prehistory through the modern era," *Very Well Health*, July 14, https://www.verywellhealth.com/longevity-throughout-history-2224054.

Batty, M. (1972) "Recent developments in land use modelling: A review of British research," *Urban Studies*, 9, 151–177.

Batty, M. (1976) *Urban Modelling: Algorithms, Calibrations, Predictions*, Cambridge University Press, Cambridge.

Batty, M. (1978) "Reilly's challenge: New laws of retail gravitation which define systems of central places," *Environment and Planning A*, 10, 185–219.

Batty, M. (1980) "Computers and design," Papers in Planning Research, 2, UWIST Department of Town Planning, Cardiff, UK.

Batty, M. (1986) *Microcomputer Graphics: Art, Design, and Creative Modelling*, Chapman and Hall, London.

Batty, M. (1989) "Technology highs," *The Guardian*, 29, June 22.

Batty, M. (1990a) "Intelligent cities: Using information networks to gain competitive advantage," *Environment and Planning B*, 17, 247–256.

Batty, M. (1990b) "Invisible cities," *Environment and Planning B*, 17, 127–130.

Batty, M. (1997) "The computable city," *International Planning Studies*, 2, 155–173.

Batty, M. (2000) "The new urban geography of the third dimension," *Environment and Planning B: Planning and Design*, 27, 483–484.

Batty, M. (2005) *Cities and Complexity*, MIT Press, Cambridge, MA.

Batty, M. (2006) "Rank clocks," *Nature*, *444*, No. 30, 592–596.

Batty, M. (2008) "Fifty years of urban modelling: Macro statics to micro dynamics," in S. Albeverio, D. Andrey, P. Giordano, and A. Vancheri (Editors), *The Dynamics of Complex Urban Systems: An Interdisciplinary Approach*, 1–20, Physica, Heidelberg, Germany.

Batty, M. (2011) "Commentary: When all the world's a city," *Environment and Planning A*, *43*, No. 4, 765–772.

Batty, M. (2013) *The New Science of Cities*, MIT Press, Cambridge, MA.

Batty, M. (2014a) "At the crossroads of urban growth," *Environment and Planning B: Planning and Design*, *41*, 951–953.

Batty, M. (2014b) "Can it happen again? Planning support, Lee's Requiem and the rise of the smart cities movement," *Environment and Planning B*, *41*, 388–391.

Batty, M. (2016a) "Big data and the city," *Built Environment*, *42*, No. 3, 321–337.

Batty, M. (2016b) "Creative destruction, long waves and the age of the smart city," in R. D. Knowles and C. Rozenblatt (Editors), *Sir Peter Hall: Pioneer in Regional Planning, Transport and Urban Geography*, 81–97, Springer Briefs on Pioneers in Science and Practice, No. 52, Springer, Mosbach, Germany.

Batty, M. (2017) "Cities as systems of networks and flows," in T. Haas and H. Westlund (Editors), *In the Post-Urban World: Emergent Transformation of Cities and Regions in the Innovative Global Economy*, 56–69, Routledge, London.

Batty, M. (2018a) "Digital twins," *Environment and Planning B: Urban Analytics and City Science*, *45*, No. 5, 817–820.

Batty, M. (2018b) "Visualizing aggregate movement in cities," *Philosophical Transactions of the Royal Society B*, *373*, July 2.

Batty, M. (2018c) *Inventing Future Cities*, MIT Press, Cambridge, MA.

Batty, M. (2019) "A map is not the territory, or is it?," *Environment and Planning B*, *19*, No. 2, 121–124.

Batty, M. (2021a) "Defining cities and growth," in E. Glaeser, P. Nijkamp and K. Kourtit (Editors), *Urban Empires: Cities as Global Rulers in the New Urban World*, 210–228, Routledge, Abingdon, UK.

Batty, M. (2021b) "Defining urban science," in J. Shi, M. Goodchild, M. P. Kwan, and A. Zhang (Editors), *Urban Informatics*, 15–28, Springer, Berlin.

Batty, M. (2022) "The post-pandemic city: Speculation through simulation," *Cities*, *124*, 103594.

Batty, M., and Hudson-Smith, A. (2005) "Urban simulacra: From real to virtual cities, back and beyond," *Architectural Design*, *75*, No. 6, 42–47.

Batty, M., and Longley, P. (1994) *Fractal Cities: A Geometry of Form and Function*, Academic Press, London, http://www.fractalcities.org/.

Batty, M., Axhausen, K., Giannotti, F., Pozdnoukhov, A., Bazzani, A., Wachowicz, M., Ouzounis, G., and Portugali, J. (2012) "Smart cities of the future," *European Physical Journal Special Topics*, *214*, 481–518.

Batty, M., Chapman, D., Evans, S., Haklay, M., Kueppers, S., Shiode, N., Smith, A., and Torrens, P. (2001) "Visualizing the city: Communicating urban design to planners and decision-makers," in Richard Brail and Dick Klosterman (Editors), *Planning Support Systems: Integrating Geographic Information Systems, Models, and Visualization Tools*, 405–443, ESRI Press, Redlands, CA.

Batty, M., Lin, H., and Chen, M. (2017) "Virtual realities, analogies and technologies in geography," in Barney Warf (Editor), *Handbook on Geographies of Technology*, 96–112, Research Handbooks in Geography Series, Edward Elgar, Cheltenham, UK.

Batty, M., Murcio, R., Iacopini, I., Vanhoof, M., and Milton, R. (2021) "London in lockdown: Mobility in the pandemic city," in A. Rajabifard, G. Foliente, and D. Paez (Editors), *COVID-19 Pandemic, Geospatial Information, and Community Resilience: Global Applications and Lessons*, 228–244, CRC Press, Boca Raton, FL.

Bauman, Z. (2005) *Liquid Life*, Polity Press, Cambridge, UK.

Beacon, G. R., and Boreham, P. G. (1978) "Computer-aided architectural design at Leeds Polytechnic," *Computer-Aided Design, 10*, 325–331.

Beckmann, M. (1957) "On the distribution of rent and residential density in cities," *Interdepartmental Seminar in Mathematical Applications in the Social Sciences, 19*, 1, Yale University, New Haven CT, and in *Journal of Economic Theory, 1*, 60–67, 1969.

Begault, L., and Khazrik, J. (2019) "Smart cities: Dreams capable of becoming nightmares," *Amnesty Tech*, June 28, https://www.amnesty.org/en/latest/research/2019/06/smart-cities-dreams-capable-of-becoming-nightmares/.

Beinhocker, E. (2006) *The Origin of Wealth: Evolution, Complexity, and the Radical Remaking of Economics*, Harvard Business Review Press, Cambridge, MA.

Benenson, I., and Torrens, P. M. (2004) *Geosimulation: Automata-Based Modeling of Urban Phenomena*, John Wiley, Chichester, UK.

Berg, N., and Knights, M. (2019) *Amazon: How the World's Most Relentless Retailer Will Continue to Revolutionize Commerce*, Kogan Page, London.

Berners-Lee, T. (1999) *Weaving the Web: The Original Design and Ultimate Destiny of the World Wide Web*, Harper One, New York.

Berry, B. J. L. (1993) "Geography's quantitative revolution: Initial conditions, 1954–1960. A personal memoir," *Urban Geography, 14*, No. 5, 434–441.

Berryman, D. R. (2012) "Augmented reality: A review," *Medical Reference Services Quarterly, 31*, No. 2, 212–218.

Best, J. (2013) "IBM Watson: The inside story of how the Jeopardy-winning supercomputer was born, and what it wants to do next," *TechRepublic*, September 9, https://www.techrepublic.com/article/ibm-watson-the-inside-story-of-how-the-jeopardy-winning-supercomputer-was-born-and-what-it-wants-to-do-next/.

Bettencourt, L. (2021) *Introduction to Urban Science: Evidence and Theory of Cities as Complex Systems*, MIT Press, Cambridge, MA.

Bettencourt, L., and West, G. (2010) "A unified theory of urban living," *Nature, 467*, 912–913.

Bhattacharya, A. (2021) *The Man from the Future: The Visionary Life of John von Neumann*, Allen Lane, Penguin Press, New York, 2021.

Bilton, N. (2013) *Hatching Twitter: A True Story of Money, Power, Friendship, and Betrayal*, Portfolio Penguin Books, New York.

Black, A. (1990) "The Chicago Area Transportation Study: A case study of rational planning," *Journal of Planning Education and Research*, 10, No. 1, 27–37.

Blondel, V., Decuyper, A., and Krings, G. (2015) "A survey of results on mobile phone datasets analysis," *European Physics Journal, Data Science*, 4, 10.

Blum, A. (2012) *Tubes: Behind the Scenes at the Internet*, Viking, New York.

Blyth, T. (2014) *Information Age: Six Networks that Changed the World*, Science Museum, London.

Boeing, G., Batty, M., Jiang, S., and Schweitzer, L. (2022) "Urban analytics: History, trajectory, and critique," in R. Franklin and S. Rey (Editors), *Handbook of Spatial Analysis in the Social Sciences*, 503–516, Edward Elgar, Cheltenham, UK.

Bogart, D., Shaw-Taylor, L., and You, X. (2018) "The development of the railway network in Britain 1825–1911," The Cambridge Group for the History of Population and Social Structure, https://www.campop.geog.cam.ac.uk/research/projects/transport/on lineatlas/railways.pdf.

Borges, J. L. (1946) "On exactitude in science," *Los Anales de Buenos Aires*, 1, No. 3, 53.

Bosch, H. van den (2022) "PlanIT Valley: The smartest city never been built," *Smart City Hub*, https://smartcityhub.com/governance-economy/planit-valley-the-smartest -city-never-been-built/.

Bostrom, N. (2014) *Superintelligence: Paths, Dangers, Strategies*, Oxford University Press, Oxford.

Bower, J. L., and Christensen, C. M. (1995) "Disruptive technologies: Catching the wave," *Harvard Business Review*, 73, No. 1, 43–53.

Box, G. E. P., and Draper, N. (1987) *Empirical Model Building and Response Surfaces*, Wiley-Blackwell, New York.

Box, G. E. P. (1979) "Robustness in the strategy of scientific model building," in R. L. Launer and G. N. Wilkinson (Editors), *Robustness in Statistics*, 201–236, Academic Press, New York.

Boyce, D., and Williams, H. (2015) *Forecasting Urban Travel: Past, Present and Future*, Edward Elgar, Cheltenham, UK.

Bradburn, J., Williams, D., Piechocki, R., and Hermans, K. (2015) *Connected and Autonomous Vehicles: Introducing the Future of Mobility*, https://www.atkinsglobal.com.

Brand, S. (1971) *The Last Whole Earth Catalog*, Random House, New York.

Brand, S. (1989) *The Media Lab: Inventing the Future at MIT*, Penguin, New York.

Brand, S. (1999) *The Clock of the Long Now: Time and Responsibility*, Doubleday, New York.

Braswell, S. (2022) "The agreement that catapulted Microsoft over IBM," *OZY News for the Disruptive*, https://www.ozy.com/flashback/the-agreement-that-catapulted-mi crosoft-over-ibm/94437/.

Braudel, F. (1982) *The Wheels of Commerce: Civilization and Capitalism: 15th– 18th Century*, Harper & Row, New York.

Brewer, G. (1973) *Politicians, Bureaucrats, and the Consultant: A Critique of Urban Problem Solving*, Basic Books, New York.

Briscoe, B., Odlyzko, A., and Tilly, B. (2006) "Metcalfe's law is wrong," *IEEE Spectrum*, July 1, https://spectrum.ieee.org/computing/networks/metcalfes-law-is-wrong.

Brömmelstroet, M. te, Pelzer, P., and Geertman, S. (2014) "Forty years after Lee's Requiem: Are we beyond the seven sins?," *Environment and Planning B*, *41*, 381–391.

Brooks, F. (1975) *The Mythical Man-Month: Essays on Software Engineering*, Addison Wesley, Reading, MA.

Brown, M. (2022) "Do the Karl Marx pub crawl," *The Londonist*, https://londonist .com/london/drink/do-the-karl-marx-pub-crawl.

Browne, R. (2018) "70% of people globally work remotely at least once a week, study says," *MakeIt*, CNBC, May 30, https://www.cnbc.com/2018/05/30/70-percent-of-people-globally-work-remotely-at-least-once-a-week-iwg-study.html.

Brynjolfsson, E., and McAfee, A. (2016) *Race Against the Machine*, Digital Frontier Press, Lexington, MA.

Burdett, R., and Philipp Rode, P. (2012) "The electric city," *Electric City Conference*, 6–7, London School of Economics, London.

Burrington, I. (2016) *Networks of New York: An illustrated Field Guide to Urban Internet Infrastructure*, Melville House, Brooklyn, NY.

Bush, V. (1945) "As we may think," *The Atlantic Monthly*, *176*, No. 1, 101–108.

Buttenfield, B. P., and McMaster, R. B. (Editors) (1992) *Map Generalization*, John Wiley, New York.

Cairncross, F. (1995) "The death of distance," *The Economist*, *336*, No. 7934, 63.

Cairncross, F. (1997) *The Death of Distance: How the Communications Revolution Is Changing Our Lives*, Harvard Business School Press, Cambridge, MA.

Calabrese, F., Di Lorenzo, G., Liu, L., and Ratti, C. (2011) "Estimating origin-destination flows using mobile phone location data," *IEEE Pervasive Computing*, *10*, No. 4, 36–44.

Calvino, I. (1974) *Invisible Cities*, Secker and Warburg, London.

Campbell-Smith, D. (2011) *Masters of the Post: The Authorised History of the Royal Mail*, Allen Lane, Penguin Press, London.

Carey, H. C. (1858) *Principles of Social Science*, 3 Volumes, J. B. Lippincott & Co., Philadelphia, PA.

Carlson, W. E. (2017) *Computer Graphics and Computer Animation: A Retrospective Overview*, Ohio State University Press, Columbus, OH.

Carpenter, L. (1982) *Star Trek II: The Wrath of Kahn*, Pixar, http://www.historyofin
formation.com/detail.php?id=3141.

Carroll, J. D. (1955) "Spatial interaction and the urban-metropolitan regional description," *Papers and Proceedings of the Regional Science Association, 1*, No. 1, 59–73.

Carroll, J. D., and Jones, G. P. (1960) "Interpretation of desire line charts made on a Cartographatron," *Highway Research Board Bulletin, 253*, Highway Research Board, Washington, DC, 86–108, at https://trid.trb.org/view/120687.

Carroll, L. (1893) *The Complete Illustrated Works*, Gramercy Books, New York, latest edition 1982.

Castells, M. (1996) *The Rise of the Network Society: Economy, Society and Culture. Volume 1: The Information Age: Economy, Society and Culture*, Blackwell, Oxford.

Castells, M., and Hall, P. (1994) *Technopoles of the World: The Making of Twenty-First-Century Industrial Complexes*, Routledge, London.

Catlett, C. (2022) Urban Center for Computation and Data (UCCD), https://www
.youtube.com/watch?time_continue=16&v=BHrsllHJHeo&feature=emb_logo.

Cerra, J. F., Weld Muller, B., and Young, F. F. (2017) "A transformative outlook on the twenty-first century city: Patrick Geddes' Outlook Tower revisited," *Landscape and Urban Planning, 166*, 90–96.

Ceruzzi, P. E., Aspray, W., and Misa, T. J. (1998) *History of Modern Computing*, MIT Press, Cambridge, MA.

Chan, J. (2019) "All you need to know about the Chinese social media landscape in 2019," *Linkfluence*, https://www.linkfluence.com/blog/chinese-social
-media-landscape-2019.

Chandler, T. (1989) *Four Thousand Years of Urban Growth*, Edwin Mellen Press, New York.

Chapin, F. S. (1974) *Human Activity Patterns in the City: What People Do in Time and Space*, John Wiley, New York.

Chapin, F. S., and Stewart, P. (1953) "Population densities around the clock," in H. H. Mayer and C. F. Kohn (Editors), *Readings in Urban Geography*, 180–182, University of Chicago Press, Chicago, IL, reprinted 1959.

Chapin, F. S., and Weiss, S. F. (1968) "A probabilistic model for residential growth," *Transportation Research, 2*, No. 4, 375–390.

Charbit, Y. (2002) "The Platonic city: History and utopia," *Population, 57*, 207–235.

Chaudhuri, G., and Clarke, K. C. (2013) "The SLEUTH land use change model: A review," *The International Journal of Environmental Resources Research, 1*, 88–104.

Chen, K. (Editor) (1972) *Urban Dynamics: Extensions and Reflections*, San Francisco Press, San Francisco, CA.

Chernavina, K. (2019) "Top 20 Chinese social media sites of 2019," *Hi-Com*, December 11, https://www.hicom-asia.com/chinese-kol-top-10-social-media-platforms-they
-use/.

Cherry, G. E. (1979) "The town planning movement and the late Victorian city," *Transactions of the Institute of British Geographers*, 4, No. 2, 306–319.

Cheshire, J., and Batty, M. (2022) "The era of the megalopolis: how the world's cities are merging," *The Conversation*, November 22, https://theconversation.com/the-era-of -the-megalopolis-how-the-worlds-cities-are-merging-193424.

Childe, G. (1942) *What Happened in History*, Penguin Books, London.

Chrisman, N. (2006) *Charting the Unknown: How Computer Mapping at Harvard Became GIS*, ESRI Press, Redlands, CA.

Christensen, C. M., Raynor, M., and McDonald, R. (2015) "What is disruptive innovation?," *Harvard Business Review*, 93, No. 12, 44–53.

Churchey, C. (2016) "Stop virtual server sprawl," *IBM Systems Magazine*, December, http://archive.ibmsystemsmag.com/power/systems-management/virtualization /stop-server-sprawl/.

Churchman, C. W. (1967) "Editorial," *Management Science*, 14, No. 4, B141–B142.

Clarke, A. C. (1962) *Profiles of the Future: An Inquiry into the Limits of the Possible*, Weidenfeld and Nicolson, London.

Clarke, A. C. (1968) *2001: A Space Odyssey*, Penguin Putnam, New York.

Clarke, H. (2019) "Leonardo the city planner: Da Vinci's New Milan," *Engineering and Technology*, May 22, https://eandt.theiet.org/content/articles/2019/05/leonardo -the-city-planner-da-vinci-s-new-milan/.

Clarke, K. C. (2012) The SLEUTH master class, https://slideplayer.com/slide/6990931/.

Clement, J. (2019a) "Number of apps available in leading apps 2019," *Statista*, October 9, https://www.statista.com/statistics/276623/number-of-apps-available-in-leading -app-stores/.

Clement, J. (2019b) "Percentage of all global web pages served to mobile phones from 2009 to 2018," *Statista*, July 22, https://www.statista.com/statistics/241462/glo bal-mobile-phone-website-traffic-share/.

Çolak, S., Alexander, L., Alvim, B. G., Mehndiretta, S., Quiros, T. P., and Gonzalez, M. (2015) "Analyzing cell phone location data for urban travel: Current methods, limitations and opportunities," *Conference of European Statisticians: Workshop on Statistical Data Collection: Riding the Data Deluge*, 29 April–1 May, Washington, DC.

Coleman, B. (2011) *Hello Avatar. Rise of the Networked Generation*, MIT Press, Cambridge, MA.

Colombo, J. (2012) "The British 'Railway Mania' bubble," *The Bubble*, April 19, http://www.thebubblebubble.com/railway-mania/.

Copeland, J. (Editor) (2006) *Colossus: The Secrets of Bletchley Park's Code Breaking Computers*, Oxford University Press, Oxford.

Coppock, J. T., and Rhind, D. W. (1991) "The history of GIS," in D. J. Maguire, M. F. Goodchild, and D. W. Rhind (Editors), *Geographical Information Systems*, 21–43, Longman, London.

Corbusier, Le (1929) *The City of Tomorrow and Its Planning*, Dover Publications, New York.

Cortada, J. W. (2019) *IBM: The Rise and Fall and Reinvention of a Global Icon*, MIT Press, Cambridge, MA.

Crawford, K. (2021) *Atlas of AI*, Yale University Press, New Haven, CT.

Cripps, E., and Foot, D. (1968) "Strategy for the South East—5: Evaluating alternative strategies," *Official Architecture and Planning*, *31*, No. 7, 928–941.

Cronon, W. (1992) *Nature's Metropolis: Chicago and the Great West*, W. W. Norton, New York.

Crooks, A., Malleson, N., Manley, E., and Heppenstall, A. (2019) *Agent-Based Modelling and Geographical Information Systems: A Practical Primer*, Sage, London.

Dahl, R. (1961) *Who Governs?: Democracy and Power in the American City*, Yale University Press, New Haven, CT.

Dawkins, O., Dennett, A., and Hudson-Smith, A. (2019) "Living with a digital twin: Operational management and engagement using IoT and mixed realities at UCL's Here East Campus on the Queen Elizabeth Olympic Park," Centre for Advanced Spatial Analysis (CASA), University College London, London.

Dechow, D. R., and Struppa, D. C. (Editors) (2015) *Intertwingled: The Work and Influence of Ted Nelson*, Springer, New York.

DeGennaro, T. (2019) "10 most popular social media sites in China," *Dragon Social*, January 31, https://www.dragonsocial.net/blog/social-media-in-china/.

Delony, D. (2013) "The laws of computing," *Technopedia*, November 15, https://www.techopedia.com/2/28205/trends/the-laws-of-computing.

Dempsey, B. (1960) *The Frontier Wage*, Loyola University Press, Chicago, IL.

Dijkstra, E. (1959) "A note on two problems in connexion with graphs," *Numerische Mathematik*, *1*, 269–271.

Dobrov, G. M. (1979) "The strategy for organized technology in the light of hard-, soft-, and org-ware interaction," *Long Range Planning*, *12*, No. 4, 79–90.

Dodge, M. (2002) "Explorations in AlphaWorld: The geography of 3D virtual worlds on the Internet," in Peter Fisher and David Unwin (Editors), *Virtual Reality in Geography*, 305–331, Taylor and Francis, London.

Dodge, M., and Kitchin, R. (2000) *Mapping Cyberspace*, Routledge, London.

Dodge, M., and Kitchin, R. (2001) *Atlas of Cyberspace*, Addison-Wesley, London.

Duany, A., Plater-Zyberk, E., and Speck, J. (2000) *Suburban Nation: The Rise of Sprawl and the Decline of the American Dream*, North Point Press, New York.

Dutton, W., Blumler, J., and Kraemer, K. (Editors) (1987) *Wired Cities: Shaping the Future of Communications*, G. K. Hall, New York.

Dyson, G. (2012) *Turing's Cathedral: The Origins of the Digital Universe*, Pantheon Books, New York.

Eames, C., and Eames, R. (1973) *A Computer Perspective: Background to the Computer Age*, Harvard University Press, Cambridge, MA, New Edition, 1990.

Earls, A. R. (2004) *Digital Equipment Corporation (MA) (Images of America)*, Portsmouth, NH.

Easton, R. (2002) *Ideal Cities: Utopianism and the (Un)Built Environment*, Thames and Hudson, London.

Eccles, R., Edmondson, A. G. C., Thyne, S., and Zuzul, T. (2013) "Living PlanIT," Harvard Business School Case 410–081, Revised November 2013, https://www.hbs.edu/faculty/Pages/item.aspx?num=38370.

Egen, S. P. (2005) "The history of avatars," *iMedia*, June 22, https://www.imediaconnection.com/articles/ported-articles/red-dot-articles/2005/jun/the-history-of-avatars/.

eMarketer Report (2022) "US time spent with media 2022 update—pivotal moments for TV, subscription OTT, digital audio, and social media," April 20, 2022, https://www.insiderintelligence.com/insights/us-time-spent-with-media/

Emerson, R. W. (1844) "The young American," A Lecture read before the Mercantile Library Association, Boston, February 7, 1844, in his *The Complete Works*, Vol. I, Nature, Addresses and Lectures, 1904.

Epstein, J. M., and Axtell, R. L. (1996) *Growing Artificial Societies: Social Science from the Bottom Up*, Bradford Books, Cambridge, MA.

Essinger, J. (2013) *Ada's Algorithm: How Lord Byron's Daughter Ada Lovelace Launched the Digital Age through the Poetry of Numbers*, Gibson Square, London.

Evans, J., and Fildes, N. (2019) "BT agrees £210m sale of historic London site: Private equity buys Newgate St headquarters that was home to GPO's telegraph office," *The Financial Times*, July 17, https://www.ft.com/content/5dc97e0a-a880-11e9-b6ee-3cdf3174eb89.

Evans, S., Liddiard, R., and Steadman, P. (2017) "3DStock: A new kind of three-dimensional model of the building stock of England and Wales, for use in energy analysis," *Environment and Planning B: Planning and Design*, *44*, No. 2, 227–255.

Evans, S., Liddiard, R., and Steadman, P. (2019) "Modelling a whole building stock: Domestic, non-domestic and mixed use," *Building Research and Information*, *47*, No. 2, 156–172.

Feigenbaum, E. A. (1983) *The Fifth Generation: Artificial Intelligence and Japan's Computer Challenge to the World*, Addison Wesley Longman, Reading, MA.

Feldman, S. (2011) "The history of web mapping," https://www.slideshare.net/stevenfeldman/history-of-web-mapping.

Ferry, G. (2003) *A Computer Called Leo: Lyons Teashops and the World's First Office Computer*, Fourth Estate, London.

Feynman, R. (1964) *The Feynman Lectures on Physics: Volume 2: Mainly Electromagnetism and Matter*, 1–6, Addison-Wesley, Reading, MA.

Finley, K. (2014) "Tech time warp of the week: Watch Andy Warhol paint Debbie Harry on an Amiga 1000 Computer," *Wired Magazine*, March 10, https://www.wired.com/2014/10/warhol-blondie-amiga/.

Fischer, C. S. (1992) *America Calling: A Social History of the Telephone to 1940*, University of California Press, Berkeley.

Fischer, C. S., and Carroll, G. (1988) "Telephone and automobile diffusion in the United States, 1902–1937," *American Journal of Sociology*, *93*, No. 5, 1153–1178.

Fleming, N. (2004) "Ten years of Cyberia," *The Telegraph*, September 1, https://www .telegraph.co.uk/news/uknews/1470674/Ten-years-of-Cyberia.html.

Fogarty, K. (2012) "Tech predictions gone wrong, A romp through some of the worst tech predictions from people who should have known better," https://www.com puterworld.com/article/2492617/tech-predictions-gone-wrong.html.

Forbes, S. (1996) "All about Eva," *Wired Magazine*, Issue 2.04, April.

Ford, M. (2015) *The Rise of the Robots: Technology and the Threat of Mass Unemployment*, Basic Books, New York.

Forester, T. (1988) "The myth of the electronic cottage," *Futures*, *3*, 227–240, June 20.

Forrester, J. W. (1961) *Industrial Dynamics*, MIT Press, Cambridge, MA.

Forrester, J. W. (1969) *Urban Dynamics*, MIT Press, Cambridge, MA.

Forrester, J. W. (1971) *World Dynamics*, Wright-Allen Press, Cambridge, MA.

Forster, E. M. (1909) *The Machine Stops*, Penguin Classics, London, reprinted 2019.

Fox, P. (2022) "Transporting bits over wires," https://www.khanacademy.org/com puting/apcomputer-science-principles/the-internet/wires-wifi-physical-network -connections/a/transporting-bits-over-wires/.

Francica, J. (2018) "Fifty years of commercial GIS—Part 1: 1969–1994," December 31, https://www.directionsmag.com/article/8410.

Frankenfield, J. (2019) "Bitcoin," *Investopedia*, October 26, https://www.investope dia.com/terms/b/bitcoin.asp.

Friedman, T. F. (2016) *Thank You for Being Late: An Optimist's Guide to Thriving in the Age of Accelerations*, Penguin Books, New York.

Gan, V., and CityLab (2015) "What telecommuting looked like in 1973: A vision of remote work from before the personal computer," *The Atlantic*, December 3, https://www.theatlantic.com/technology/archive/2015/12/what-telecommuting -looked-like-in-1973/418473/.

Ganapati, S. (2020) "Using mobile apps in government," *IBM Center for the Business of Government*, http://www.businessofgovernment.org/sites/default/files/Using%20 Mobile%20Apps%20in%20Government.pdf.

Gardner, M. (1970) "Mathematical games: The fantastic combinations of John Conway's new solitaire game 'Life,'" *Scientific American*, *223*, 120–123.

Gardner, M. (1985) *Wheels, Life and Other Mathematical Amusements*, W. H. Freeman, San Francisco, CA.

Garreau, J. (1991) *Edge Cities: Life on the New Frontier*, Anchor Books, Doubleday, New York.

Gayford, M. (2010) "David Hockney's iPad art," *The Telegraph*, October 20, https:// www.telegraph.co.uk/culture/art/art-features/8066839/David-Hockneys-iPad-art .html.

Geddes, P. G. (1906) "A suggested plan for a civic museum and its associated studies," *Sociological Papers*, *3*, 197–240.

Geddes, P. G. (1915) *Cities in Evolution: An Introduction to the Town Planning Movement and to the Study of Civics*, Williams and Norgate, London.

Geere, D. (2011) "How the first cable was laid across the Atlantic," *Wired Magazine*, January 18, https://www.wired.co.uk/article/transatlantic-cables.

Gertner, J. (2013) *The Idea Factory: Bell Labs and the Great Age of American Innovation*, Allen Lane, Penguin Press, New York.

Gholipour, B. (2018) "What is a normal heart rate?," *Live Science*, January 12, https://www.livescience.com/42081-normal-heart-rate.html.

Gibson, D., Kozmetsky, G., and Smilor, R. (1992) *The Technopolis Phenomenon: Smart Cities, Fast Systems, Global Networks*, Rowman and Littlefield, Lanham, MD.

Gibson, W. (1984) *Neuromancer*, Victor Gollanz, London.

Gilder, G. (1993) "Telecosm: Metcalfe's law and legacy," *Forbes*, September 13.

Gilder, G. (2000) *Telecosm: The World After Bandwidth Abundance*, Simon & Schuster, New York.

Gillies, J., and Cailliau, R. (2000) *How the Web Was Born: The Story of the World Wide Web*, Oxford University Press, Oxford.

Glaeser, E. (2013) *Triumph of the City*, Macmillan, London.

Glaeser, E., and Cutler, D. (2021) *Survival of the City: Living and Thriving in an Age of Isolation*, Basic Books, London.

Gleick, J. (2000) *Faster: The Acceleration of Just about Everything*, Random House, New York.

Gonzalez, M., Hidalgo, C., and Barabasi, A.-L. (2008) "Understanding individual human mobility patterns," *Nature*, *453*, 779–782.

Goodchild, M. (2007) "Citizens as sensors: The world of volunteered geography," *GeoJournal*, *69*, No. 4, 211–221.

Goodchild, M. F. (2018) "Reimagining the history of GIS," *Annals of GIS*, *24*, No. 1, 1–8.

Goodwin, T. (2015) "The battle is for the customer interface," *TechCrunch*, March 3, https://techcrunch.com/2015/03/03/in-the-age-of-disintermediation-the-battle-is-all-for-the-customer-interface/.

Graham-Cumming, J. (2012) "The great railway caper: Big data in 1955," https://www.youtube.com/watch?v=pcBJfkE5UwU.

Gray, J. (2022) "Sixteen laws of computing," https://jimgray.azurewebsites.net.

Gray, S., O'Brien O., and Hügel, S. (2016) "Collecting and visualizing real-time urban data through City Dashboards," *Built Environment*, *42*, No. 3, 498–509.

Green, J. (2018) "Charles-Joseph Minard: A legacy of beautiful data-based maps," in *The Dirt: Uniting the Built and Natural Environments*, October 3, https://dirt.asla.org/2018/10/03/charles-joseph-minard-a-legacy-of-beautiful-data-based-maps/.

Green, M. (2018) "The fascinating history of Fleet Street," *The Daily Telegraph*, May 21, https://www.telegraph.co.uk/travel/destinations/europe/united-kingdom/england /london/articles/the-history-of-fleet-street-and-british-newspaper-industry/.

Greenberg, J. (2014) *Gordon Welchman: Bletchley Park's Architect of Ultra Intelligence*, Frontline Books, London.

Greening-Jackson, T. (2012) "LEO I and the BR job," https://blog.jgc.org/2012/10/the -great-railway-caper-big-data-in-1955.html.

Gruen, V. (1964) *The Heart of Our Cities: The Urban Crisis. Diagnosis and Cure*, Thames and Hudson, London.

Gunawardene, N. (2003) "Humanity will survive information deluge," *OneWorld South Asia*, December 5, https://bazaarmodel.net/phorum/read.php?1,462.

Hafner, K., and Lyon, M. (1996) *Where Wizards Stay Up Late: The Origins of The Internet*. Touchstone, Simon & Schuster, New York.

Haklay, M., and Weber, P. (2008) "OpenStreetMap: User-generated street maps," *Pervasive Computing*, October–December, 12–17.

Hall, J., Palsson, C., and Price, J. (2018) "Is Uber a substitute or complement for public transit?," *Journal of Urban Economics*, *108*, 36–50.

Hall, M., and Barry, J. (1990) *Sunburst: The Ascent of Sun Microsystems*, Contemporary Books, Bel Air, CA.

Hall, P. G. (1981) "The geography of the Fifth Kondratieff Cycle," *New Society*, *26*, March, 535–537.

Hall, P. G. (1998) *Cities in Civilization: Culture, Innovation and Urban Order*, Weidenfeld and Nicolson, London.

Hall, P. G., and Pain, K. (Editors) (2006) *The Polycentric Metropolis: Learning from Mega-city Regions in Europe*, Earthscan, London.

Han, B., Cook, P., and Baldwin, T. (2014) "Text-Based Twitter user geolocation prediction," *Journal of Artificial Intelligence Research*, *49*, 451–500, https://people.eng.uni melb.edu.au/paulcook/live-4200-7781-jair.pdf.

Hansell, S. (2008) "Zuckerberg's law of information sharing," *The New York Times, Bits*, November 6, https://bits.blogs.nytimes.com/2008/11/06/zuckerbergs -law-of-information-sharing/.

Hansen, W. G. (1959a) "How accessibility shapes land use," *Journal of the American Institute of Planning*, *25*, No. 2, 73–76.

Hansen, W. G. (1959b) "Accessibility and residential growth," MCP degree thesis, MIT, Cambridge, MA.

Harari, Y. (2015) *Sapiens: A Brief History of Humankind*, Vintage, London.

Harari, Y. (2017) *Homo Deus: A Brief History of Tomorrow*, Vintage, London.

Hardy, D. (1991) *From Garden Cities to New Towns: Campaigning for Town and Country Planning 1899–1946*, Spon, London.

Harrington, J. L. (2009) *Relational Database Design, Clearly Explained*, 3rd Edition, Morgan Kaufmann, New York.

Harris, B. (1963) "Linear programming and the projection of land uses," PJ Paper 20, The PJTS Commission, Philadelphia, PA.

Harris, B. (1965) "New tools for planning," *Journal of the American Institute of Planners*, *31*, 90–95.

Harris, B. (1968) "Quantitative models of urban development: Their role in metropolitan policy making," in Harvey S. Perloff and Lowdon Wingo, Jr. (Editors), *Issues in Urban Economics*, 363–410, Resources for the Future, Johns Hopkins University Press, Baltimore, MD.

Harris, R. (1995) *Enigma*, Random House, London.

Harvey, D. (1990) "Between space and time: Reflections on the geographical imagination," *Annals of the Association of American Geographers*, *80*, No. 3, 418–434.

Heathman, A. (2019) "iPhone through the ages: How Apple smartphones have evolved from 1st generation to the new iPhone 11," *Evening Standard*, September 10, https://www.standard.co.uk/tech/apple-iphone-2019-through-the-ages-iphone-to -iphone-xs-a4232841.html.

Heckbert, P. (1999) "History of computer graphics," https://www.cs.cmu.edu/~ph/nyit /masson/history.htm.

Hepner, G. (2022) "An art history of Andy Warhol," https://guyhepner.com/an-art -history-of-andy-warhol/.

Heppenstall, A., Crooks, A., Malleson, N., Manley, E., Ge, J., and Batty, M. (2021) "Agent-based models for geographical systems: A review," *Geographical Analysis*, *53*, 76–91.

Herbert, J. D., and Stevens, B. H. (1960) "A model for the distribution of residential activity in urban areas," *Journal of Regional Science*, *2*, No. 2, 21–36.

Herrmann, J. (2015) "The failure of London's Tech City," *Spectator Life*, September 26, https://life.spectator.co.uk/articles/the-failure-of-londons-tech-city/.

Hevner, A. R., and Berndt, D. J. (2000) "Eras of business computing," *Advances in Computers*, *52*, 1–90.

Higgins, C. (2017) "Seeing 'sights' that don't exist: Karl Marx in the British Museum Round Reading Room," *Library & Information History*, *33*, No. 2, 81–96.

Hiltzik, M. A. (1999) *Dealers of Lightning: Xerox PARC and the Dawn of the Computer Age*, Harper Collins, New York.

Hodges, A. (1983) *Alan Turing: The Enigma*, Hutchinson Publishing Co., London.

Hodges, A. (2022) "The Alan Turing Internet Scrapbook," https://www.turing.org .uk/scrapbook/oracle.html.

Hoffmann, J. (1997) "The history of the browser wars: When Netscape met Microsoft," *The History of the Web*, October 1, https://thehistoryoftheweb.com/browser-wars/.

Holland, J. (2015) *The Times History of Britain's Railways*, Times Books, London.

Horni, A., Nagel, K., and Axhausen, K. (Editors) (2017) *The Multi-Agent Transport Simulation MATSim*, Ubiquity Press, London.

Howard, E. (1898) *To-Morrow: A Peaceful Path to Reform*, Routledge, London, revised 2009.

Hudson, P. (2022) Colouring London, CASA, UCL, http://colouringlondon.org/.

Hudson-Smith, A., Batty, M., Crooks, A., and Milton, R. (2009) "Mapping for the masses: Accessing Web 2.0 through crowdsourcing," *Social Science Computer Review*, *27*, No. 4, 524–538.

Hudson-Smith, A., Milton, R., Dearden, J., and Batty, M. (2007) "Virtual cities: Digital mirrors into a recursive world," Working Paper 125, Centre for Advanced Spatial Analysis, University College London, London.

Hudson-Smith, A., Milton, R., Dearden, J., and Batty, M. (2008) "The neogeography of virtual cities: Digital mirrors into a recursive world," in Marcus Foth (Editor), *Handbook of Research on Urban Informatics: The Practice and Promise of the Real-Time City*, 270–290, Information Science Reference, IGI Global, Hershey, PA.

Ingram, G. K., Kain, J., and Royce Ginn, J. (1972) *The Detroit Prototype of the NBER Urban Simulation Model*, National Bureau of Economic Research, Columbia University Press, New York.

Iqbal, M. (2022) "App download and usage statistics," *BusinessofApps*, November 19, 2019, https://www.businessofapps.com/data/app-statistics/.

Isaacson, W. (2014) *The Innovators: How a Group of Hackers, Geniuses and Geeks Created the Digital Revolution*, Simon & Schuster, London.

Isaacson, W. (2015) *Steve Jobs: The Exclusive Biography*, Little, Brown Book Group, New York.

Isaacson, W. (2018) *Leonardo Da Vinci*, Simon & Schuster, New York.

Isard, W. (1956) *Location and Space-economy: A General Theory Relating to Industrial Location, Market Areas, Land Use, Trade, and Urban Structure*, MIT Press, Cambridge, MA.

Jackson, A. (1969) *London's Termini*, David & Charles, London, new revised edition, 1984.

Jacobs, J. (1961) *The Death and Life of Great American Cities*, Random House, New York.

Jacobs, J. (1969) *The Economy of Cities*, Random House, New York.

Jang, M., and Suh, S-T (2010) "U-City: New trends of urban planning in Korea based on pervasive and ubiquitous geotechnology and geoinformation," in David Taniar, Osvaldo Gervasi, Beniamino Murgante, Eric Pardede, and Bernady Apduhan (Editors), *Computational Science and Its Applications—ICCSA 2010. ICCSA Lecture Notes in Computer Science*, 6016, 262–270, Springer, Berlin, Heidelberg, Germany.

Jankel, A., and Morton, R. (1984) *Creative Computer Graphics*, Cambridge University Press, Cambridge.

Jerde, S. (2019) "The colossal amount of time spent consuming media may finally be flatlining, according to Nielsen Report," *Adweek40*, March 19, https://www.adweek

.com/tv-video/the-colossal-amount-of-time-spent-consuming-media-may-finally-be
-flatlining-according-to-nielsen-report/.

Johnson, J., Nowak, A., Ormerod, P., Rosewell, B., and Zhang, Y.-C. (Editors) (2017) *Non-Equilibrium Social Science and Policy: Introduction and Essays on New and Changing Paradigms in Socio-Economic Thinking*, Springer, New York.

Johnson, S. (2007) *The Ghost Map: The Story of London's Most Terrifying Epidemic—And How It Changed Science, Cities, and the Modern World*, Riverhead Books, New York.

Judge, B. (2022) "11 March 1702—the world's first daily newspaper published," https://moneyweek.com/383504/11-march-1702-elizabeth-mallet-daily-courant -first-daily-newspaper/.

Kadanoff, L. (1971) "From simulation model to public policy: An examination of Forrester's 'urban dynamics,'" *Simulation, 16*, 261–268.

Kadanoff, L., Voss, J. R., and Bouknight, W. J. (1970) "Computer display and analysis of urban information through time and space," *Technological Forecasting and Social Change, 2*, No. 1, 77–103.

Kagan, J. (2019) "Financial technology—Fintech," *Investopedia*, June 25, https:// www.investopedia.com/terms/f/fintech.asp.

Karlgaard, R., and Malone, M. (1995) "City vs. country: Tom Peters and George Gilder debate the impact of technology on location," *Forbes ASAP Technology Issue*, February 27, *155*, No. 5, 56–61.

Kasik, D., and Senesac, C. (2013) "Visualization: Past, present, and future at Boeing," http://on-demand.gputechconf.com/gtc/2013/presentations/S3440-Boeing -Visualization-Past-Present-Future.pdf.

Keegan, V. (2017) *Lost London 8: Hidden History of Holborn Viaduct*, https://www.on london.co.uk/vic-keegans-lost-london-8-hidden-history-of-holborn-viaduct/.

Kelly, K. (1998) *New Rules for the New Economy*, Viking Penguin, New York.

Kelly, K. (2016) *The Inevitable: Understanding the 12 Technological Forces That Will Shape Our Future*, Penguin Viking Books, New York.

Kemeny, J. G., and Kurtz, T. E. (1964) *Basic: A Manual for BASIC, the Elementary Algebraic Language Designed for Use with the Dartmouth Time Sharing System*, Dartmouth College, Hanover, NH.

Kennedy, C. A., et al. (2015) "Energy and material flows of megacities," *Proceedings of the National Academy of Sciences, 112*, No. 19, 5985–5990.

Kenyon, K. M. (2022) "Jericho Town, West Bank," *Encyclopaedia Britannica*, https:// www.britannica.com/place/Jericho-West-Bank.

Kerrigan, D. (2017) *Life as a Passenger: How Driverless Cars Will Change the World*, CreateSpace, New York.

Khurana, R. (2017) "Why is the UK so hostile to innovation?," *CAPX*, April 3, https://capx.co/why-is-the-uk-so-hostile-to-innovation/.

Kidder, T. (1981) *The Soul of a New Machine*, Little, Brown and Co., New York.

Kim, L. (2022) "What is PPC? Learn the basics of pay-per-click (PPC) marketing," https://www.wordstream.com/ppc.

King, J. L., and Kraemer, K. L. (Editors) (1978) *Computers and Local Government: Volume 2*, Praeger, New York.

Kirkpatrick, D. (2010) *The Facebook Effect: The Real Inside Story of Mark Zuckerberg and the World's Fastest Growing Company*, Simon & Schuster, New York.

Kirstein, P. (2022) "Early experiences with the ARPANET and Internet in the UK," http://nrg.cs.ucl.ac.uk/internet-history.html.

Kitchin, R. (1997) *Cyberspace: The World in the Wires*, Wiley-Blackwell, London.

Kitchin, R. (2014) *The Data Revolution: Big Data, Open Data, Data Infrastructures, and Their Consequences*, Sage, London.

Kitchin, R., and Dodge, M. (2014) *Code/Space: Software and Everyday Life*, MIT Press, Cambridge, MA.

Kitchin, R., Lauriault, T., and McArdle, G. (2015) "Knowing and governing cities through urban indicators, city benchmarking and real-time dashboards," *Regional Studies, Regional Science, 2*, 1–28.

Kitchin, R., Lauriault, T., and McArdle, G. (Editors) (2017) *Data and the City*, Routledge, London.

Kitson, K., and Michie, J. (2019) "The deindustrial revolution: The rise and fall of UK manufacturing, 1870–2010," Working Paper No. 459, Centre for Business Research, University of Cambridge, Cambridge, UK.

Kleinrock, L. (1964) *Communication Nets: Stochastic Message Flow and Delay*. McGraw-Hill, New York.

Klemmers, H. (2013) "A city that refused to die, a study of blitzed and fire bombed Plymouth in 1941"; also "The making of a modern city, Plymouth City Museum and Art Gallery," April 27–June 29, 2013, https://plymhearts.org/.

Klosterman, R. E. (1994) "Large-scale urban models: Retrospect and prospect," *Journal of the American Planning Association, 60*, No. 1, 3–6.

Kondratieff, N. D. (1925) *The Major Economic Cycles*, republished as *The Long Wave Cycle*, translation by G. Daniels, E. P. Dutton, New York, 1984.

Kondratieff, N. D. (1926, 1935) "The long waves in economic life," *The Review of Economics and Statistics, 17*, 105–115.

Koolhaas, R., Boeri, S., Kwinter, S., Tazi, N., and Obrist, H.-U. (2001) *Mutations*, ActarD, New York.

Korzybski, A. (1931) "A non-Aristotelian system and its necessity for rigor in mathematics and physics," *American Association for the Advancement of Science*, New Orleans, LA, reprinted in *Science and Sanity*, International Non-Aristotelian Library, Oxford, 747–761, 1933.

KPMG (2015) "Connected and autonomous vehicles—The UK economic opportunity," https://www.smmt.co.uk/wp-content/uploads/sites/2/Connected-and-Autonomous-Vehicles-–-The-UK-Economic-Opportu. . . . pdf.

Kraemer, K. L. (1977) *Computers and Local Government*, Praeger, New York.

Kraemer, K. L., and Perry, J. L. (1980) *Technological Innovation in American Local Governments: The Case of Computing*, Pergamon Press, New York.

Krugman, P. (1996) *The Self-Organizing Economy*, Blackwell, Oxford.

Kuenne, R. E. (1954) "Walras, Leontief and the interdependence of economic activities," *Quarterly Journal of Economics, 68*, 323–354.

Kuhn, T. S. (1962) *The Structure of Scientific Revolutions*, University of Chicago Press, Chicago, IL.

Kurzweil, R. (2000) *The Age of Spiritual Machines: When Computers Exceed Human Intelligence*, Penguin Books, New York.

Kurzweil, R. (2006) *The Singularity Is Near: When Humans Transcend Biology*, Penguin Books, New York.

Kurzweil, R, (2022) Accelerating Intelligence weblog, https://www.kurzweilai.net/the -law-of-accelerating-returns.

Lahiri, S. (2000) "Professor Wassily W. Leontieff 1905–1999," *The Economic Journal, 110*, F695–F707.

Land, F. (2012) "Remembering LEO," London School of Economics and Political Science, January, https://www.researchgate.net/publication/292055297.

Landau, V. (2018) "How Douglas Engelbart invented the future," *Smithsonian Magazine*, January, https://www.smithsonianmag.com/innovation/douglas-engelbart-invented-future-180967498/.

Landow, G. P. (2014) "Railway mania," *The Victorian Web*, March, http://www .victorianweb.org/technology/railways/fraud.html.

Lanier, J. (2017) *Dawn of the New Everything: A Journey Through Virtual Reality*, Picador, London.

Laposky, B. (1976) "Oscillons: Electronic abstractions," in Ruth Leavitt (Editor), *Artist and Computer*, 21–22, Creative Computing Press, Morristown, NJ.

Larijani, A. N., Olteanu-Raimond, A.-M., Perret, J., Brédif, M., and Ziemlicki, C. (2015) "Investigating the mobile phone data to estimate the origin destination flow and analysis; case study: Paris region," *Transportation Research Procedia, 6*, 64–78.

Lavington, S. (1986) *Early British Computers*, Manchester University Press, Manchester, UK.

Lavington, S. (2019) *Early Computing in Britain: Ferranti Ltd. and Government Funding, 1948–1958*, Springer, Berlin, Germany.

Lean, T. (2016) "Prestel: The British Internet that never was," August 23, https:// www.historytoday.com/history-matters/prestel-british-internet-never-was.

Lee, C. (2008) "Where have all the Gophers gone? Why the web beat Gopher in the battle for protocol mind share," https://ils.unc.edu/callee/gopherpaper.htm.

Lee, D. B. (1973) "Requiem for large-scale models," *Journal of the American Institute of Planners, 39*, 163–178.

Lee, D. B. (1994) "Retrospective on large-scale urban models," *Journal of the American Planning Association*, *60*, No. 1, 35–40.

Lee, K.-F. (2018) *AI Superpowers: China, Silicon Valley, and the New World Order*, Houghton Mifflin Harcourt, Boston, MA.

Lee, T. B. (2014) "Forty maps that explain the internet," *Vox*, June 2, https://www.vox.com/a/internet-maps.

Leonhardt, D. (2000) "John Tukey, 85, statistician; coined the word 'software,'" *The New York Times*, July 28, Section A.

Levitt, T. (1983) "The globalization of markets," *Harvard Business Review*, May, https://hbr.org/1983/05/the-globalization-of-markets.

Levy, S. (2010) *Hackers: Heroes of the Computer Revolution*, O'Reilly Media, Sebastopol, CA.

Levy, S. (2020) *Facebook: The Inside Story*, Penguin Books, New York.

Lewis, M. (n.d.) "Killer apps explained—history, examples, impacts & future applications," *Money Crashers*, https://www.moneycrashers.com/killer-apps-explained-history-examples-impacts-future/.

Lin, H., and Batty, M. (Editors) (2009) *Virtual Geographic Environments*, Science Press, Beijing, China.

Lin, H., Chen, M., Lu, L., Zhu, Q., Gong, J., You, X., Wen, Y., Xu, B., and Hu, M. (2013) "Virtual geographic environments (VGEs): A new generation of geographic analysis tools," *Earth-Science Reviews*, *126*, 74–84.

Liu, Y., Batty, M., Wang, S., and Corcoran, J. (2012) "Modelling urban change with cellular automata: Contemporary issues and future research directions," *Progress in Human Geography*, *45*, 3–24.

Longley, P. A., Goodchild, M. F., Maguire, D. J., and Rhind, D. W. (2015) *Geographic Information Science and Systems*, 4th Edition, John Wiley, New York.

Lowry, I. S. (1963) "Location parameters in the Pittsburgh model," *Papers and Proceedings of the Regional Science Association*, *11*, 145–165.

Lowry, I. S. (1964) *A Model of Metropolis*, Memorandum RM-4035-RC, The RAND Corporation, Santa Monica, CA.

Luebke, D., Reddy, M., Cohen, J. D., Varshney, A., Watson, B., and Huebner, R. (2002) *Level of Detail for 3D Graphics*, Morgan Kaufmann, San Francisco, CA.

McAfee, A., and Brynjolfsson, E. (2017) *Machine, Platform, Crowd: Harnessing Our Digital Future*, W. W. Norton, New York.

McBride, S. (2019) "Uber's nightmare has just begun," *Forbes*, September 4, https://www.forbes.com/sites/stephenmcbride1/2019/09/04/ubers-nightmare-has-just-started/#42be598b7e03.

McDonald, J. (2007) "William Alonso, Richard Muth, resources for the future, and the founding of urban economics," *Journal of the History of Economic Thought*, *29*, 67–84.

McEwan, I. (2019) *Machines Like Me, And People Like You*, Jonathan Cape, London.

McGrath, R. G. (2011) "The world is more complex than it used to be," *Harvard Business Review*, August 31, https://hbr.org/2011/08/the-world-really-is-more-compl.

McHarg, I. (1967) *Design with Nature*, Doubleday, Garden City, NY.

MacKay, J. (2019) '"Here's how much you use your phone during the workday,"' *Rescue Time Blog*, March 21, https://blog.rescuetime.com/screen-time-stats-2018/.

McKay, S. (2013) *The Lost World of Bletchley Park: The Illustrated History of the Wartime Codebreaking Centre*, Aurum Press, London.

McKenzie, R. (1924) "The ecological approach to the study of the human community," *American Journal of Sociology*, *30*, No. 3, 287–301.

McLoughlin, J. B. (1973) *Control and Urban Planning*, Faber and Faber, London.

McLuhan Galaxy (2011) "Any Warhol and Marshall McLuhan: The artist and the visionary," https://mcluhangalaxy.wordpress.com/2011/12/08/andy-wharhol-marshall-mcluhan-the-artist-the-visionary/.

McNish, J., and Silcoff, S. (2015) *Losing the Signal: The Untold Story Behind the Extraordinary Rise and Spectacular Fall of BlackBerry*, Flatiron Books, New York.

Macrae, N. (1992) *John Von Neumann: The Scientific Genius Who Pioneered the Modern Computer, Game Theory, Nuclear Deterrence, and Much More*, Pantheon Books, New York.

Madhani, A. (2016) "Chicago begins building 'fitness tracker' to check its vitals," *USA Today*, https://www.usatoday.com/story/news/2016/08/29/chicago-begins-building-fitness-tracker-check-its-vitals/89434620/.

Mahon, B. (2004) *The Man Who Changed Everything: The Life of James Clerk Maxwell*, John Wiley, Chichester, UK.

Mailland, J., and Driscoll, K. (2017) *Minitel: Welcome to the Internet*, MIT Press, Cambridge, MA.

Malleson, N., Heppenstall, A., and Crooks, A. (2018) "Place-based simulation modelling: Agent-based modelling and virtual environments," *Oxford Research Encyclopaedia of Criminology and Criminal Justice*, Oxford University Press, Oxford.

Marchant, J. (2016) "A journey to the oldest cave paintings in the world," *Smithsonian Magazine*, December, https://www.smithsonianmag.com/history/journey-oldest-cave-paintings-world-180957685/.

Marchetti, C., and Ausubel, J. (2012) "Quantitative dynamics of human empires," *International Journal of Anthropology*, *27*, No. 1–2, 1–62, http://phe.rockefeller.edu/docs/empires_booklet/.

Markoff, J. (2022) *Whole Earth: The Many Lives of Stewart Brand*, Penguin Press, New York.

Marsh, A. (2018) "Crumbling Britain: The slow death of the high street," *New Statesman*, July 11, https://www.newstatesman.com/politics/uk/2018/07/crumbling-britain-slow-death-high-street.

Marshall, A. (1890) *Principles of Economics*, 8th Edition, Macmillan, London.

Marshall, S. (2004) *Streets and Patterns: The Structure of Urban Geometry*, Routledge, London.

Martin, J. (1978) *The Wired Society: A Challenge for Tomorrow*, Prentice-Hall, Englewood Cliffs, NJ.

Martin, J., and Norman, A. (1970) *The Computerized Society*, Prentice-Hall, Englewood Cliffs, NJ.

Marvin, S., and Luque-Ayala, A. (2017) "Urban operating systems: Diagramming the city," *International Journal of Urban and Regional Research*, *41*, No. 1, 84–103.

Marx, K. (1857) "The fragment on machines," *Grundrisse*, 690–712, http://thenew objectivity.com/pdf/marx.pdf.

Marx, Karl (1867). *Das Kapital: Kritik der politischen Oekonomie*. Vol. 1: *Der Produktionsprozess des Kapitals*, Verlag von Otto Meissner, Hamburg, DE.

Marx, K. (1973) *Grundrisse, Foundations of the Critique of Political Economy, 1861*, Penguin Books in association with New Left Review, London.

Marx, K., and Engels, F. (1848) *Manifesto of the Communist Party*, February 1848, in Karl Marx, *Selected Works, Vol. 1*, Progress Publishers, Moscow.

Mattern, S. (2021) *A City Is Not a Computer: Other Urban Intelligences*, Princeton University Press, Princeton, NJ.

Maxwell, J. C. (1870) "On hills and dales," *The London, Edinburgh, and Dublin Philosophical Magazine and Journal of Science*, Series 4, *40*, 421–427.

Mayer-Schönberger, V., and Cukier, K. (2013) *Big Data: A Revolution That Will Transform How We Live, Work, and Think*, John Murray, London.

Meadows, D. H., Randers, J., Meadows, D. L., and Behrens, W. W. (1972) "The limits to growth: A report for the Club of Rome's project on the predicament of mankind," Earth Island, London.

Menn, J. (2003) *All the Rave: The Rise and Fall of Shawn Fanning's Napster*, Crown Business, New York.

Merchant, B. (2017) *The One Device: The Secret History of the iPhone*, Bantam Press, New York.

Messerschmitt, D. G. (2022) "The future of computer-telecommunications integration," Department of Electrical Engineering and Computer Sciences, University of California, Berkeley, CA, https://pdfs.semanticscholar.org/c6b0/dd0532cfa194f4581 ca3149a279b0bd7e88b.pdf.

Metcalfe, R. M. (2007) "It's all in your head," *Forbes*, April 20, https://www.forbes .com/forbes/2007/0507/052.html#504565b047d3.

Metz, C. (2012) "How the Queen of England beat everyone to the internet," *Wired Magazine*, December, https://www.wired.com/2012/12/queen-and-the-internet/.

Metz, D. (2013) "Peak car and beyond: The fourth era of travel," *Transport Reviews*, *33*, 255–270.

Milton, R. (2022) MapTube, CASA, UCL, http://www.maptube.org/.

Mitchell, M. (2022) Complexity Explorer, Santa Fe Institute, Santa Fe, NM, https:// www.complexityexplorer.org.

Modelski, G. (2003) *World Cities: –3000 to 2000*, Faros 2000, Washington, DC.

Moore, G. E. (1965) "Cramming more components onto integrated circuits," *Electronics*, *38*, No. 8, April 19, 114–117.

Morphet, J., and Morphet, R. (2019) "New urban agenda: New urban analytics," Centre for Advanced Spatial Analysis, The MacArthur Project Research Report, University College London.

Morris, W. (1893) "Town and country," *The Journal of Decorative Art*, April 13, https://www.marxists.org/archive/morris/works/1892/town.htm.

Moser, W. (2016) "The Array of Things is coming to Chicago (and the world)," https://www.chicagomag.com/city-life/September-2016/The-Array-of-Things-Is-Coming-to-Chicago-and-the-World/.

Moskvitch, K. (2011) "Smart cities get their own operating system," *BBC News Technology*, September 30, https://www.bbc.co.uk/news/technology-15109403.

Mumford, L. (1938) *The Culture of Cities*, Martin Secker and Warburg, London.

Mumford, L. (1961) *The City in History*, Harcourt, Brace and World, New York.

Mumford, L. (1970) *The Myth of the Machine*, Harcourt Brace Jovanovich, New York.

Murcio, R., Milton, R., and Batty, M. (2021) "The impact of lockdowns on mobility in city systems," *Scienze Regionali*, *3*, 373–390.

Namazi-Rad, M., Padgham, L., Perez, P., Nagel, K., and Bazzani, A. (Editors) (2017) "Agent based modelling of urban systems," First International Workshop, ABMUS 2016, Notes in Computer Science, Springer, Berlin, Germany.

Negroponte, N. (1969) "Toward a theory of architecture machines," *Journal of Architectural Education*, *23*, No. 2, 9–12.

Negroponte, N. (1979) "The return of the Sunday Painter," in Michael L Dertouzos and Joel Moses (Editors), *The Computer Age: A Twenty-Year View*, 21–37, MIT Press, Cambridge, MA.

Negroponte, N. (1995) *Being Digital*, Vintage Books, New York.

Neumann, J. von (1956) *The Computer and the Brain*, Yale University Press, New Haven, CT.

Neumann, J. von (1966) *Theory of Self-Reproducing Automata*, University of Illinois Press, Champaign-Urbana, IL.

Nguyen, J. (2018) "Archaeologists figured out that some of the world's oldest cave drawings don't just depict animals—They're constellations of stars," *Business Insider*, December 20, https://www.businessinsider.com/ancient-cave-drawings-are-constellations-of-stars-2018-12?r=US&IR=T.

Nielsen, K. H. (2017) "From monstrosity to laptop: The story of the personal computer," https://sciencenordic.com/computers-denmark-forskerzonen/from-monstrosity-to-laptop-the-story-of-the-personal-computer/1452258.

Nikolic, D. (2017) "Headsets, caves, AR and VR—What's the reality for construction?," *BIM+*, February 8, http://www.bimplus.co.uk/technology/hea8dsets-cav6es-ar-and-vr-w7hats-reality/#.

Nilles, J., Carlson, R., Gray, P., and Hanneman, G. (1976) *The Telecommunications-Transportation Tradeoff: Options for Tomorrow*, Booksurge, New York.

NRF (2022) Virtual Singapore, Prime Minister's Office, Government of Singapore, https://www.nrf.gov.sg/programmes/virtual-singapore.

NSF (2022) A brief history of NSF and the internet, https://www.nsf.gov/news/news _summ.jsp?cntn_id=103050.

O'Brien, O., Cheshire, J., and Batty, M. (2014) "Mining bicycle sharing data for generating insights in sustainable transport systems," *Journal of Transport Geography*, *34*, 262–273.

Odlyzko, A. (2015) "The forgotten discovery of gravity models and the inefficiency of early railway networks," School of Mathematics University of Minnesota, Minneapolis, MN, http://www.dtc.umn.edu/~odlyzko/doc/mania09.pdf.

O'Regan, G. (2013) *Giants of Computing: A Compendium of Select, Pivotal Pioneers*, Springer, London.

Orcutt, M. (2020) "An elegy for cash: The technology we might never replace," *MIT Technology Review*, January 3, https://www.technologyreview.com/s/614998/an-elegy -for-cash-the-technology-we-might-never-replace/.

Ozkul, B. D. (2015) "Von Thünen revisited," *Built Environment*, *41*, No. 1, 99–111.

Pace, S., Frost, G., Lachow, I., Frelinger, D., Fossum, D., Wassem, D. K., and Pinto, M. (1995) *The Global Positioning System Assessing National Policies*, Critical Technologies Institute, The Rand Corporation, Santa Monica, CA, https://www.rand.org/content /dam/rand/pubs/monograph_reports/MR614/MR614.pdf.

Panko, R. (2018) "Mobile app usage statistics 2018," https://themanifest.com/app -development/mobile-app-usage-statistics-2018.

Park, R., and Burgess, E. (1925) *The City*, University of Chicago Press, Chicago, IL.

Parker, G., Van Alstyne, M., and Choudary, S. (2016) *Platform Revolution: How Networked Markets Are Transforming the Economy and How to Make Them Work for You*, W. W. Norton, New York.

Pfanner, E. (2004) "A decade of internet cafes 'The world's first', Cafe Cyberia in London, takes a bow," *International Herald Tribune*, September, https://www.evapascoe .com/a-decade-of-internet-cafes-the-worlds-first-cafe-cyberia-in-london-takes-a-bow/.

Phillips, D., and Yeh, A. (Editors) (1987) *New Towns in East and South-east Asia: Planning and Development*, Oxford University Press, Hong Kong.

Piaget, J. (1971) *Structuralism*, Routledge, London.

Plato (1976) *The Republic*, Penguin Classics, London, 1976 (translation from the original c. 390 BCE).

Plummer, A. V. (2006) "The Chicago Area Transportation Study: Creating the first plan (1955–1962): A narrative," http://www.surveyarchive.org/Chicago/cats_1954-62.pdf.

Poo, L. (2018) "Sleepy in Songdo, Korea's smartest city," *CityLab*, https://www.city lab.com/life/2018/06/sleepy-in-songdo-koreas-smartest-city/561374/.

Popper, K. (1937, 1959) *The Logic of Scientific Discovery*, Routledge and Kegan Paul, London.

Popper, K. (1957) *The Poverty of Historicism*, Routledge and Kegan Paul, London.

Popper, K. (1963) *Conjectures and Refutations: The Growth of Scientific Knowledge*, Routledge and Kegan Paul, London.

Popper, K. (1995) *The Myth of the Framework: In Defence of Science and Rationality*, edited by M. A. Notturno, Routledge, London.

Quantified Self (2022) Quantified Self Institute, https://qsinstitute.com/about/what-isquantified-self/.

Quigley, R. (2010) "Kitty: One of the first-ever computer animations," March 22, https://www.themarysue.com/kitty-computer-animation-russia-1968-video/.

Rabari, C. (2013) PhD thesis, UCLA, Los Angeles, CA.

Rabari, C., and Storper, M. (2014) "The digital skin of cities: Urban theory and research in the age of the sensored and metered city, ubiquitous computing and big data," *Cambridge Journal of Regions, Economy and Society*, 8, No. 1, 27–42.

Ramasamy, B., Chakrabarty, A., and Cheah, M. (2004) "Malaysia's leap into the future: An evaluation of the multimedia super corridor," *Technovation*, 24, No. 11, 871–883.

Rashid, T. (n.d.) *Make Your Own Neural Network*, Amazon Media, London.

Ratcliffe, J. (2019) "How many CCTV cameras are there in London 2019?," May 29, https://www.cctv.co.uk/how-many-cctv-cameras-are-there-in-london/.

Ravenstein, E. G. (1885, 1889) "The laws of migration," *Journal of the Statistical Society*, 48, 167–227, 1885, and 52, 214–301, 1889.

Reades, J., Zhong, C., Manley, E., R. Milton, and Batty, M. (2016) "Finding pearls in London's oysters," *Built Environment*, 42, No. 3, 365–381.

Reba, M., Reitsma, E., and Seto, K. C. (2016) "Spatializing 6,000 years of global urbanization from 3700 BC to AD 2000," *Scientific Data*, 3, 160034.

Redding, A. C. (2018) *Google It: A History of Google*, Feiwel & Friends, New York.

Reilly, W. J. (1931) *The Law of Retail Gravitation*, Pilsbury, New York, republished 1953.

Reismann, D. (1990) *Alfred Marshall's Mission*, Palgrave Macmillan, London.

Reps, J. (1965) *The Making of Urban America: A History of City Planning in the United States*, Princeton University Press, Princeton, NJ.

Reynard, C. (2018) "Death of the high street," *Forbes*, April 26, https://www.forbes.com/sites/cherryreynard/2018/04/26/death-of-the-high-street/#7793247b45cf.

Rheingold, H. (1985) *Tools for Thought: The History and Future of Mind-Expanding Technology*, Simon & Schuster, Prentice Hall, Englewood Cliffs, NJ.

Rheingold, H. (1992) *Virtual Reality: The Revolutionary Technology of Computer-Generated Artificial Worlds—and How It Promises to Transform Society*, Simon & Schuster, New York.

Rheingold, H. (2000) *The Virtual Community: Homesteading on the Electronic Frontier*, MIT Press, Cambridge, MA.

Rhind, D. (1977) "Computer-aided cartography," *Transactions of the Institute of British Geographers, New Series, 2*, No. 1, 71–97.

Richards, O. (2015) "Best idea wins: How Pixar grew up," *The Telegraph*, November 21, https://www.telegraph.co.uk/film/what-to-watch/pixar-history-good-dinosaur -toy-story/.

Richter, F. (2017) "Smartphones cause photography boom," *Statista*, August 31, https://www.statista.com/chart/10913/number-of-photos-taken-worldwide/.

Rigby, S. (2019) "James Clerk Maxwell: the great scientist with a profound impact on modern physics," *Science Focus*, February 4, https://www.sciencefocus.com/science /james-clerk-maxwell-the-most-important-physicist-you-havent-heard-of/.

Ring, J. (2018) *We Were Yahoo!: From Internet Pioneer to the Trillion Dollar Loss of Google and Facebook*, Post Hill Press, New York.

Ritchie, H., and Roser, M. (2018) "Urbanization," *Our World in Data*, September, https://ourworldindata.org/urbanization.

Rittel, H., and Webber, M. (1973) "Dilemmas in a general theory of planning," *Policy Sciences, 4*, No. 2, 155–169.

Roberts, J. (2017) *Lorenz: Breaking Hitler's Top Secret Code at Bletchley Park*, History Press, Stroud, UK.

Roberts, M., and Turner, C. (2005) "Conflicts of liveability in the 24-hour city: Learning from 48 hours in the life of London's Soho," *Journal of Urban Design, 10*, No. 2, 171–193.

Robinson, A. H. (1955) "The 1837 maps of Henry Drury Harness," *The Geographical Journal, 121*, 440–450.

Rodrigue, J.-P. (2017) *The Geography of Transport Systems*, Routledge, New York.

Ronda, J. P. (2008) "How railroads forever changed the frontier," *American Heritage, 58*, No. 4, https://www.americanheritage.com/how-railroads-forever-changed-frontier.

Rosenau, H. (1974) *The Ideal City: Its Architectural Evolution*, Studio Vista, London.

Rossi, R. (2015) "The Uber-fication of everything: how Uber changed the world," *Information Age*, August 19, https://www.information-age.com/uber-fication-everyth ing-how-uber-changed-world-123460024/.

Saadia, M. (2019) "How 'Blade Runner' and sci-fi made everything dystopian. Science fiction, especially Blade Runner, has spawned so many dystopias that dystopias itself has become banal. We need a new utopianism that embraces the city," *CityLab*, November 1, https://www.citylab.com/perspective/2019/11/dystopian-cities-science -fiction-blade-runner-books-movies/598624/.

Salustri, F. (2022) "A brief history of computer graphics," *DesignWIKI MEC222*, Engineering Graphical Communications, https://deseng.ryerson.ca/dokuwiki/mec222 :brief_history_of_computer_graphics.

Samuelson, P. A. (1983) "Thünen at two hundred," *Journal of Economic Literature, 21,* No. 4, 1468–1488.

Saxenian, A. (1996) *Regional Advantage—Culture and Competition in Silicon Valley and Route 128,* Harvard University Press, Cambridge, MA.

Scales, L. (2017) "How television was invented in London," *Londonist,* https://londonist.com/2016/01/how-television-was-invented-in-london.

Schlager, K. (1965) "A land use plan design model," *Journal of the American Institute of Planners, 31,* 103–111.

Schmidt, A. (2004) ESRI Users Conference, San Diego, CA, https://www.youtube.com/watch?v=aySwJKK6i2s.

Schmidt, E. (1999) Internet World Trade Show, New York, November 18, https://www.oxfordreference.com/view/10.1093/acref/9780191826719.001.0001/q-oro-ed4-00017947.

Schmidt, E., and Jonathan Rosenberg, J. (2014) *How Google Works,* Grand Central Publishing, New York.

Schumacher, E. F. (1973) *Small Is Beautiful: A Study of Economics as if People Mattered,* Vintage, London.

Schumpeter, J. (1939) *Business Cycles 1: A Theoretical, Historical, and Statistical Analysis of the Capitalist Process,* McGraw Hill, New York.

Schwab, K. (2016) *The Fourth Industrial Revolution,* World Economic Forum, Portfolio Penguin, Geneva, Switzerland.

Schwab, K., and Davis, N. (2018) *Shaping the Future of the Fourth Industrial Revolution: A Guide to Building a Better World,* Penguin Books, London.

Selyukh, A. (2016) "The big internet brands of the '90s—Where are they now?," https://www.npr.org/sections/alltechconsidered/2016/07/25/487097344/the-big-internet-brands-of-the-90s-where-are-they-now?t=1572597015091.

Shapiro, C., and Varian, H. (1998) *Information Rules: A Strategic Guide to the Network Economy,* Harvard Business School Press, Cambridge, MA.

Shaw, K. (2018) "The OSI model explained: How to understand (and remember) the 7 layer network model," *Network World,* October 22, https://www.networkworld.com/article/3239677/the-osi-model-explained-how-to-understand-and-remember-the-7-layer-network-model.html.

Shirky, C. (2008) *Here Comes Everybody: The Power of Organizing Without Organizations,* Penguin Books, New York.

Shurkin, J. N. (1996) *Engines of the Mind: The Evolution of the Computer from Mainframes to Microprocessors,* W. W. Norton, New York.

Sidewalk Labs (2022) "Reimagining cities to improve quality of life," https://www.sidewalklabs.com/ and in Toronto https://www.sidewalktoronto.ca/.

Simon, P. (2011) *The Age of the Platform: How Amazon, Apple, Facebook, and Google Have Redefined Business,* Motion Publishing, Henderson, NV.

Singleton, A., Spielman, S., and Folch, D. C. (2018) *Urban Analytics*, Sage, London.

Skaburskis, A. (2008) "The origin of 'wicked problems,'" *Planning Theory and Practice*, *9*, No. 2, 277–280.

Skidmore, Owings & Merrill (1984) "A look back at the early days of 3D visualization," https://www.som.com/news/this_som_archive_video_offers_a_look_back_at_the_early_days_of_3d_visualization, and on Vimeo at https://vimeo.com/93315120.

Smith, A. (1776) *The Wealth of Nations*, Vols. 1 and 2, Strahan, Penguin Classics, 1982 Printing, London.

Smith, D. A. (2022) CityGeographics: Urban Form, Dynamics and Sustainability weblog, http://luminocity3d.org/WorldPopDen/.

Snow, J. (1848) *On the Mode of Communication of Cholera*, Cosimo Classics, London, 2020 edition.

Sobel, D. (1995) *Longitude: The True Story of a Lone Genius Who Solved the Greatest Scientific Problem of His Time*, Fourth Estate, London.

Solow, R. M., and Vickrey, W. S. (1971) "Land use in a long narrow city," *Journal of Economic Theory*, *3*, No. 4, 430–447.

Southorn, G. (2014) "Why are some people better at map reading than others?," https://360.here.com/2014/10/06/people-better-map-reading-others/.

Standage, T. (1998) *The Victorian Internet: The Remarkable Story of the Telegraph and the Nineteenth Century's On-Line Pioneers*, Weidenfeld and Nicolson, London.

Steinitz, C. (2013) *A Framework for Geodesign: Changing Geography by Design*, ESRI Press, Redlands, CA.

Steinitz, C., Parker, P., and Jordan, L. (1976) "Hand-drawn overlays: Their history and prospective uses," *Landscape Architecture*, *66*, 444–455.

Sterling, B., Wild, L., and Lunenfeld, P. (2005) *Shaping Things*, MIT Press, Cambridge, MA.

Sterman, J. (2002) "All models are wrong: Reflections on becoming a systems scientist," *Systems Dynamics Review*, *18*, No. 4, 501–531.

Stewart, J. Q. (1947) "Empirical mathematical rules concerning the distribution and equilibrium of population," *Geographical Review*, *37*, 461–485.

Stone, B. (2017) *The Upstarts: How Uber, Airbnb, and the Killer Companies of the New Silicon Valley are Changing the World*, Little, Brown and Company, New York.

Strohmeyer, R. (2008) "The 7 worst tech predictions of all time," *PC World*, December 31, https://www.pcworld.com/article/155984/worst_tech_predictions.html.

Sulis, P., Manley, E., Zhong, C., and Batty, M. (2018) "Mobility data as proxy for measuring urban vitality," *Journal of Spatial Information Science*, *16*, 137–162.

Sullivan, L. (1896) "The tall office building artistically considered," *Lippincott's Magazine*, March 23, 403–409.

Sullivan, M. (2012) "A brief history of GPS," *PC World*, August 9, https://www.pcworld.com/article/2000276/a-brief-history-of-gps.html.

Sutherland, I. E. (1963) "Sketchpad: A man-machine graphical communication system," PhD Thesis, MIT, 1963, Technical Report, Number 574, UCAM-CL-TR-574, Computer Laboratory, University of Cambridge, UK, https://www.cl.cam.ac.uk/tech reports/UCAM-CL-TR-574.pdf.

Swaine, M., and Freiberger, P. (2014) *Fire in the Valley: The Birth and Death of the Personal Computer*, 3rd Edition, Pragmatic Bookshelf, Raleigh, NC.

Sweney, S. (2019) "Britons hang up the landline as call volumes halve: Rise in mobile usage causes rapid decline of traditional telephones over last six years," *The Guardian*, January 5, https://www.theguardian.com/business/2019/jan/05/britons -hang-up-landline-call-volumes-halve.

Taleb, N. (2007) *The Black Swan: The Impact of the Highly Improbable*, Random House, New York.

Tarafdar, K. (2022) "What was the relationship between James Clerk Maxwell and Michael Faraday?," https://www.quora.com/What-was-the-relationship-between-James -Clerk-Maxwell-and-Michael-Faraday.

Taylor, F. (2017) "First review of new Google Earth," *Google Earth Blog*, April 18, https:// www.gearthblog.com/blog/archives/2017/04/first-review-new-google-earth.html.

Taylor, J. (1997) "The emerging geographies of virtual worlds," *Geographical Review*, 87, No. 2, 172–192.

Taylor, T. (2017) "The cycles of cities," *Conversable Economist*, July 20, https:// conversableeconomist.blogspot.com/2017/07/the-cycles-of-cities.html.

Tennekes, H. (2019) "Karl Popper and the accountability of scientific models," in J. Grasman and G. van Straten (Editors), *Predictability and Nonlinear Modelling in Natural Sciences and Economics*, 6–11, Springer Nature, Dordrecht, The Netherlands.

Tetlock, P., and Gardner, D. (2015) *Superforecasting: The Art and Science of Prediction*, Random House, New York.

Tewdwr-Jones, M., Sookhoo, D., and Freestone, R. (2020) "From Geddes' city museum to Farrell's urban room: Past, present, and future at the Newcastle City Futures exhibition," *Planning Perspectives*, 35, 277–297.

Thibodeau, P. (2016) "Envisioning a 65-story data center," *Computer World*, https:// www.computerworld.com/article/3054603/a-65-story-data-center-design-that-soars -with-ideas.html.

Thompson, P. (2018) "John von Neumann, the last great polymath," https://www .sothebys.com/en/articles/john-von-neumann-the-last-great-polymath.

Thünen, J. H. von (1826) *Der Isolierte Staat in Beziehung auf Landwirtschaft und Nation-alökonomie*, Wirtschaft & Finan, Hamburg, Germany, translated by Carla M. Wartenberg and published as *Von Thünen's Isolated State*, Pergamon Press, Oxford, 1966.

Tobler, W. (1970) "A computer movie simulating urban growth in the Detroit region," *Economic Geography*, 46 (Supplement), 234–240.

Toffler, A. (1970) *Future Shock*, Bantam, New York.

Toffler, A. (1979) *The Third Wave*, Bantam, New York.

Tomlinson, R. (2003) *Thinking About GIS: Geographic Information Systems Planning for Managers*, ESRI Press, Redlands, CA.

Tomlinson, R. (2012) "Origins of the Canada geographic information system," *ArcUser Fall 2012*, ESRI, Redlands, CA, https://www.esri.com/news/arcnews/fall12ar ticles/origins-of-the-canada-geographic-information-system.html.

Tomlinson, R., and Toomey, M. A. G. (1999) "GIS and LIS in Canada," in G. McGrath and L. Sebert (Editors), *Mapping a Northern Land: The Survey of Canada 1947–1994*, 462–490, McGill-Queen's University Press, Montreal, Canada.

Townsend, A. M. (2013) *Smart Cities: Big Data, Civic Hackers, and the Quest for a New Utopia*, W. W. Norton, New York.

Tufte, E. (1983) *The Visual Display of Quantitative Information*, Graphics Press USA, New Haven, CT.

Tukey, J. W. (1962) "The future of data analysis," *The Annals of Mathematical Statistics*, *33*, No. 1, 1–67.

Turing, A. M. (1937) "On computable numbers, with an application to the Entscheidungsproblem, 1936," *Proceedings of the London Mathematical Society*, *s2–42*, No. 1, 230–265.

Turing, A. M. (1950) "Computing machinery and intelligence," *Mind: A Quarterly Review of Psychology and Philosophy*, *LIX*, No. 236, October, 433–460.

Turing, A. M. (1952) "The chemical basis of morphogenesis," *Philosophical Transactions of the Royal Society of London B*, *237*, No. 641, 37–72.

Ulam, S. (1958) "John von Neumann, 1903–1957," *Bulletin of the American Mathematical Society*, *64*, 1–49.

Ulam, S. (1976) *Adventures of a Mathematician*, Charles Scribner's Sons, New York.

Vinci, L. da (1519) *Notebooks of Leonardo da Vinci 1452–1519*, translated and arranged by Edward MacCurdy, Braziller, New York, 1955, https://archive.org/details/noteboo 00leon.

Virtual Reality Society (2017) "History of virtual reality," https://www.vrs.org.uk/vir tual-reality/history.html.

Vlaskovits, P. (2011) "Henry Ford, innovation, and that 'Faster Horse' quote," *Harvard Business Review*, August 29, https://hbr.org/2011/08/henry-ford-never-said-the-fast.

Voorhees, A. M. (2013) "A general theory of traffic movement," *Proceedings*, Institute of Traffic Engineers, New Haven, CT, 46–56, reprinted in *Transportation*, *40*, 1105–1116.

Waddell, P. (2002) "UrbanSim: Modeling urban development for land use, transportation, and environmental planning," *Journal of the American Planning Association*, *68*, No. 3, 297–314.

Waddell, P. (2011) "Integrated land use and transportation planning and modeling: Addressing challenges in research and practice," *Transport Reviews*, *31*, No. 2, 209–229.

Wakefield, J. (2013) "Tomorrow's cities: Do you want to live in a smart city?," *BBC News Technology*, August 19, http://www.bbc.co.uk/news/technology-22538561.

Wall, M. (2019) "Virtual cities: Designing the metropolises of the future," *BBC Technology of Business*, January 18, https://www.bbc.co.uk/news/business-46880468.

Warntz, W. (1959) *The Geography of Price: A Study in GeoEconometrics*, University of Pennsylvania Press, Philadelphia, PA.

Warntz, W. (1964) "A new map of the surface of population potentials for the United States," *Geographical Review*, *54*, No. 2, 170–184.

Wasse, L.A. (2020) "The basics of LiDAR," January, https://www.neonscience.org/lidar-basics.

Watson, B., Müller, P., Veryovka, O., Fuller, A., Wonka, P., and Sexton, C. (2008) "Procedural urban modeling in practice," *IEEE Computer Graphics and Applications*, *28*, No. 3, 18–26.

Watson, J. (1998) *The Double Helix*, Touchstone Books, New York.

Watts, D. (2003) *Six Degrees: The Science of a Connected Age*, Random House, New York.

Watts, S., and Raza, M. (2019) "SaaS vs PaaS vs IaaS: What's the difference and how to choose," *BmcBlogs*, June 15, https://www.bmc.com/blogs/saas-vs-paas-vs-iaas-whats-the-difference-and-how-to-choose/.

Webber, M. M. (Editor) (1964) *Explorations into Urban Structure*, University of Pennsylvania Press, Philadelphia, PA.

Weinberger, M. (2016) "The worst things Bill Gates ever said," *Business Insider*, April 18, https://www.independent.co.uk/news/people/the-worst-things-bill-gates-ever-said-a6990046.html.

Weiser, M. (1991) "The computer for the 21st century," *Scientific American, 265*, No. 3, 94–104.

Weizenbaum, J. (1976) *Computer Power and Human Reason: From Judgment to Calculation*, W. H. Freeman, New York.

Wellar, B. (Editor) (2012) *Foundations of Urban and Regional Information Systems and Geographic Information Systems and Science*, Special Publication in Celebration of URISA's 50th Anniversary Conference 1963–2012, https://www.urisa.org/clientuploads/directory/Documents/Books%20and%20Quick%20Study/Foundations_FINAL.pdf.

Wells, H. G. (1902) "The probable diffusion of great cities," in *Anticipations*, 33–65, Chapman and Hall Ltd., London.

West, G. (2017) *Scale: The Universal Laws of Life and Death in Organisms, Cities and Companies*, Weidenfeld and Nicolson, London.

Wheen, F. (1999) *Karl Marx*, Fourth Estate, London.

White, B. A. (2007) *Second Life: A Guide to Your Virtual World*, Que, Hoboken, NJ.

White, R., Engelen, G., and Uljee, I. (2015) *Modeling Cities and Regions as Complex Systems: From Theory to Planning Applications*, MIT Press, Cambridge, MA.

Whyte, W., Bello, F., Freedgood, S., Seligman, D., and Jacobs, J. (1958) *The Exploding Metropolis*, Doubleday, Garden City, New York.

Wiener, N. (1948) *Cybernetics: Or Control and Communication in the Animal and the Machine*, MIT Press, Cambridge, MA.

Wilkes, M. V. (1975) "Early computer developments at Cambridge: The EDSAC," *Radio and Electronic Engineer, 45*, No. 7, 332–335.

Williamson, L. (2013) "Tomorrow's cities: Just how smart is Songdo?," *BBC News Technology*, http://www.bbc.co.uk/news/technology-23757738.

Wilson, A. G. (1967) "A statistical theory of spatial distribution models," *Transportation Research*, 1, No. 3, 253–269.

Wilson, A. G. (1970) *Entropy in Urban and Regional Modelling*, Pion Press, London.

Winchester, S. (2002) *The Map That Changed the World: A Tale of Rocks, Ruin and Redemption*, Viking Penguin, London.

Wolfram, S. (1994) *Cellular Automata and Complexity: Collected Papers*, CRC Press, Taylor and Francis, Boca Raton, FL.

Wood, G. (2002) *The American Revolution: A History*, Weidenfeld and Nicolson, London.

Wooldridge, M. (2021) *The Road to Conscious Machines: The Story of AI*, Penguin Books, London.

Wright, A. (2017) "Mapping the Internet of Things," *Communications of the ACM, 60*, No. 1, January, 16–18.

WUPEH (2019) Wuhan Urban Planning Exhibition Hall, *16th International Conference on Computers in Urban Planning and Urban Management*, July 8–12, http://cupum2019 .aconf.org/news/2606.html.

Wurster, C. B. (1963) "The form and structure of the future urban complex," in Lowdon Wingo, Jr. (Editor), *Cities and Space: The Future Use of Urban Land*, 73–102, Johns Hopkins University Press, Baltimore, MD.

Wycherley, R. (1949) *How the Greeks Built Cities*, Macmillan, New York.

Zahavi, Y. (1976) *The Unified Mechanism of Travel (UMOT) Model*, The World Bank, Washington, DC, http://www.surveyarchive.org/Zahavi/TheUMOTModel.pdf.

Zuboff, S. (2019) *The Age of Surveillance Capitalism*, Profile Books, London.

Zuboff, S. (2020) *#BlockSidewalk*, January, https://www.blocksidewalk.ca/supporters.

ILLUSTRATION CREDITS

1   The General Post Office and the Royal Mail in 1830.
From the Alamy Photo Stock, 2E3PP92. Permission granted. Originally published by Thomas McLean, 26 Haymarket, April 19, 1830. Original painting by James Pollard.

2   The information mile, walking west from the GPO at St. Paul's.
Author's own figure.

3   King's College London, where Maxwell first presented his wave equations.
Wellcome copyright. Permission granted. Wellcome Library and Wikimedia, London Copyright holder. Copyrighted work available under Creative Commons Attribution license CC BY 4.0. https://wellcomecollection.org/works/e3yjqd34/images?id=cg4jaz7f; https://commons.wikimedia.org/wiki/File:King%27s_College,_London;_the_interior _of_the_theatre_on_prize_Wellcome_L0007544.jpg.

4   Turing worked here next to the Manchester Baby in the early 1950s.
Author's own photograph.

5   Colossus: One of the first digital computers at Bletchley Park in 1943–1944.
Permission granted by the National Archives Image Library, Kew, London. Image library reference: FO 850/234. https://images.nationalarchives.gov.uk/assetbank-nation alarchives/action/viewHome.

6   John von Neumann and the IAS machine at the Institute for Advanced Study, Princeton, 1951.
Permission granted by Shelby White and the Leon Levy Archives Center, Institute for Advanced Study in Princeton, NJ. Credit to Alan Richards, photographer.

7   The first nodes in the ARPANET (internet)—UCSB, UCLA, Utah, and SRI—1969.
Unknown attribution. Len Kleinrock, one the ARPANET pioneers involved in the construction of the first node at UCLA in 1969, writes: "The bottom line is we don't know who drew that original figure. We have had group email discussions about it

including folks like myself, Steve Crocker, Bob Kahn, and others and there has been no agreement for ownership among us; some of us claim possible ownership but no one is sure. It's likely to remain one of those unknowns of the early Arpanet." (personal communication, December 22, 2022)

8   Bill Gates and Paul Allen at Lakeside School, Seattle, 1968.
Permission granted from the family of Bruce R. Burgess, photographer, the offices of Bill Gates and of Paul Allen, and Leslie Schuyler, school archivist. https://www.lakesideschool.org/schoolarchives.

9   Queen Elizabeth II sends her first email in 1976 from the Royal Signals Research Establishment in Malvern, UK.
Photograph by Peter Kiersten, Computer Science, UCL. C. Metz, "How the Queen of England beat everyone to the internet," *Wired Magazine*, December 25, 2012, https://www.wired.com/2012/12/queen-and-the-internet/.

10   Thomas Kurtz and John Kemeny with BASIC on PCs in 1984.
Permission granted from Office of the General Counsel, Dartmouth, Hanover, NH. https://www.dartmouth.edu/legal/.

11   Johann Heinrich von Thünen and his model of concentric land-use rings.
Portrait from an original painting by Wilhelm Ternite (1786–1871) in 1840 at the Thünen-Museum-Tellow in Mecklenberg, Germany; and https://www.thuenen.info/?page_id=2009&lang=en and CC https://commons.wikimedia.org/wiki/Category:Johann_Heinrich_von_Thünen and http://www.personal.umich.edu/~copyrght/thunen/thunen/. The rings are reproduced from the author's *Inventing Future Cities*, MIT Press, Cambridge MA, 2018, p. 80.

12   Marx, Engels, and their families.
Wikimedia Commons, photographer unknown, 1864. https://commons.wikimedia.org/wiki/File:Marx%2BFamily_and_Engels.jpg.

13   Jane Jacobs, vocal critic of architecture and planning, in NYC in 1961.
Jane Jacobs, chair of the Committee to Save the West Village, holds up documentary evidence at a press conference at the Lion's Head restaurant at Hudson and Charles Streets, NYC. This work is from the New York World-Telegram and Sun collection at the Library of Congress. There are no known copyright restrictions on the use. https://commons.wikimedia.org/wiki/Category:Jane_Jacobs#/media/File:Jane_Jacobs.jpg.

14   An interactive dashboard capturing traffic data.
Author's own figure with Oliver O'Brien, CASA, UCL, January 2016, and also from the author's *Inventing Future Cities*, MIT Press, Cambridge MA, 2018, p. 113.

15   Capturing demand for travel linked to the supply of trains on the London subway.
Author's own figure, with Jonathan Reades and Richard Milton, CASA, UCL, March 2017.

16   Wassily Leontieff, inventor of input–output analysis.
Permission given by Svetlana Alpers, executor of Leontieff's estate. https://www.the
famouspeople.com/profiles/wassily-leontief-299.php.

17   Wilbur's simultaneous equation solver used by Leontieff for his input–output
model in 1936.
The MIT History Collection, https://mitmuseum.mit.edu/collections/object/GCP
-00029516, and reproduced in J. H. H. Speak, *Robot Mathematician Solves Nine Simultaneous Equations*, University of Groningen, Faculty of Science and Engineering, Groningen,
The Netherlands, 2017.

18   The CATS Cartographatron, 1959.
J. D. Carroll, and G. P. Jones, "Interpretation of desire line charts made on a Cartographatron," *Highway Research Board Bulletin*, *253*, Highway Research Board, Washington, DC, 1960, 86–108. http://onlinepubs.trb.org/Onlinepubs/hrbbulletin/253/253
-004.pdf.

19   Hansen's accessibility model, 1959.
MIT archives: https://dspace.mit.edu/bitstream/handle/1721.1/74869/32597665-MIT.
pdf. W. G. Hansen, "Accessibility and residential growth," MCP degree thesis, MIT,
Cambridge, MA, 1959, and his paper "How accessibility shapes land use," *Journal of
the American Institute of Planners*, *25*, 73–76, 1959.

20   Ivan Sutherland's Sketchpad, 1961.
At the MIT Lincoln Laboratory, Sutherland, in 1961, sitting at the console of TX-2
driving Sketchpad. Permission granted by Ivan Sutherland and the MIT Lincoln Laboratory Archives.

21   Kadanoff's interactive graphical simulations of urban development, 1970.
Permission granted by Elsevier Copyright Clearance Center, license number
5462520844058, January 5, 2023, from L. Kadanoff, J. R. Voss, and W. J. Bouknight,
"Computer display and analysis of urban information through time and space,"
Technological Forecasting and Social Change, *2*, No. 1, 77–103, 1970.

22   Early graphics screen dumps from a DEC VAX GIGI terminal, 1982.
Author's own figures. The DEC VAX machine was used at the University of Mel-
bourne Architecture and Planning School in October 1982. https://vt100.net/dec/vk
100in2.pdf.

23   Early line-printer graphics, 1960s.
Author's own line-printer-like drawing of Mickey Mouse produced on a laser printer.
Permission from Walt Disney Company granted in 1986 for this image to be used.
See M. Batty, *Microcomputer Graphics*, Chapman and Hall, London, 1987.

24   Andy Warhol draws Debbie Harry, 1985.
Permission agreed by Thomas Manzi, Artist Management, NYC, for the group Blondie.
Photograph from Getty Images and granted by photographer Allan Tannenbaum.
Image # 525549358. Rights managed, Premium Archive.

25 A simple 3D rendition of an English village using AutoCAD, 1980.
Permission granted by Peter Boreham for reproduction in the author's "Computers and design," Papers in Planning Research, 2, UWIST Department of Town Planning, Cardiff, UK, 27 pages, 1980; from G. R. Beacon and P. G. Boreham, "Computer-aided architectural design at Leeds Polytechnic," *Computer-Aided Design, 10*, 325–331, 1978.

26 Agent-based modeling of mobility.
Author's own figure from his book *Cities and Complexity*, MIT Press, Cambridge, MA, 2005.

27 Urban dynamics with Forrester-like graphics, 1970.
Author's own figure from his book *Urban Modelling*, Cambridge, UK, 1976.

28 Desktop urban simulation of land use and transport integrated with Google Earth for 3D rendition of map data.
Images compiled and collated from the author's paper "Visually-driven urban simulation: Exploring fast and slow change in residential location, *Environment and Planning A, 45*, 532–552, 2013.

29 Patrick Geddes' Outlook Tower: The idea in 1910 and the reality in 2013.
This first picture is from P. Geddes, *Cities in Evolution*, Norgate and Williams, London, 1915. Permission for the second picture of the Outlook Tower is granted by David Martin of Geograph and is available at https://www.geograph.org.uk/photo/3452202.

30 John Paton-Watson and Patrick Abercrombie discuss the Plymouth Plan in 1943.
From the 10th Abercrombie Lecture, University of Liverpool, May 16, 2006, delivered by the author and published in M. Batty, "Model cities," *Town Planning Review, 78*, No. 2, 125–151, 2007.

31 Wuhan Citizens Home: An entire building for participation using city models, 2017.
Author's own photograph. See also M. Batty, "At the crossroads of urban growth," *Environment and Planning B, 41*, 951–953, 2014.

32 The first 3D digital city models by Skidmore, Owings & Merrill, 1984.
Nine cities by Skidmore, Owings & Merrill, 1984. Scanned from the original 16-mm film by Peter Little, with the frame reproduced here extracted by the author. https://vimeo.com/93315120.

33 Embedding Virtual London into a virtual room where avatars act as participants in the design process.
Figure from the Virtual London model produced by the author with Andy Hudson-Smith, CASA, UCL, 2010.

34 Models constructed from web-based services, big data, and fast computation.
Author's own figure with Richard Milton, CASA, UCL, 2022.

35 Singapore: Building an intelligent island, 1988.
Author's own photograph, Changi, Singapore.

36 Geospatial server farm at Harwell, UK.
Author's own photograph, *The Jasmine Supercomputer Installation*, Centre for Environmental Analysis, Rutherford-Appleton Laboratory, https://www.ceda.ac.uk/services/jasmin

37 Cooling pipes inside Google's data center in Douglas County, Georgia.
https://www.google.com/about/datacenters/gallery/#!#gallery. The picture is in the public domain. See https://www.google.com/about/datacenters/faq/.

38 Modeling riots and policing using the London touch table.
Author's own photograph with Hannah Fry and Thomas Evans, CASA, UCL.

39 Augmenting models of physical plans with computable touch-table technologies.
Iconic models augmented with information technologies for public participation. https://pipersmodelmakers.co.uk/; https://programme.openhouse.org.uk/listings/2653.

40 Ideal cities: Le Corbusier and his plan for Paris, 1925.
https://commons.wikimedia.org/wiki/File:Plan_Voisin_model.jpg; https://en.wikipedia.org/wiki/Plan_Voisin#/media/File:Plan_Voisin_model.jpg.

41 Ebenezer Howard's garden city, 1901.
Republished and quoted in the author's *Inventing Future Cities*, MIT Press, Cambridge, MA, 2018, and in E. Howard, *To-Morrow: A Peaceful Path to Real Reform*, republished by Routledge, London, 2003.

42 Masdar Smart City UAE, 2011.
The master plan. https://masdarcity.ae; https://transsolar.com/projects/abu-dhabi-masterplan-masdar-city.

43 Greenbelt, Maryland, 1938.
Image from the Rare Book and Manuscript Collections, Greenbelt, MD Collection, 1939, Clarence Stein papers, #3600, Cornell University Library, Repository: ID Number: RMC2010_0339 Catalog Record: 2068664.0, Collection Number: 3600, Clarence S. Stein (American architect and urban planner, 1882–1975). https://digital.library.cornell.edu/catalog/ss:5/4381.

44 The end: The last of the physical Post Office: Demolition of the BT HQ at St. Paul's, 2023.
Author's own photograph, The demolition and refurbishment of the BT HK from the original 1985 building; the original building is at https://commons.wikimedia.org/wiki/Category:BT_Centre. From the Geograph Project. Photograph by Martin Addison and licensed for reuse under the Wikimedia Creative Commons License.

45 The future: Social networking.
Author's own photograph: At a rail station complex in Guangzhou, China, December 2019.

NAME INDEX

SUBJECT INDEX